AF598895

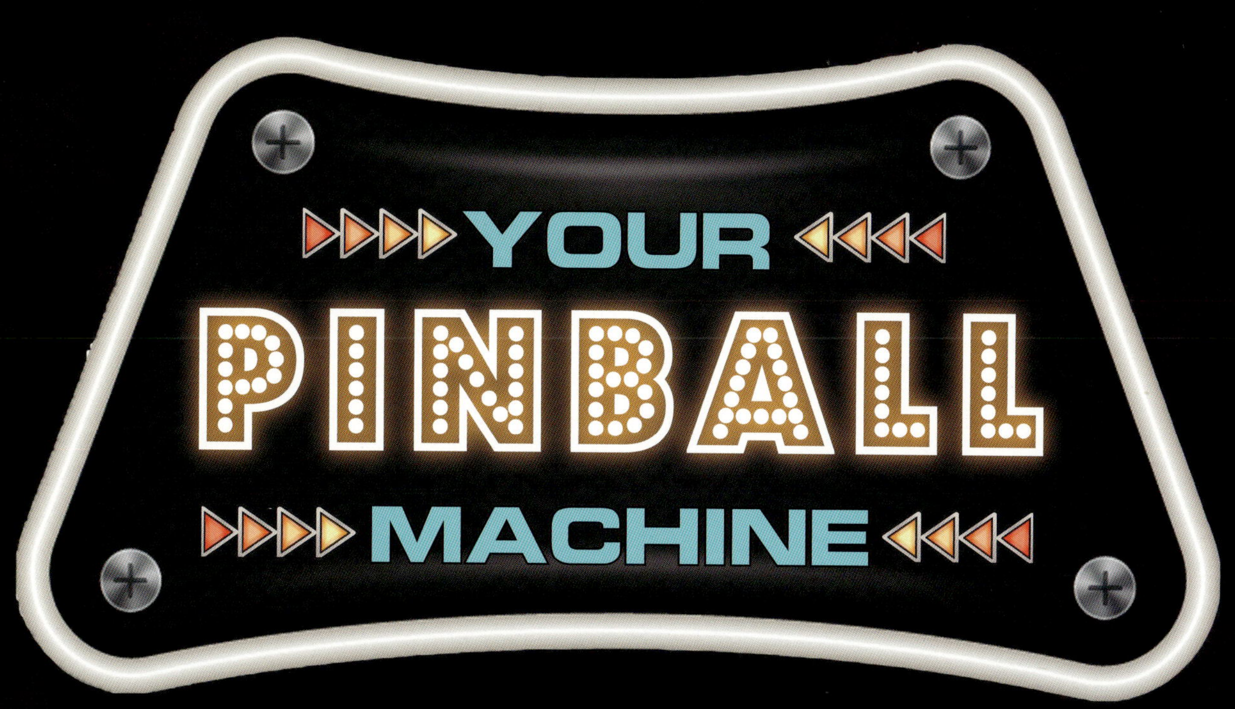

HOW TO PURCHASE, ADJUST, MAINTAIN, AND REPAIR YOUR OWN MACHINE

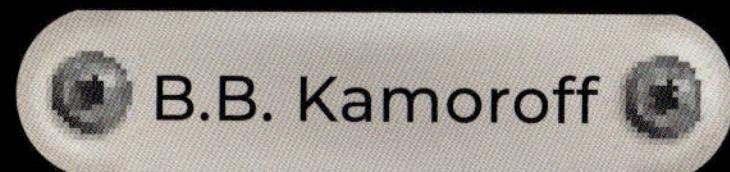

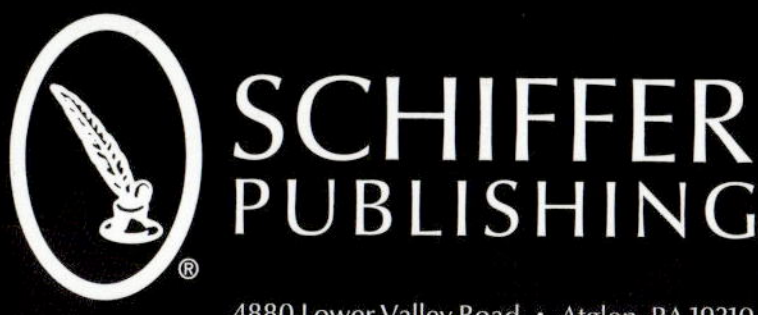

4880 Lower Valley Road • Atglen, PA 19310

Library of Congress Control Number: 2020943658

Edited by Ian Robertson
Designed by Jack Chappell
Cover design by Jack Chappell
Type set in Budmo/Micro Technic Extended/Pixelmix/Montserrat/Chaparral

ISBN: 978-0-7643-6180-7
Printed in China

Published by Schiffer Publishing, Ltd.
4880 Lower Valley Road
Atglen, PA 19310
Phone: (610) 593-1777; Fax: (610) 593-2002
E-mail: Info@schifferbooks.com
Web: www.schifferbooks.com

For our complete selection of fine books on this and related subjects, please visit our website at www.schifferbooks.com. You may also write for a free catalog.

Schiffer Publishing's titles are available at special discounts for bulk purchases for sales promotions or premiums. Special editions, including personalized covers, corporate imprints, and excerpts, can be created in large quantities for special needs. For more information, contact the publisher.

We are always looking for people to write books on new and related subjects. If you have an idea for a book, please contact us at proposals@schifferbooks.com.

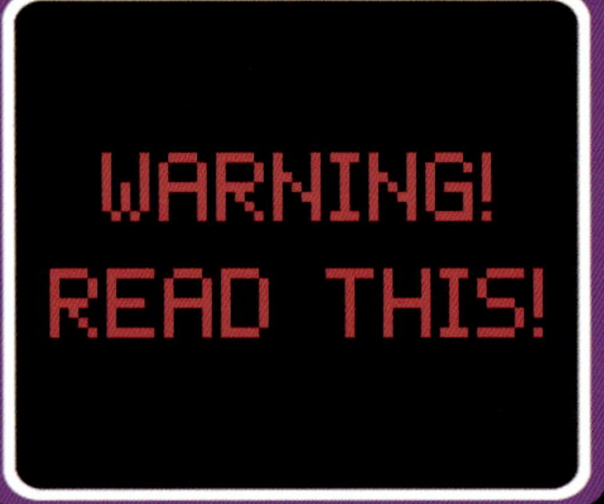

Be Careful: A plugged-in pinball machine has 120 lethal volts in it and exposed wires. You can kill yourself. Unplug the machine before you work on it. Use only the correct fuses. Do not touch the large capacitors in electronic machines even with the machine unplugged; the capacitors store high voltage. Read "Safety First" in the "Repairs: Basics" chapter before opening and working on a machine.

You are responsible for your own pinball machine and your own safety. This manual is not an electrician's guide, and I am not in a position to tell you how to be an electrician. If you are unsure of your wiring, get a professional electrician to help you.

DEDICATED TO . . .

Crystal Rose: 289,840 score on Flip Flop

Julia: 93,180 on Lawman

Colleen: 8,875,770 on Jokerz

Corrina: 167,070 on Hi-Deal

Sharon: Jackpot, Special, and Double Bonus

THANK YOU

Special thanks to Jim and Judy Tolbert, Steve Young, Peter Schiffer, and Sharon Kamoroff.

Thank you Michael Schiess, Founding Director of the Pacific Pinball Museum, for the photographs, for keeping the book alive, and all around support. Very special thanks to Kevin Tiell, Melissa Harmon, and Rob Perica for the photography.

Thank you Kelly Altemueller, Tim Arnold, Chuck Artigues, Lou Brooks, Dick Bueschell, David Carter, Richard Conger, Joel Cook, Tim Ferrante, Gary Flower, Joel Goldoor, Tim Hanna, Jim Hewett, Don Highley, Russ Jensen, Jack Lee, Marc and Nancy Mandeltort, Carey Massimini, Harry McKeown, Dave Mercer, Lyle Morris, Tony Page, Jonathan Quandt, Ian Robertson, John Robertson, Terry Ruffridge, Jim Schelberg, Michael Scott, Kevin Steele, Brad Walton, Yogi Taylor, Daniel Zavaro. And Sam Leandro.

Thanks also to the Pacific Pinball Museum (pacificpinball.org), the International Pinball Database (ipdb.org), Pinball Resource (pbresource.com), Marco Specialties (marcospecialties.com), and the long gone Gottlieb, Bally, and Williams companies.

Michael Schiess, Pacific Pinball Museum:
I would like to thank my wife, Melissa Harmon, for tolerating and then joining me in my fascination of pinball. Larry Zartarian and the PPM Board for joining me in the quest for creating the "Smithsonian of Pinball." d'Arci Bruno for making our museum so beautiful and accessible for all. Kevin Tiell for helping capture the soul of pinball in imagery. My good friends Mark McDonald, Jem Gruber, and Oakland Pinball Mafia Emmet Cadigan and Joseph Bansuelo for supporting the museum effort from the start. Chris Kuntz, The Pinball Pirate, mentor to me and all the repair volunteers who keep everything running. Wade Krause, Jim Dietrich, Clay Harrell, Steve Young, Gordon Hasse, Richard Conger, and Mark Mandeltort for helping me out in so many ways. Bear Kamoroff for letting me be part of it and sharing the proceeds with the Pacific Pinball Museum. And last, my dear friend, Ed Cassel, who played pinball 'round the world with me and captured their beauty in his murals.

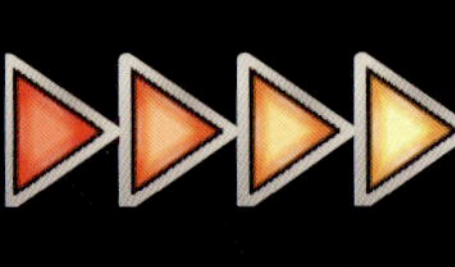

CONTENTS

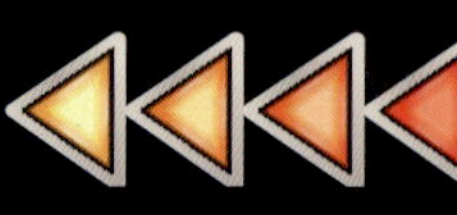

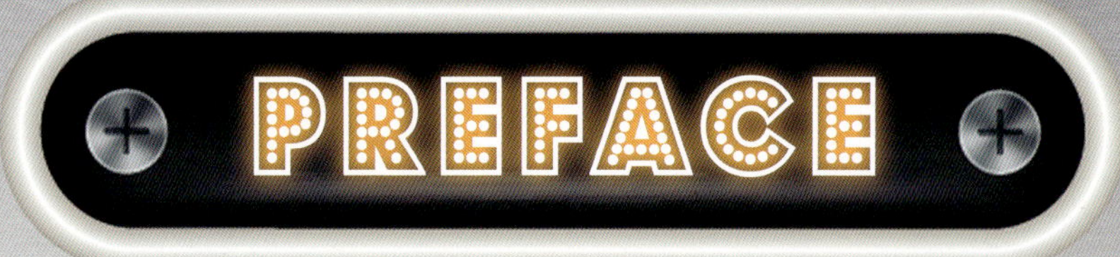

Preface

The Story of a Book

> Even if you don't intend to do your own repairs, knowing what makes bumpers bump and lights flash can give you a new feel for the game. You'll be more respectful of your 'opponent.' You may even begin to feel like another part in the charged circuit.
>
> —**Jim Tolbert**, *Tilt: The Pinball Book*

The original concept for this book came out of a completely different idea. A good friend of mine, Jim Tolbert—who humbly referred to himself as the "High Priest of Pinball"—had written the very first "home" pinball repair guide in 1978, *Tilt: The Pinball Book*. Jim owned and operated a pinball sales and repair business, For Amusement Only in Berkeley, California, for more than thirty years. Jim is a living pinball history. He knew the Gottlieb family, and he was friends with Harry Williams. He knew the stories behind every strange pinball machine ever made, personally told to him by the people who created the machines. Jim taught me how to diagnose and repair pinball machines, he traced schematics with me, and he patiently (well, semi-patiently) answered every question I had.

Jim's book *Tilt* was long out of print, and copies were fetching more than $100 on eBay (it originally sold for $4.50). Other than Jim's hard-to-find book, there was very little information available to help home pinball owners. I suggested to Jim that he update his book. I'd help edit it and we'd get it re-published. Jim immediately liked the idea, but he never followed through on it. About a year later, I asked Jim if he would mind if I wrote my own book. Jim gave me his blessing (I think he said, "Why not?"), as he admitted that he wasn't going to get around to updating *Tilt*.

I had a background in how-to book writing and editing. I had been fixing pinball machines for a few years. I had a successful pinball sales and repair business and had worked on maybe 200 machines. My business was truly on-the-job training. I was learning how to repair pinball machines while I was getting paid to do it.

When I started on my book, I wrote down just about everything I could think of, a process that took a full year, and brought out the first printing of *Pinball Machine Care & Maintenance*, the title I originally gave the book (500 copies). There's a famous old saying in business (actually from Henry Kaiser in the 1950s), "Find a need and fill it." There was obviously a need. The 500 copies sold out almost immediately.

The criticisms, comments, and suggestions from various pinball "experts," and many questions from my readers, also came almost immediately. I started making notes about more details to add and more repair tips. Almost every pinball machine I worked on revealed yet another interesting item that I wanted to add to the book.

Your Pinball Machine is all of that book and much more. Fully twice as much information as the original edition, with more questions answered and more pinball "tips and tricks" for everyone from the first-time pinball owner to the experienced pinball wizards working on their tenth or twentieth machine.

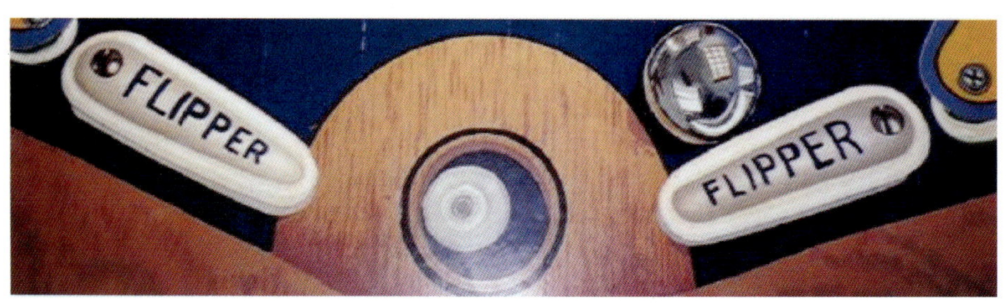

The showroom floor of Jim Tolbert's pinball shop For Amusement Only was a mural of an old pinball machine. Mural painted by Dan Fontes of San Rafael, California

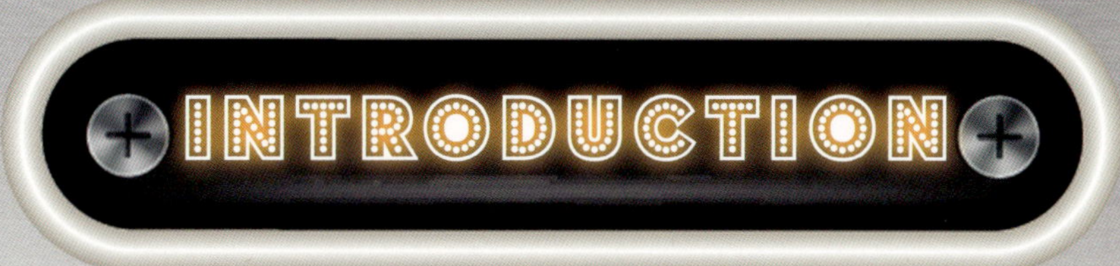

Introduction

> No one can escape the transforming fire of machines.
>
> —**Kevin Kelly**, *Out of Control*

A pinball machine will give you many years of enjoyment and good play if you take a few steps to clean, protect, and maintain it. Pinball machines will go up in value over the years if you keep them properly cared for. The difference between a $25 parts machine and a $2,500 collector's treasure is often only the condition of the backglass, the playfield, and the cabinet.

This manual will help you choose a pinball machine, help you evaluate its condition, and determine its value. This manual will show you how to set up a machine and how to protect it and keep it clean. This manual will explain the easiest and most common repairs and adjustments you can make at home with common tools.

Pinball, like everything else in the world, has its own jargon, its own terminology. Descriptions of different machines, features, parts, adjustments, and repairs involve words and expressions unique to pinball. If you come across a word or expression you don't know, look it up in the back of this manual in the "Terminology" chapter.

Like most how-to manuals, this book may sound confusing, not always making sense until you are actually working on a pinball machine. After you stick your head inside your machine, trace wires, and observe components at work—or not at work—you'll understand the book much better.

Fixing a pinball machine is often a matter of experimentation and trial and error. What works for some machines does not work for others. Almost every pinball machine I've worked on has had at least one thing that was different than any other machine I've repaired. Like most everything else in the world, you learn by doing.

There is nothing more satisfying than having a machine that you can care for yourself, that you can repair, maintain, and keep in fine-tuned running order. This manual is, hopefully, a start in that direction.

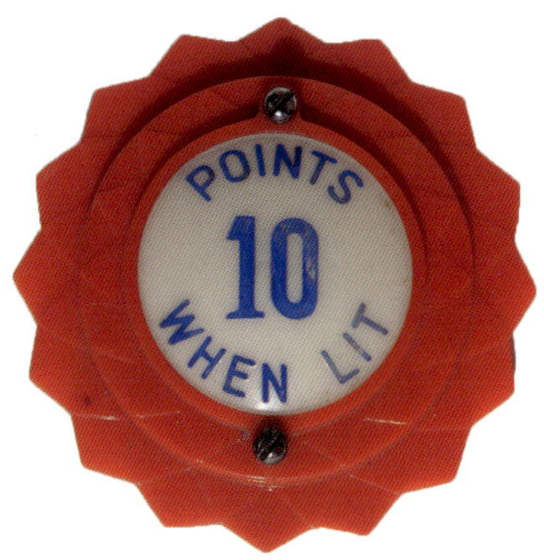

> Those pinball designers were no dummies. I am constantly amazed at the new things I discover while working on games. New troubleshooting techniques are developed by necessity. That old adage "experience is the best teacher" is certainly true. There is no substitute for experience.
>
> —**Russ Jensen**, *Pinball Troubleshooting Guide*

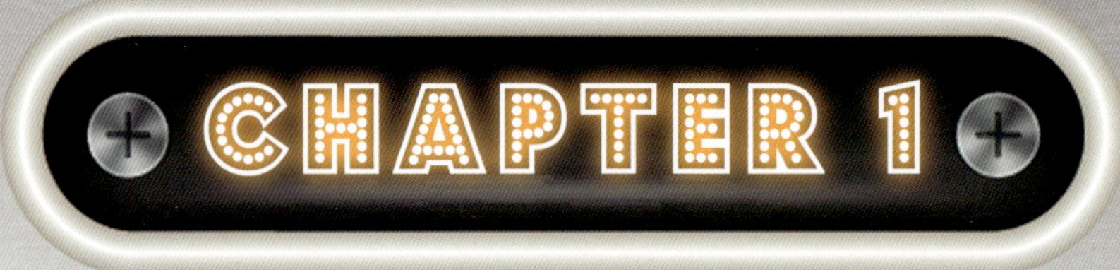

THE HISTORY OF PINBALL

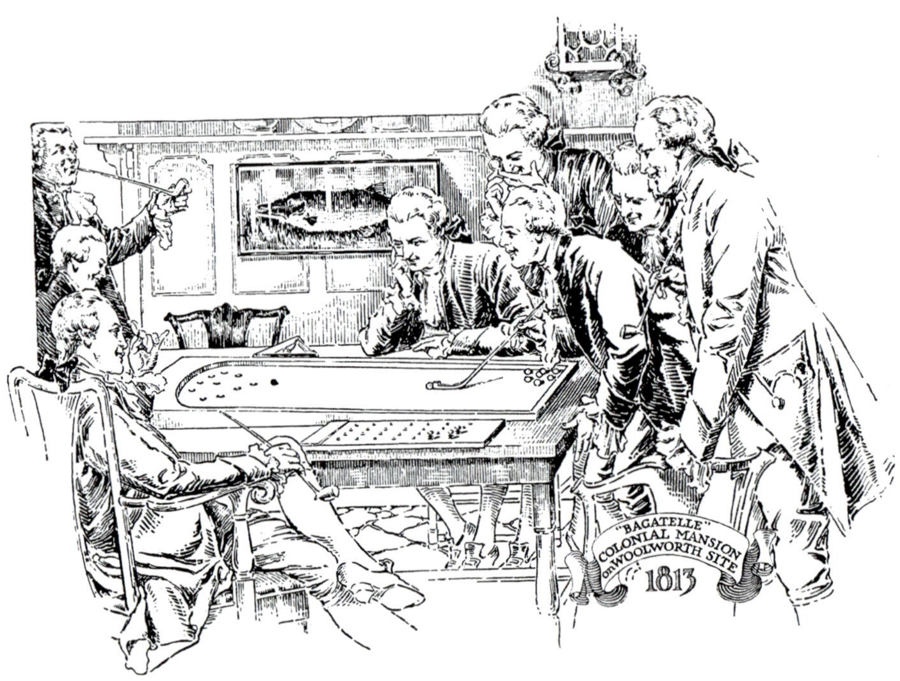

The beginning of pinball goes back to France in the late 1700s, to a table game where a ball was hit by a cue stick and could land in different holes on a playfield, each hole earning a different score. The game was called bagatelle, named after Chateau Bagatelle, the mansion where the game was invented and first played.

About 1900, bagatelles migrated to the United States. A game that was basically unchanged for more than 100 years suddenly came under the scrutiny of Yankee mechanical ingenuity, and by 1920, the evolution of pinball was underway.

Someone invented a spring-loaded plunger to shoot the ball. Someone came up with a clever idea to surround the holes with little nails or pins sticking up from the playfield to make it more easy, or difficult, to get the ball in the holes. Those pins gave the game its new name—pinball—the name still in use more than eighty years after the pins disappeared.

In the 1930s pinballs started their march to greatness. The first coin mechanisms, electric lights, electric coils, and the first bumpers appeared. The distinct and dramatic upright backbox (the head) showed up in 1938, and pinballs began looking much like today's pinball machines.

And then, in 1947, a genius named Harry Mabs of the Gottlieb Pinball Company in Chicago invented the flipper, and pinball as we know it arrived.

The first flipper machines, manufactured from 1947 until about 1959, look like antique museum pieces from your great grandfather's day. The machines were elegant in their artwork. The cabinets and the legs were varnished wood, and there is almost a delicateness to them. Yet these "woodrails," as they are called—referring to the all-wood railing around the playfield glass—were built for and played mostly by teenage boys who were just as roughhouse with them as kids are today.

About 1960, the elegant woodwork of the woodrails was replaced by brightly painted cabinets, steel rails around the glass, and steel legs. The playfields started getting a more streamlined, more modern look. By the 1970s, pinball machines were faster, flashier, more sophisticated, and more fun to play. But throughout the evolution of the game, pinball machines kept the same basic electric circuitry originally created in the 1930s: switches that opened and closed using electro-magnetic coils. These machines are known as "electro-mechanical" or "EM." People sometimes refer to them simply as vintage machines.

In 1976, the first electronic, digital scoring pinball machines were invented, and the day of reel-turning, chime-ringing, electro-mechanical pinball machines gave way to the digital era. The earliest electronic machines looked almost identical in design and playfield layout as their electro-mechanical predecessors, except for their new digital scoring displays.

By the 1980s, electronic machines were evolving right along with the sophisticated, computer programmed electronic circuitry. Most machines built since the mid-1980s have rocket-fast ramps, flashing strobe lights, complex scoring, and several balls flying around the playfield all at the same time.

Today, more than seventy years after Harry Mabs and the Gottlieb Company gave us the flipper, pinball machines are still being manufactured and are still being played. And despite the huge advances in pinball technology and design, pinball machines still use the same concept: You and the flippers against a little steel ball. Human intelligence, skill, and determination against gravity and the laws of physics. No programmer sat in front of a monitor deciding how a pinball will react when whacked by a flipper, when it blasts off a slingshot, or boomerangs off a pop bumper. You never know what that ball is gonna do.

> Pinball history is a rich broth of pop culture, plus a constant and dazzling array of new technologies.
>
> —Richard M. Bueschel, *Pinball 1*

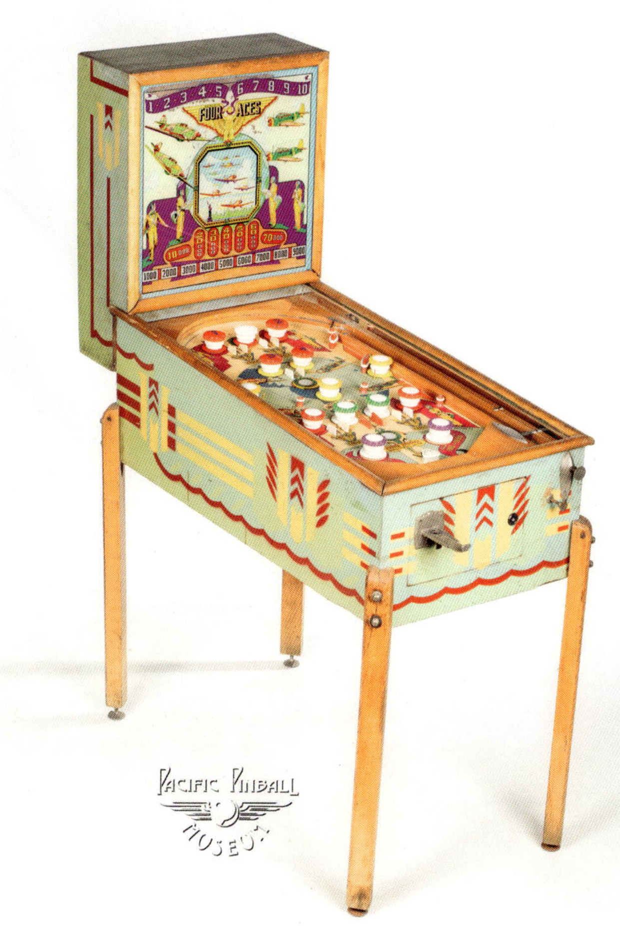

Genco **Four Aces**, 1942. Before score reels were invented, scoring was indicated by painted numbers on the back-glass. Donated to the PPM by Michael Schiess. Photography by RobPerica.com

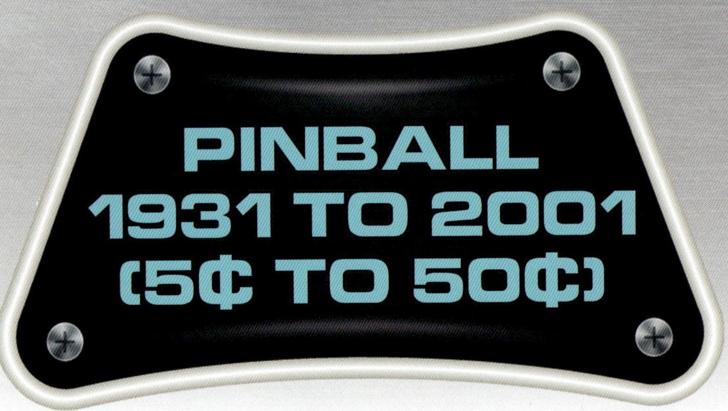

PINBALL 1931 TO 2001 (5¢ TO 50¢)

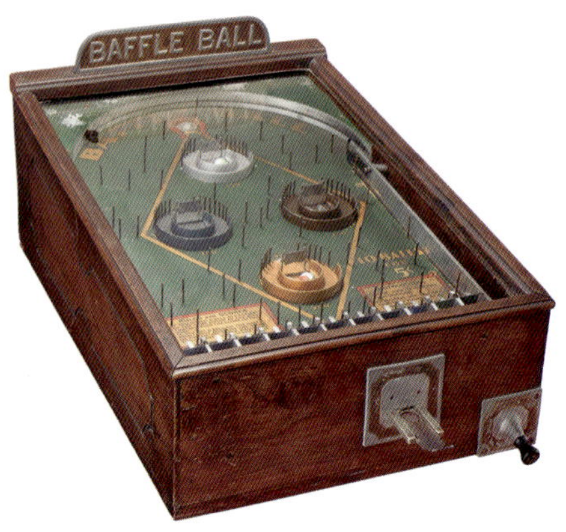

Gottlieb **Baffle Ball**, 1931. The "Elvis Presley of Pinball," a game that became a phenomenon. What Elvis was to rock & roll, Baffle Ball was to pinball.

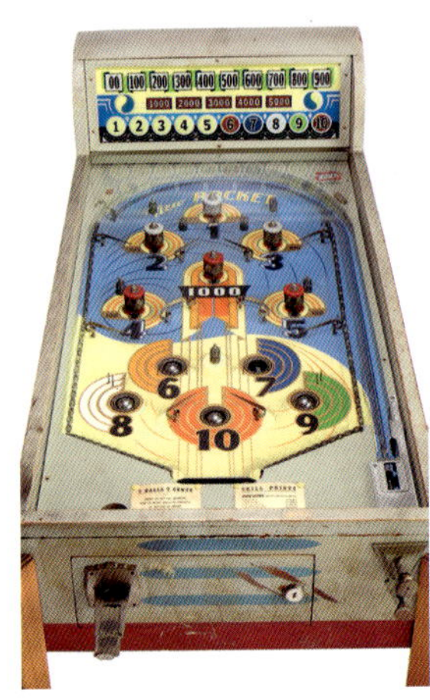

Bally **New Rocket**, 1938. Bally's introduction of the upright backbox.

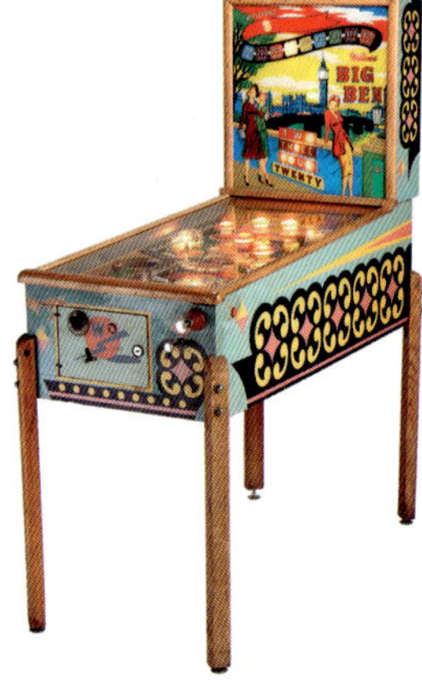

Williams's classic woodrail **Big Ben**, 1954.

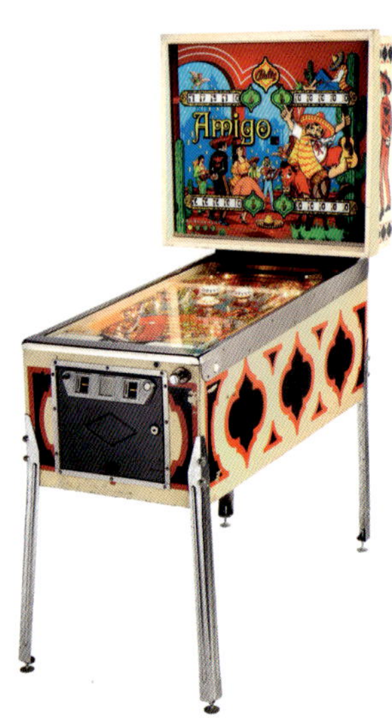

Bally **Amigo,** 1974. The last electro-mechanical machines were streamlined, chromed, fast-paced, and solidly made beauties.

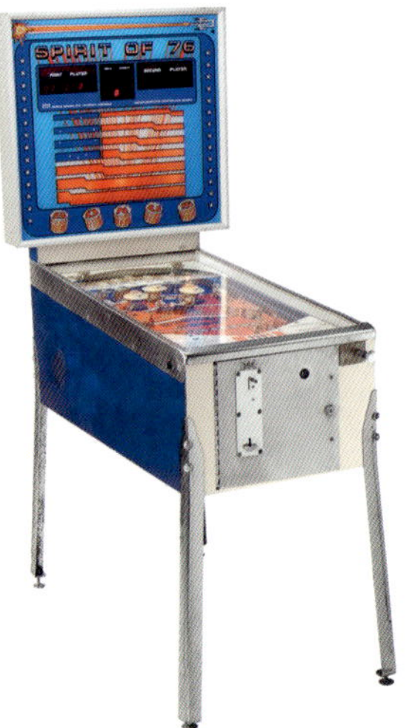

Micro Games **Spirit of 76,** 1975. The first electronic pinball machine ever manufactured. Micro, in Phoenix, Arizona, built only two pinball machines, this one and 1978's Lucky Draw, before closing its doors.

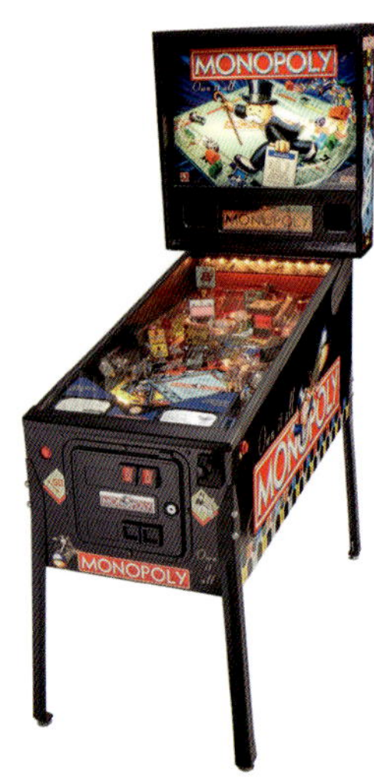

Stern **Monopoly,** 2001. The Mr. Monopoly character, created in 1936, was "updated" in 1985 by Lou Brooks, an artist from Mendocino County, California (who is a friend—and a pinball collector).

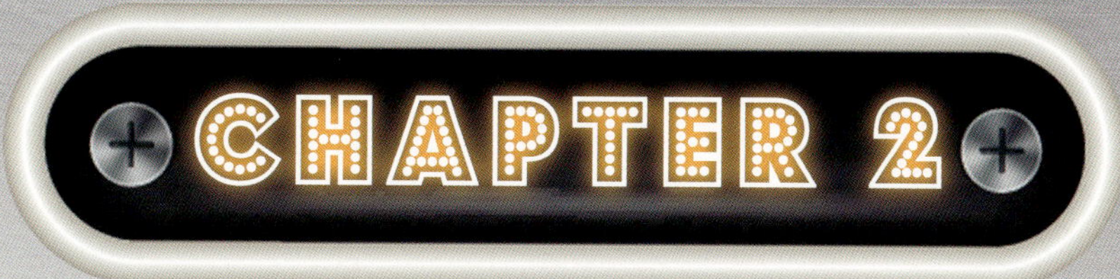

BUYING A PINBALL MACHINE

> Make friends with a pinball machine. And the best way to do that is to invite it into your home.
>
> —**Gary Flower**, *The Lure of the Silver Ball*

Some people know exactly, or almost exactly, what kind of pinball machine they are looking for, but most people aren't really sure, other than it would be fun to own one. Many people played a certain game when they were kids and are looking for that particular game. Some people want electro-mechanical machines only, some people want electronic machines only, others just want to find a pinball. Some have generous budgets and some can barely afford the gas to go check one out. Whether you are looking for a particular game or any game, here are some tips.

DIFFERENT KINDS OF PINBALL MACHINES

While everyone is familiar with what a pinball machine looks like (at least I hope so), there have been several variations over the years.

PRE-1947: NO FLIPPERS

Pinball machines made before 1947 did not have flippers. They looked and scored like regular pinball machines, but other than shaking the machines to nudge the ball, players had no control of the action.

When the flipper was invented in 1947, interest in the pre-flipper machines dropped to zero. Operators with the old machines were suddenly earning no money. Some clever manufacturers offered flipper conversion kits for the old machines. Pinball businesses could drill holes in the playfields to install flippers, drill holes in the cabinets to install flipper buttons, wire it all up, and rescue their investment.

In the twenty years I've run a pinball business, I have had only one pre-flipper machine in the shop. I've never seen a converted machine.

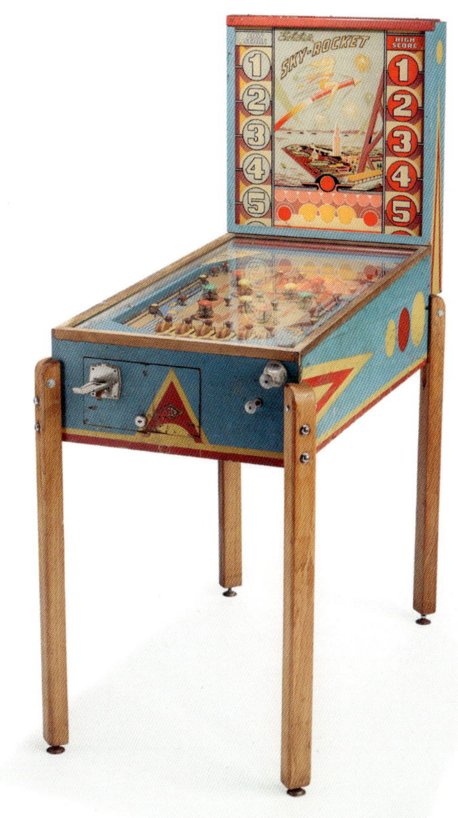

Exhibit Supply **Sky-Rocket,** circa 1939, before flippers were invented. Illustration is of the 1939 Golden Gate International Exposition Worlds Fair, held on Treasure Island, in the San Francisco Bay. In the foreground is the brand new Bay Bridge. Exhibit Supply in Chicago made pinball machines from 1932 until 1957.

BINGOS

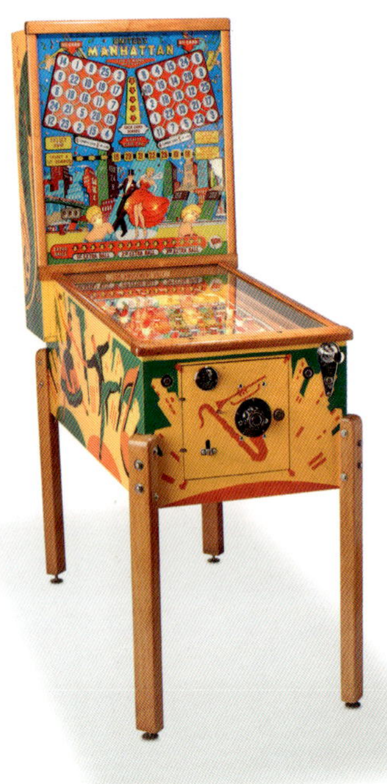

United **Manhattan** bingo pinball machine, 1955. Almost all bingo machines were made in the 1950s by Bally or United. United also made a pinball machine called Manhattan in 1948. United, in Chicago, made machines from 1942 until 1962.

In the 1950s and 1960s, a different type of pinball machine was manufactured without flippers called a bingo. You shot a ball up into the playfield just like a pinball, but the playfield had twenty or more holes where the ball landed and stayed. Bingos are gambling games and do not have the play and action of regular pinball machines. Bingos look like pinball machines, but there is a huge difference. Don't buy a bingo by mistake.

COCKTAIL TABLE MODELS

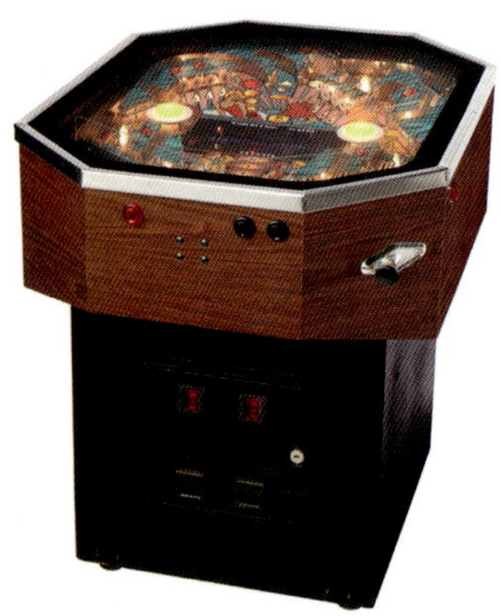

International Concepts **Night Moves,** 1989, was actually manufactured by Premier/Gottlieb, who also built International Concepts' only other pinball machine, Caribbean Cruise.

In the 1970s, some electronic pinball machines were designed to resemble cocktail tables. The machines are smaller and lower than regular pinball machines. People sat down, set their drinks right on top of the glass, and played. A company called Allied Leisure made dozens of these machines.

HOME MODELS

Most pinball machines for sale, both new and used, are commercial games designed to take coins. A few electronic machines were made just for the home market. These machines were produced without coin mechanisms, without the match feature, and often without some of the other features found on commercial games. Home models were less expensive to produce than commercial coin-operated models.

The Bally Company made a series of four home pinball machines in the late 1970s they called "Professional Home Models." I don't know what made the machines professional, other than they did have the same size playfield as Bally's commercial games, though with smaller and lighter cabinets. Bally's home models used the same coils and many of the same playfield components, such as flippers and spinners, but used more rudimentary (cheaper) circuit boards than their commercial machines.

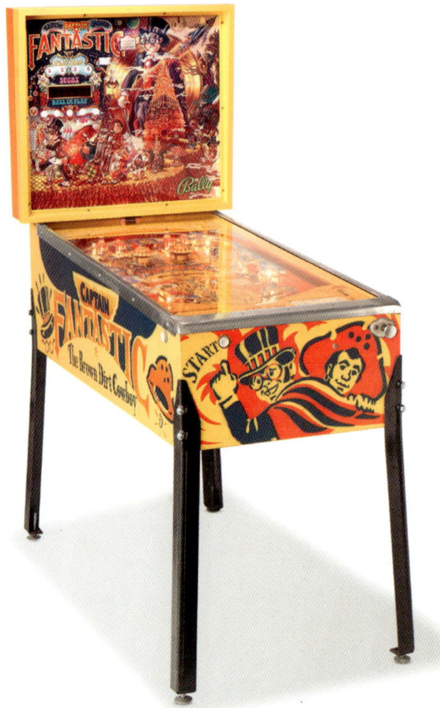

Bally "Professional Home Model" **Captain Fantastic,** 1977. Looks nothing like Bally's commercial Captain Fantastic from 1976, except for Elton John. Note that the front panel of the cabinet has no coin door, the backbox is thin and made of plastic, and there is only one digital scoring display.

Three of the four home model machines that Bally made—Fireball, Captain Fantastic, and Evel Knievel—were also the names of well known, and well loved, Bally full-size commercial machines (so don't confuse them!). Bally's fourth machine, Galaxy Ranger, was only made as a home model. And most interesting, all four models used the exact same playfield layout. They were identical except for the artwork.

The Brunswick Company made some home model machines in the 1970s, though I've never seen one, so I guess they're rare. In the last few years the Stern Pinball Company manufactured some home models with smaller displays than their coin machines and, to quote Stern, "A more economical cabinet."

People looking for a modern pinball machine that is easier to move and that takes up less space than a full side commercial machine may want to consider a home model if you can find one. They are not common. Home models usually fetch a lower price than full size commercial machines.

If you are looking at a pinball machine that obviously never had a coin slot, that's a sure sign it was a home model.

Don't confuse home models with toy pinball machines. The toy machines, even those with legs, have smaller size playfields and use cheap components. Home models look very much like their coin-operated big brothers. Toy pinball machines look like toys.

PACHINKOS

There is a kind of pinball machine called a pachinko (pa-chink-o), a wall mounted Japanese gambling game where you shoot many small balls up into a playfield full of little pins. You try to land the balls in scoring holes.

Pachinkos are small, only about two feet by three feet, and lightweight. Most of the pachinkos imported into the US have been converted to 120 volts, and some have been mounted into free-standing cabinets.

When someone tells you they have a miniature pinball machine or a Japanese pinball machine, they usually are referring to a pachinko.

> Space is a real commodity in Japan. In America, pinball machines grew larger and larger. In Japan, there is just not enough space for the proportions of pinball machines. This precipitated the first major Japanese innovation: The game was set upright to conserve valuable space. The word Pachinko originally meant a slingshot or the ball shot from a slingshot.
>
> —Eric C. Sedensky, *Winning Pachinko*

VIDEO PINBALLS

One day a couple years ago, a guy walked into my shop. There were about twenty pinball machines on display. He looked around, his eyes opened wide, and he said, "Video games!" Well, *no*.

Video games are not pinball machines. Video games have no physical action and no moving parts other than the joysticks and buttons. Video monitors on the old machines were cathode ray tubes just like old television screens. Repairing an old video game requires the same knowledge as repairing an old television. If your video flipper isn't working, I have no idea what to do.

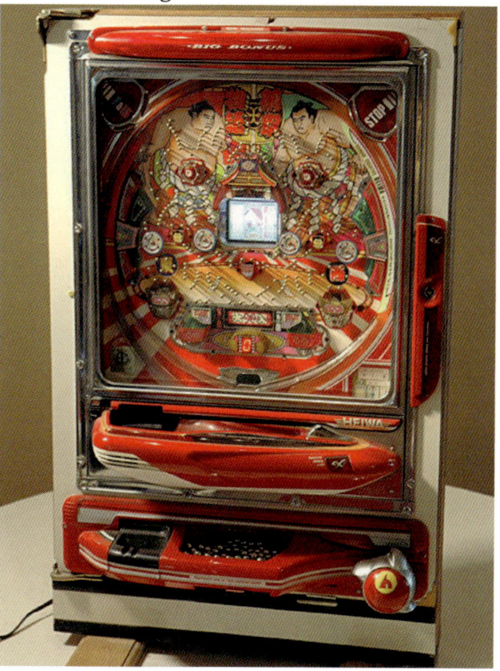

An electro-mechanical pachinko from the 1970s (*left*) and an electronic pachinko from the 1980s (*right*).

In 1999, two pinball machines were manufactured as pinball video games. The combined Williams/Bally Company designed what was going to be a series of hybrid machines called Pinball 2000. The machines had a pinball playfield with real flippers and a real ball, but had a video monitor in the upright backbox. A hidden mirror made the monitor reflect onto the playfield. To the player, it appeared that the ball (the real ball) was interacting with video animation on the playfield. The two games—Revenge From Mars and Star Wars Episode 1—were popular, but they were expensive to produce and weren't profitable. Sadly, those two games not only were the only Pinball 2000 games produced, they were Williams's (and Bally's) swan song. Neither Williams nor Bally, after 67 years in business, made another pinball machine.

And then, in 2020, Stern brought out Stranger Things, a pinball machine with a video display on the playfield that interacted with the ball. Virtual pyrotechnics, real ball.

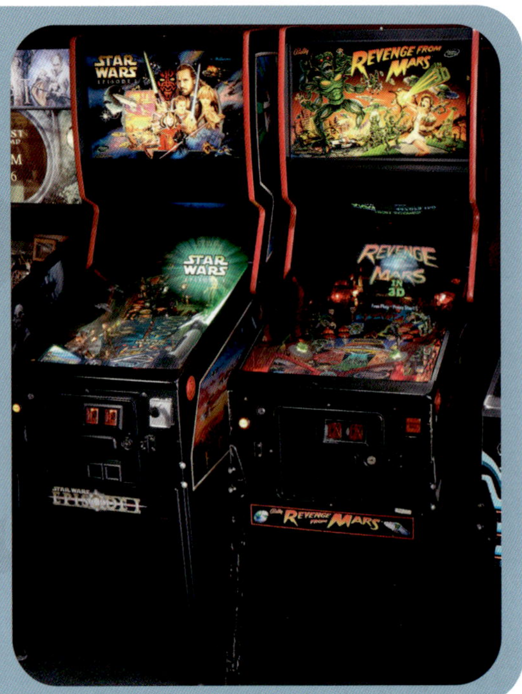

ARCADE GAMES

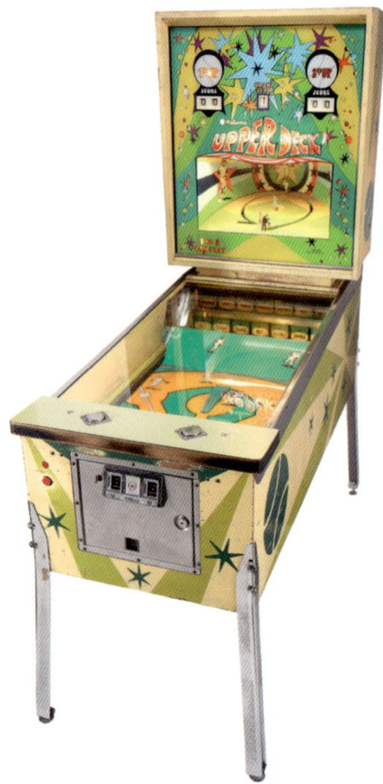

Williams **Upper Deck,** 1973. Pitch and bat arcade game. Note the baseball players inside the backbox, circling the bases.

Arcade games are physical action games and amusements: bowling games with real bowling balls, target games with real rifles, driving games with steering wheels and foot pedals, pitch-and-bat baseball games, cranes, fortune tellers, miniature airplanes flying in circles, and all kinds of other novelties designed to catch your attention and your quarters.

Arcade games are obviously not pinball machines. But some of the mechanical action parts and electro-mechanical circuits in arcade games are similar enough to pinball machines that someone who knows how to fix a pinball machine can probably fix an arcade game. Arcade games and pinball machines are soul mates.

Commercial video games are sometimes called arcade games, but here I am referring to games that have actual physical movement, not to video games.

CHOOSING A PINBALL MACHINE

I could fill an entire book describing the amazing variety of pinball machines that have been manufactured since the flipper was invented. In fact, my publisher Schiffer has produced a dozen books on pinball machines—big full-color beauties that will be a huge help in deciding what kind of pinball machine, what vintage, what design, and what features you'd like to own.

One thing common to all pinball machines, other than cocktail table machines, is their size. Most pinball machines are about 4½ feet long, two feet wide, and nearly six feet tall, and weigh, depending on their design, 100 to 200 pounds.

VINTAGE OR MODERN?

The terms vintage and modern mean different things to different people. For pinball machines (those with flippers), there were two distinct periods in the evolution of the machines: the electro-mechanical era (1947–1976) and the electronic era (1976 to today). This is the dividing point between what many people call vintage (electro-mechanical) and modern (electronic). Still, many lovers of early electronic machines consider them to be "vintage." Vintage, like beauty, is in the eye of the beholder. And confusion over terminology is never helpful. So we'll refer to pinball machines as either electro-mechanical or electronic. The pinball machines of the two eras are similar in many ways, but the technology of the machines are dramatically different.

ELECTRO-MECHANICAL MACHINES

Up until 1976, all pinball machines were electro-mechanical. The operation of the game, the scoring, the bonuses—everything—was accomplished by electro-magnetic coils that opened and closed switches that performed the tasks needed to make the machine work. The coils themselves were activated by switches on the playfield, switches that the ball hit or rolled over or landed on; or the coils were activated by switches on other coils. Real bells and chimes would be struck by a metal arm or plunger activated, like everything else in the machine, by a coil.

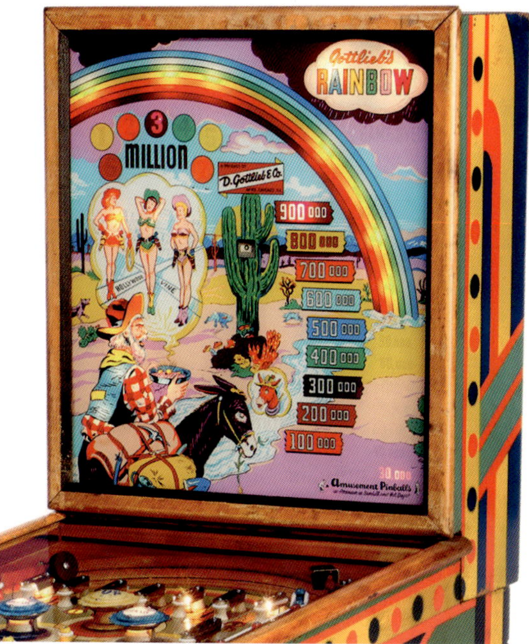

Gottlieb **Rainbow**, 1956, with Gottlieb's old trademark on the bottom of the glass: "Amusement Pinballs, as American as Baseball and Hot Dogs!"

Electro-mechanical scoring was accomplished by score reels, with the numbers 0 through 9 on each reel. Three or four reels were mounted next to each other in the backbox (the head, the upright part of the cabinet). On very old machines from the 1950s and earlier, scoring was accomplished by lights behind the backglass. The lights illuminated painted score numbers on the backglass. On these old machines most of the backglass was taken up with painted numbers.

The last electro-mechanical pinball machine was made in 1978. Although these machines are quite old, the machines can and do run almost forever if they are maintained. Non-working electro-mechanical machines can usually be resuscitated if they haven't been damaged or cannibalized for parts.

1940S AND 1950S MACHINES

From 1947—when the first flipper machines were being manufactured—and through the 1950s, pinball manufacturers were experimenting with different designs for flippers, relays, rotating units, and other electrical and mechanical components. I don't know if "experimenting" is the right word. The manufacturers would try out components and designs that turned out to be unreliable, or clumsy, or difficult to keep in good working order, or too expensive to produce. So the manufacturers would redesign the components, and redesign them again, trying to create a reliable, cost effective product. Finally, about 1960, components were being manufactured that worked, were easy to maintain, and were not expensive to produce. From then on, until the new electronic era started the trial-and-error process all over again, all manufacturers used similar components. Relays, rotating units, and flippers all worked very much alike. A lot of pinball parts were interchangeable, even from different manufacturers.

For these 1940s and 1950s machines, the basic circuits and how components operated are similar to machines manufactured in the 1960s and 1970s. But adjusting and repairing 1940s and 1950s components—"first generation" relays, rotating units, and flippers—can present challenges that require some genuine creativity on your part, patching or rebuilding components that not only are no longer being produced, they never worked well when they were new; or possibly adapting newer generation components to work in the old machines.

I have worked on machines from the 1940s and 1950s. I find it fun to restore these old beauties, but boy, it can be an adventure. A lot of trial and error. Still, if you take your time, use this manual to learn how circuits are connected and how components work, and are prepared to experiment a bit, you can give new life to old machines—even a better life than they originally had.

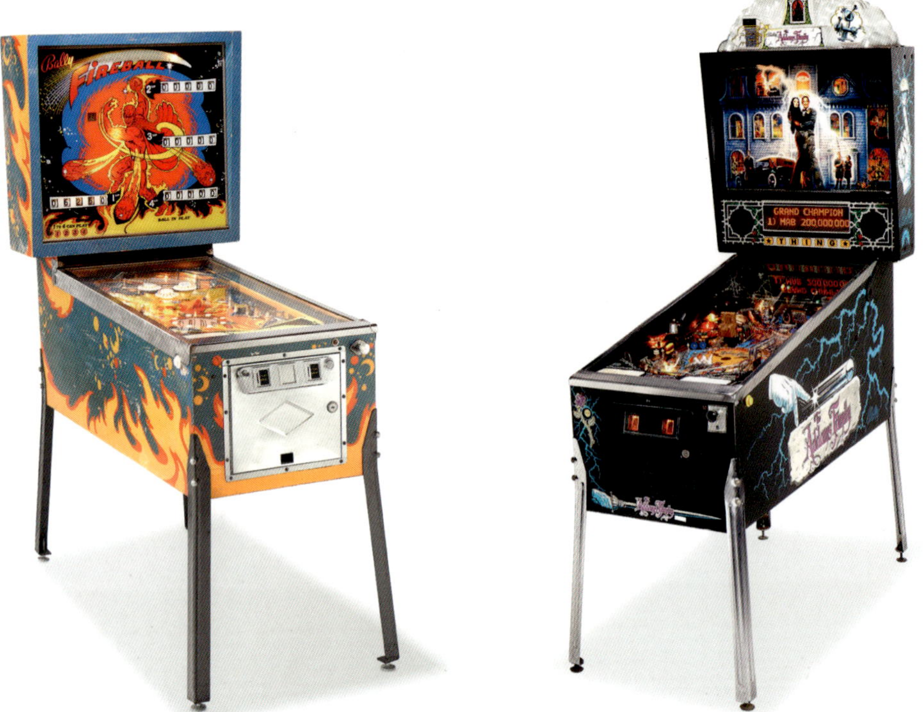

The most popular electro-mechanical pinball machine ever made, Bally **Fireball**, and the most popular electronic machine ever made, Bally **Addams Family**, created 20 years apart (1972 and 1992).

ELECTRONIC MACHINES

Electronic machines—also called digital and solid state—first appeared in 1976. The operation and scoring is controlled by computer chips (ICs), transistors, and other solid state components. Scores are shown on electronic displays. The first electronic machines still used real chimes, but it wasn't long before all electronic machines offered digital sounds and music and, eventually, talking.

A more important difference between electro-mechanical and electronic machines is that electronic machines are usually much faster paced than electro-mechanical machines. Electronic flippers, pop bumpers, and slingshots are more powerful than those on most electro-mechanical machines. The extra power is because electronic machines use DC (direct current) power to operate the flippers, bumpers, and other playfield action, while the electro-mechanical machines used AC (alternating current) power. Without getting into electrical theory, DC power is stronger than AC power—more kick per volt.

Some of the last electro-mechanical machines used DC power for flippers and pop bumpers, but still retained the old relays-and-switches technology for everything else in the machine. You can convert AC powered flippers and bumpers on old electro-mechanical machines to the more powerful DC voltage. This is covered under "Converting to DC Power" in the "Repairs: Flippers" chapter.

Electronic components are much more delicate and much less forgiving than electro-mechanical components. The transistors, diodes, and ICs that control electronic machines can be damaged very easily. Electro-mechanical machines are darn near indestructible.

Time has not always been kind to older electronic machines. Digital components, such as capacitors and diodes, give out over time. You cannot depend on a thirty-year-old circuit board. There is a good chance it will need rebuilding, or replacement if anyone is manufacturing new boards for your old machine.

Be especially wary of the first electronic machines made by the Gottlieb Company, machines made in the late 1970s and early 1980s, and known as System One or System 80 machines. These machines have a reputation for failure and are considered difficult to repair. New and better designed CPU boards are available for many of these machines.

Most important, troubleshooting and repairing the electronics in modern machines requires a knowledge of digital circuitry. Electro-mechanical machines have circuits that often can be diagnosed and repaired by just about anyone who can operate a soldering gun.

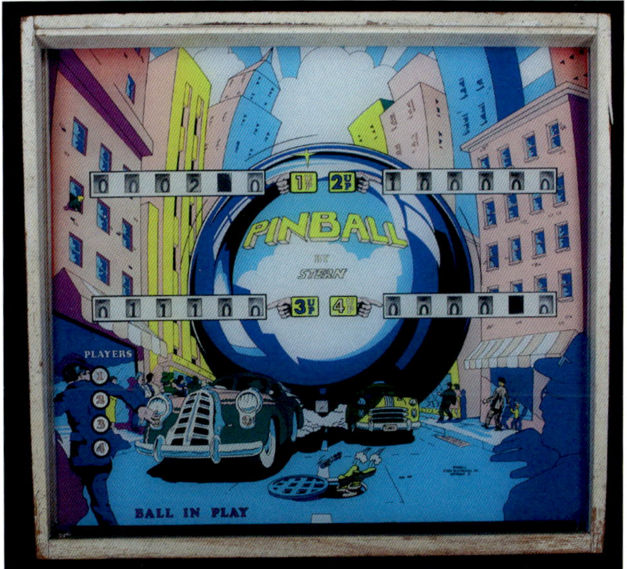

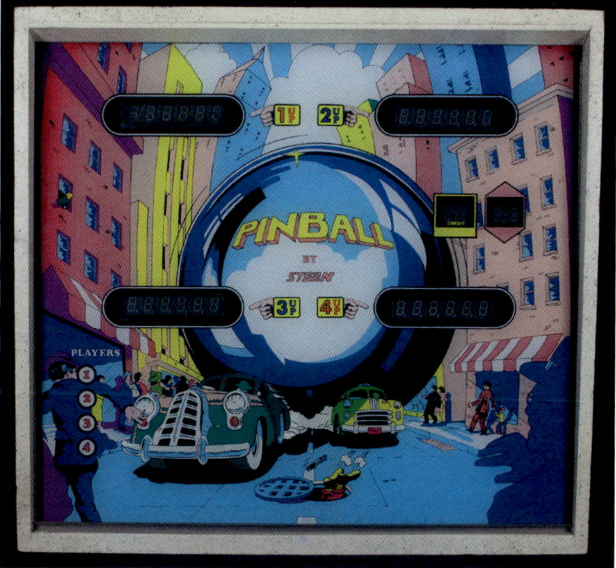

Stern **Pinball**, 1977, was manufactured as both an electro-mechanical and an electronic machine. Same game, different electronics. From 1976 through 1978, several manufacturers made two version of some machines, mostly electronic, but also a few electro-mechanicals for the "old school" holdouts.

You can tell at a glance if a machine is electro-mechanical or electronic simply by looking at the scoring mechanism visible in the backglass. All electro-mechanicals used score reels or, if made in the 1940s and 1950s, painted numbers that light up the score. All electronic machines use digital scoring.

NEW OR USED?

Obviously you can't buy a new electro-mechanical pinball machine—they haven't made any in more than forty years—so this question is for people considering an electronic machine. There are four major considerations:

Cost: For starters—and I imagine for many people, finishers—new pinball machines are expensive. They start at about $5,000 and can go up to $12,000! The cost of a used pinball machine typically runs from a few hundred dollars to a few thousand dollars, depending on age and condition.

Condition: A new pinball machine is a sparkling clean beauty. A used machine has been played by hundreds of people and may have taken some hard knocks along the way. The years may show, or may not. There are many fine looking old machines out there.

Complexity: New pinball machines are high tech "computers with flippers" running on sophisticated software. The typical pinball owner cannot repair these machines, although there are adjustments and maintenance that anyone can do. Older machines, even some of the older electronic machines, are easier to work on, and the inner workings easier to understand and repair.

Design: Almost all new pinball machines have licensed themes: a famous movie, or television show, or rock star, complete with music from the performers or spoken dialogue recorded by the actors. Older machines also had themes, but the subjects were unlimited in their variety and less likely to look dated as the years, and the popularity of the personalities, passed along.

> Some people insist that new is always best. The connoisseur with a big entertainment budget can have the newest, hottest pinball in town right in his or her own living room. But in addition to the high cost, a new pinball is inherently lacking in the appealing character of an older machine. That classic quality comes from responding to a thousand different hands, from winking back at countless pairs of eyes that have followed the steel ball in its swift adventure.
>
> —Jim Tolbert, *Tilt*

THEME AND DESIGN

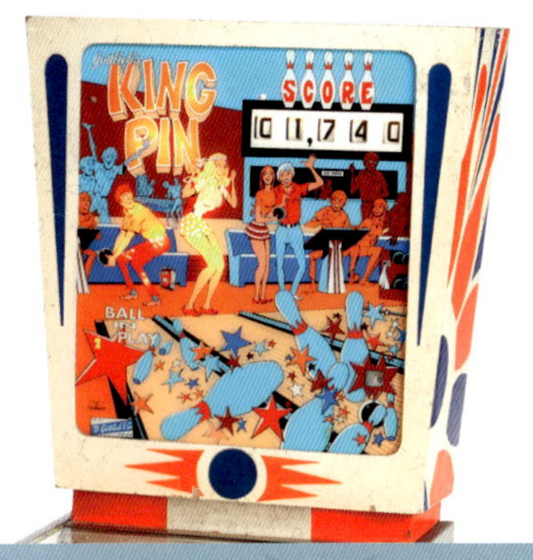

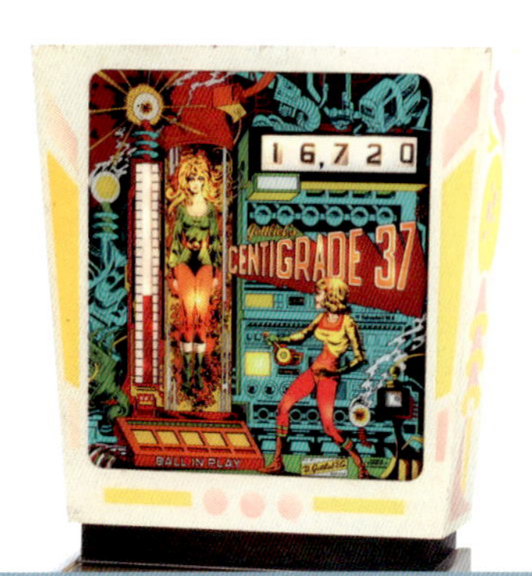

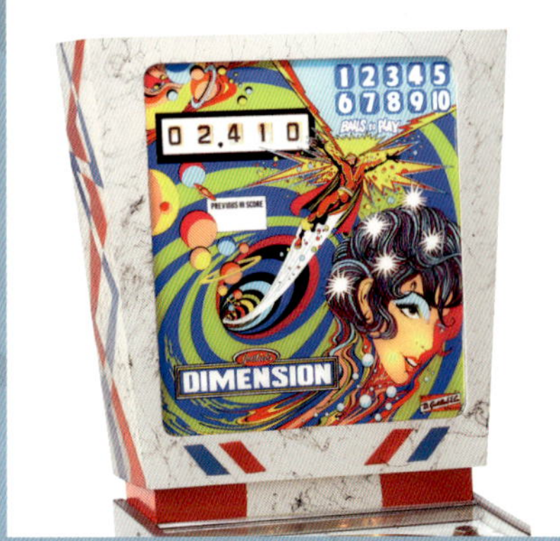

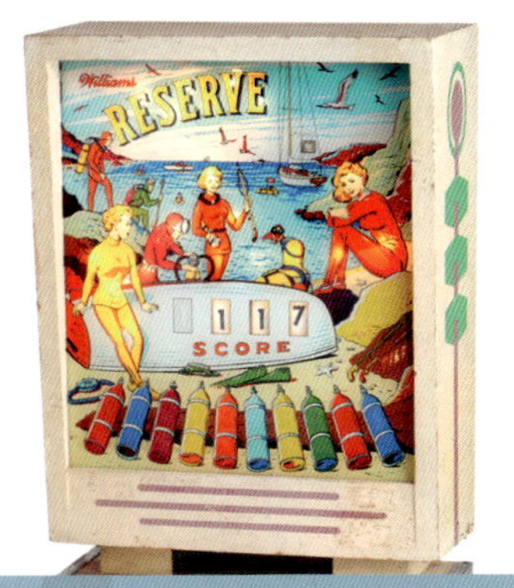

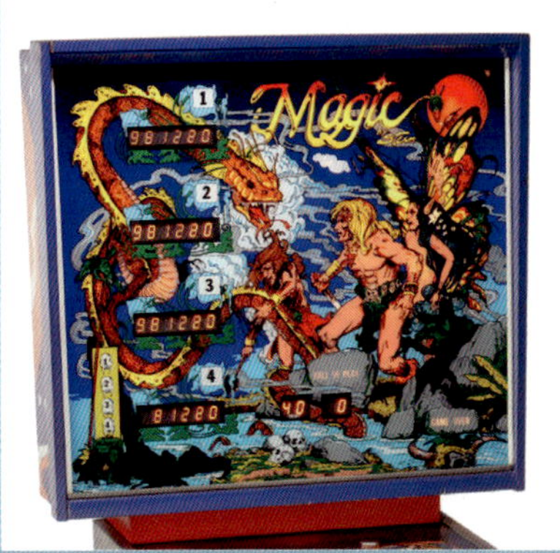

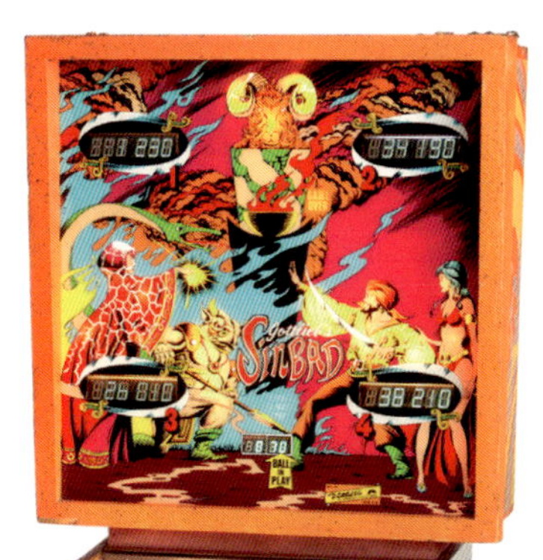

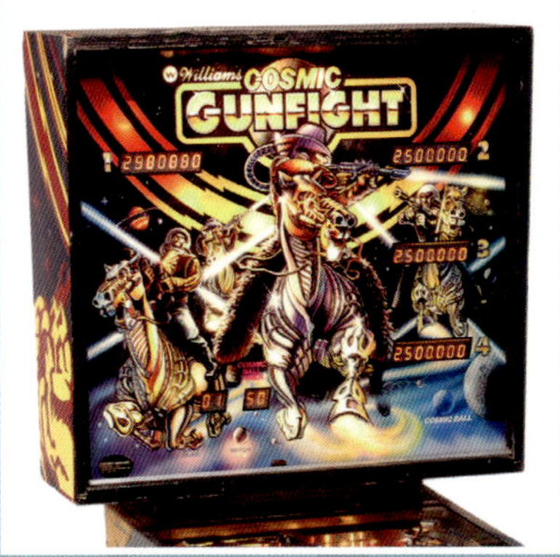

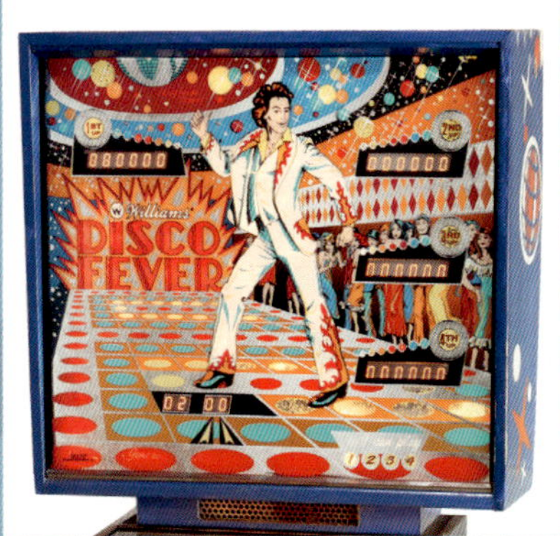

Gottlieb **King Pin**, 1973. Gottlieb **Centigrade 37**, 1977. Gottlieb **Dimension**, 197. Williams **Reserve**, 1961. Stern **Magic**, 1979. Gottlieb **Sinbad**, 1978. Williams **Cosmic Gunfight**, 1982. Williams **Disco Fever**, 1978.

Are you looking for—or looking to avoid—a particular theme or type of artwork? Do you want a machine that features space ships, or car races, or people playing pool, or pretty girls, or a famous rock and roll band, or Indiana Jones? You'll be staring at the machine for a long time, so pick one that is pleasing to your eye and taste, and to that of your family.

Thousands of games have been manufactured over the course of pinball history, each one a little or a lot different than the next. Don't bring home a game you don't like the look of.

THREE BALL, FIVE BALL, AND MULTI-BALL

Through the 1960s, pinball machines had five balls per game. Starting about 1970, many machines offered the option of a five-ball or a three-ball game. If the operator was greedy, he could simply move a plug and your quarter would suddenly only get three balls. By the 1980s all games went to three-ball play, although many electronic machines have an adjustment that allows five-ball play if you want it.

Most newer electronic machines offer multi-ball play, where you can have more than one ball on the playfield at a time: You hit a bonus or jackpot and suddenly you're trying to keep two, three, or four balls on the playfield at the same time. Some people love the multi-ball feature and some people find it no fun to have that many balls in play all at once. Most electro-mechanical and early electronic machines don't offer multi-ball.

SINGLE PLAYER VS. MULTI-PLAYER

"It's more fun to compete," the Gottlieb Company used to advertise on its multi-player pinballs. Multi-player machines allow one to four players to play a game, each player getting a ball in turn. The first player gets a first ball; then the second, third, and fourth players (if there are additional players) each get their first ball before the first player gets a second ball.

Electro-mechanical machines were manufactured as single-player, two-player, or four-player games. A few machines were six-player machines, but they are very uncommon. Almost all electronic machines except the "home" models are four-players. On electro-mechanical machines and early electronic machines, you can easily tell the number of players by looking at the number of player scoring windows or displays on the backglass. On newer electronic machines that have one combined display, all of the machines are four-player.

If you are buying an electro-mechanical machine, you will have to decide whether you want a single player or multi-player machine. While multi-player machines do in fact make it "more fun to compete," single player electro-mechanicals often had a more interesting game play. On single-player machines, building up bonuses, or trying to hit all of the targets, or light all the numbers could progress from ball to ball. You had all five balls to try to complete a sequence and "beat" the game. On most

A ONE-OF-A-KIND PINBALL MACHINE

Every pinball machine I've ever seen was either 5 balls per game or 3 balls per game, with one exception.

In 1972, Williams produced a machine called Travel Time (also called Summer Time in a slightly different version). Travel Time was a timed game, with the length of play controlled by a playfield clock. You had an unlimited number of balls, but you had a limited time in which to play. Certain targets would stop the clock from ticking down and certain targets would extend the time, adding an extra 10 or 25 seconds.

The challenge to "beat the clock" and rack up maybe ten or twenty balls in one game was a fun innovation. But most players, expecting five balls, were confused by Travel Time. The game did not earn a lot of money for route operators. Alas, the "timed" game was never repeated. Travel Time became a once-ever "no-other-like-it" pinball machine.

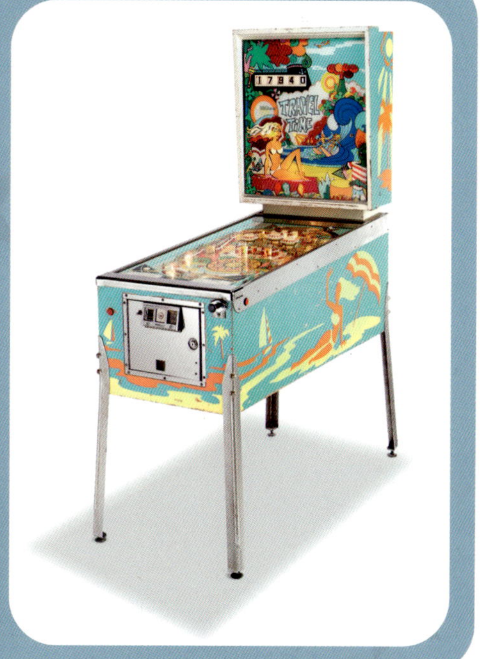

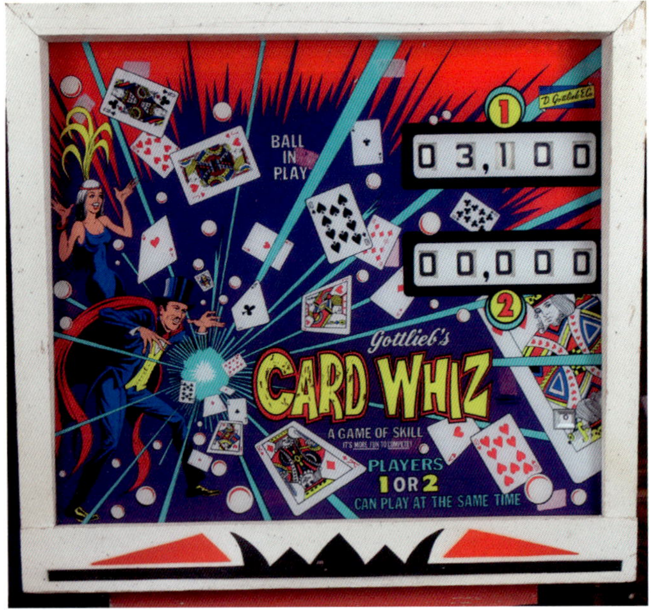

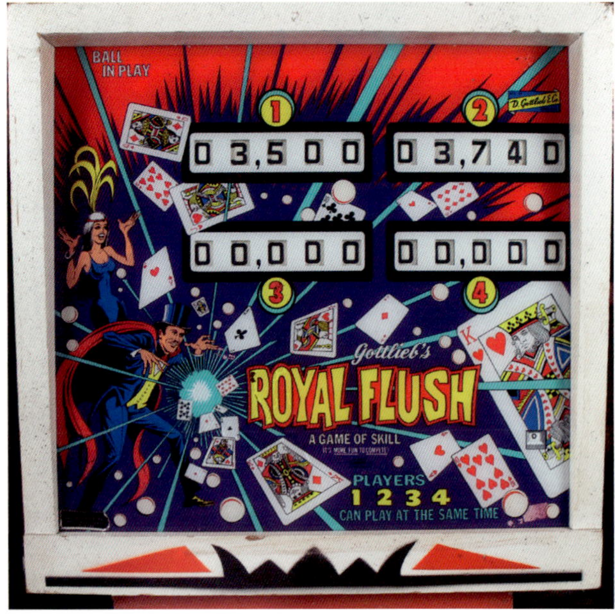

Gottlieb **Card Whiz** and **Royal Flush**, 1976. Many electro-mechanical machines were made in two-player and four-player versions. The machines were identical except for the number of players—and the loss of backglass art to make room for the additional score reels. The machines always had different but similar names.

multi-player electro-mechanical machines, the bonuses and targets reset after every ball, which is fun, too: you have to accomplish *a lot* with just one ball.

Single player electro-mechanical machines are easier to work on than two-player and four-player machines. Single-player machines do not have a player step-up unit, a complex apparatus that controls which player is up and which ball is in play—and that often needs fixing. Single player machines have only one set of score reels: three or four reels instead of eight or twelve or sixteen reels to clean and adjust.

ADD-A-BALL

Some machines are "add-a-ball" games. When you get up to a certain score or light the Special, instead of getting a free game, you get an extra ball. Instead of racking up a bunch of free games, you might get six, seven, eight, or more balls in one game.

When you own your own machine you are already getting all the free games you want, and winning a free game sometimes loses its excitement. Getting extra balls instead of free games can add tremendously to the fun of playing.

Don't confuse the add-a-ball feature with the more common "extra ball" feature. Many pinball machines award an extra ball if you hit a certain target or complete a series of plays, but when you hit the Special, you still get a free game, not another extra ball. If your machine has the add-a-ball feature, you get an extra ball not just when you win the "extra ball," but also when you hit the Special.

Most (but not all) electronic machines have an adjustment that converts the game to add-a-ball. Your machine's instruction manual will tell you if this option is available.

Some electro-mechanical machines have an add-a-ball feature and some don't. For electro-mechanicals, there are two kinds of add-a-balls: what we call "genuine" add-a-ball games, and games that have an option to convert to add-a-ball.

Genuine add-a-ball games, which are not common and are often difficult to find, allow you to get more than one extra ball per ball in play. You can rack up two, three, four, or more extra balls while playing a single ball. You can usually tell from the backglass if you have a genuine add-a-ball game. If the number of balls shown goes up to nine or ten instead of five, it's a genuine add-a-ball. Genuine add-a-ball games indicate "Balls Left to Play" instead of "Ball In Play," although some replay machines also indicate "Balls Left to Play." (The manufacturers referred to machines that award a free game as "replay" machines to distinguish them from add-a-ball machines.)

Electro-mechanical machines that offer the option to convert to add-a-ball, as opposed to a genuine add-a-ball, will only give you one extra ball per ball in play. You can tell if you have a machine that converts to add-a-ball by looking inside the backbox, or on the inside bottom of the cabinet for a plug with two options—usually marked "Extra Ball" and "Replay"—that converts the game to add-a-ball (the "Extra Ball" option). Your machine's instruction manual will also let you know if the game offers an add-a-ball feature.

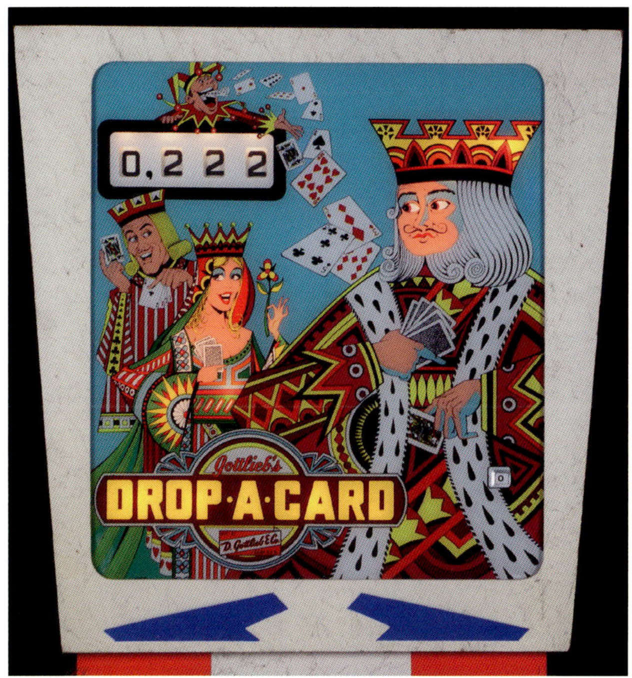

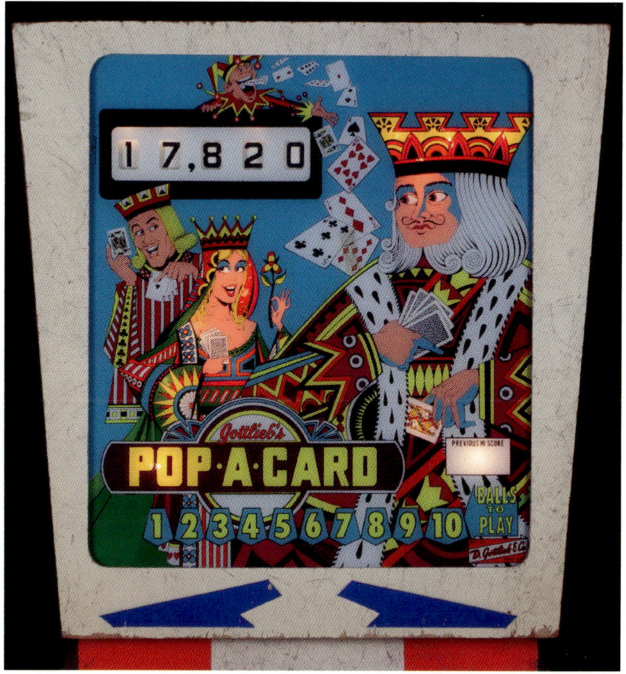

In some cities, overzealous city fathers decided that pinball machines were gambling games (because you could win a free game!) and outlawed them. So the ever-clever pinball manufacturers created add-a-ball versions that awarded extra balls instead of those illegal free games. These are Gottlieb's **Drop A Card**, 1971, that awarded free games, and **Pop A Card**, 1972, that awarded extra balls. Note the numbers 1–10 at the bottom of the backglasss on Pop A Card.

Add-a-ball is a very desirable feature in a game. Many people who own add-a-ball games don't even know they have this option. Rarely does the add-a-ball feature increase the cost of a machine, although it adds greatly to its value.

WHERE TO FIND A PINBALL MACHINE

Old pinballs are everywhere. They're in barns and garages, and storage units and basements and bedrooms. They're in antique stores and second hand stores. They're in the back rooms of arcades and bowling alleys and amusement companies that put machines out on routes. Some machines are in great shape and some are in just horrible shape. Some are expensive, some are very affordable, and some are free—"Just haul this thing away."

You are likely to find a much better price on a machine if you buy it from an individual than if you buy it from a business. Many individuals who are selling a pinball machine have played it for several years and have lost interest, or the kids went off to college and it's sitting there, or they need the space, or they're moving. These machines are often well taken care of and usually go for a reasonable price. However, there is a good chance that something won't be working on the machine, and that it will need repair or adjustment. After you read this manual, you may discover that it is a minor repair, one you can do yourself.

If you are looking at a machine that needs repair and you can tell what's wrong and how to fix it, lucky you. A non-working game sells for a whole lot less than a working game, even if it's a five-minute adjustment.

If the machine isn't working and you don't know how to fix it, you have a problem. Pinball repair people, if there is one in your area, are not inexpensive. You can pay several hundred dollars to get a machine repaired. Repairs are often more expensive than buying a working machine.

Antique store owners think pinballs are like other antiques: valuable because they are old. Antique stores tend to ask high prices for pinballs, and chances are the machines aren't working either. I've never found a good deal on a pinball in an antique store, but it's still fun to look. And antique store owners often know someone who has a pinball machine for sale.

Companies and individuals who regularly sell and repair pinballs also charge high prices for pinball machines, but when you buy a machine from these people, you should be getting a machine that has been "shopped": gone through entirely, repaired, adjusted, cleaned, waxed, new balls and rubber rings, and all the lights working. You should also get some sort of guarantee—at least 30 days. By the way, be aware that the term "shopped" means different things to different people. If someone tells you a machine has been shopped ask what "shopped" means.

Arcades and route operators often sell older machines—some working, some not, sometimes guaranteed, but usually not. These people will sometimes remove the coin mechanism from the machine before selling it to be sure no one will buy the game and then compete with them! (Don't worry if the coin mechanism is gone; you can easily set games for free play.)

You can find pinballs at auctions, though this is risky, because you may not be allowed to turn on the machine or even look inside it.

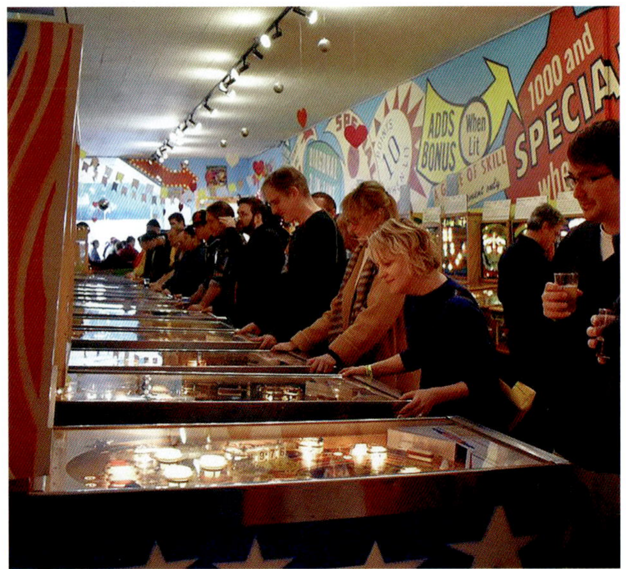

A great place to find pinball machines is at a pinball swap meet or pinball show. All over the country there are pinball gatherings where people bring machines to show off, to play, and to sell. Usually held at some community hall or county fairgrounds over a weekend, there may be anywhere from a few dozen to several hundred machines, all of them set up for free play. For a daily admission of $20 or $25, you can play every machine in the place as often as you like, find one you love, negotiate the price with the owner, and take your prize home with you. Pinball shows are also a great place to find people who repair machines and sell parts.

PINBALL MACHINES ON THE INTERNET

There are many pinball machines for sale on the internet. People do buy machines sight unseen and have them shipped from clear across the country. Well, as Queen Isabella told Christopher Columbus, *good luck*. It's hard to tell what a machine looks like and what condition it is in without seeing it in person. Photographs are often misleading; they always look better than the machines themselves. Sellers are often misleading; they do not accurately describe the condition of the machines they are selling.

Shipping is risky. Machines get damaged in transit. Machines that were working when they left Iowa are not working when they arrive in California. In my own shop we've often repaired what I call "eBay machines," pinball machines purchased over the internet that not only were not working, but they obviously were not working before they were shipped (missing parts, burned fuses), despite the seller's assurance that the machines were in perfect working condition.

You can use the internet to get an idea of what people are asking for certain machines, but you'll find that prices on the internet, like prices in the real world, are all over the board.

You can use the internet to find out what different machines look like. There are a few internet pinball sites that have photographs of hundreds of pinball machines. One site, the International Pinball Database (ipdb.org), probably has the most complete listing of pinball machines.

WORKING OR NOT WORKING?

Most people want a pinball machine that is working. But any used pinball machine, unless it is a refurbished one from a pinball shop, is likely to have something wrong with it. There are so many facets to a pinball machine, it's almost guaranteed that something is not going to be working, or not working well. And it is very likely something you can fix yourself.

Find out for yourself if the pinball machine works. Don't take the seller's word for it. I mention this elsewhere in the book, but be aware that sellers often tell buyers that a machine is working perfectly when in fact the machine isn't working right, or isn't working at all. The seller is always "completely puzzled" at this discovery. I can almost write the script I've heard it so many times: "It never did that before." "It was playing great yesterday." Then, always, "It's got to be something minor." Which it often was, but the seller has no idea what's wrong, or what to do about it. And suddenly the price comes waaaayyy down. While working machines can fetch big bucks, non-working machines often sell for next to nothing, and sometimes for nothing at all. I've often paid $50 or $100 for machines that need work. Over the years I've been offered enough free non-working machines to fill the entire shop!

Always turn on a machine and play it before going any further.

HOW MUCH WILL IT COST?

Asking how much a pinball machine costs is like asking how much a car costs. So much depends on age, condition, how many buyers are out there, and how eager the seller is to sell. You can pay as much as $12,000 for a brand new pinball machine right out of the crate from the manufacturer. You can find junker pinball machines for as little as $25. Decent looking, older used machines in good working order, particularly electro-mechanical machines, typically sell for anywhere from $300 to $1,000. Electronic machines from the 1990s and newer can fetch anywhere from $1,000 to $5,000. A lot depends on the age and condition of the machine, but a lot also depends on the seller.

There is an old expression in the rural area where I live: "You can always buy a goat, but you can't always sell one." There is not a huge market for goats. Or for pinball machines (I've owned both).

And like goat owners, people who are selling pinball machines often need to get rid of them. The people are moving, or remodeling the house, or maybe the kids have moved away and the machine is sitting there unused, taking up valuable space. Sellers are often very eager to get the machine out of there, and the sellers may soon discover that there are no immediate buyers.

So if the price is not right, you don't have to haggle like a used car salesman, and you don't have to make a counter offer unless you like that kind of give-and-take. Just walk away from the deal. Tell the seller something like, "Thanks, but that's way out of my price range," and give the seller a way to contact you if anything changes. There's a very good chance you'll get a call in a week or so with a much better offer. And if you lose out and someone else gets the machine that you now wish you bought, don't worry; it wasn't the right machine. There's another machine out there waiting for you, and better and less expensive, too.

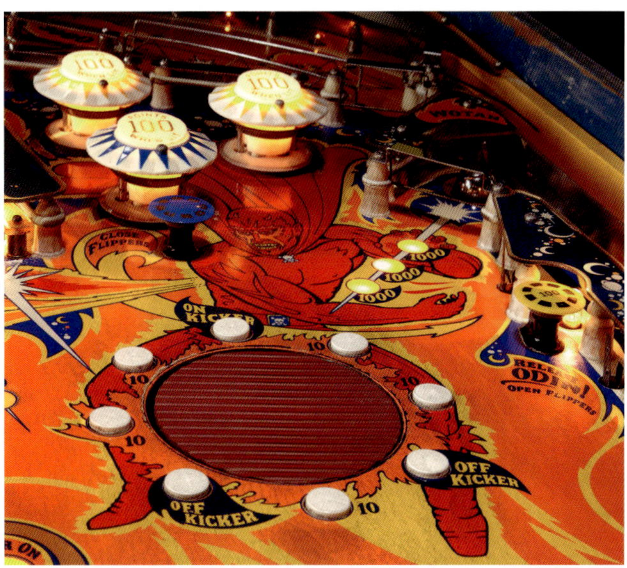

Bally **Fireball**, 1972. The large disk on the playfield, a feature of several Bally machines from the 1970s, spins constantly the entire game, causing havoc when a ball tries to cross it. And totally irritating if you have to work on the machine in play mode. I wired an off-on toggle switch to the disk on my machine.

EXAMINING A MACHINE

It's real easy to get excited about a pinball machine and miss some important detail or problem that you should consider before you buy it.

PLAY A GAME

Before you go any farther, turn on the machine and play it several times. Is it fun or is it boring? Is it challenging? Every machine is different. Some are fast paced, some slower. Some require a lot of skill. Some are loaded with gimmicks. Some have complex scoring, some very simple scoring. Some have the extra ball feature, some don't.

You are going to be playing this machine probably hundreds of times, so look for a game that has long term appeal.

PHYSICAL CONDITION OF THE MACHINE

The cosmetic condition of the machine—the backglass, playfield, and the cabinet—affects the value of a machine more than anything else. Most old machines will have a little missing paint, some faded colors, and minor scratches and dings. But some machines are badly worn and should be avoided, unless you don't really care and the price is too good to pass up.

What you want to avoid is someone's amateur touch-up job. Touching up the paint on an old playfield, or cabinet, or backglass is a real challenge. It requires not just skill, but the ability to mix the correct paints to achieve matching colors. Too often the results make the machine look worse.

PHYSICAL CONDITION: BACKGLASS

Most important is the backglass. It is the most dramatic and most visual part of every pinball machine. All machines until the late 1980s had painted (silkscreened) backglasses.

Backglasses may be difficult or impossible to replace, and they are next to impossible to repair. If the backglass is broken or missing, the machine has little resale value.

If the paint is partly scraped away or missing in spots the machine may still be attractive and valuable. But if the paint is bubbling off or peeling off, the damage is irreversible, the value is decreased. You can't unbubble bubbled paint.

Starting late in the 1980s, pinball manufacturers switched from painted backglasses to translites: flexible printed plastic sheets mounted behind clear glass. The translites can fade if left in bright sunlight and they can be damaged if a hot backglass lamp makes contact. Other than that, translites show excellent lasting power. There is no paint to fall off. For more information see "Backglasses" and "Translites" in the "Things to Know" chapter.

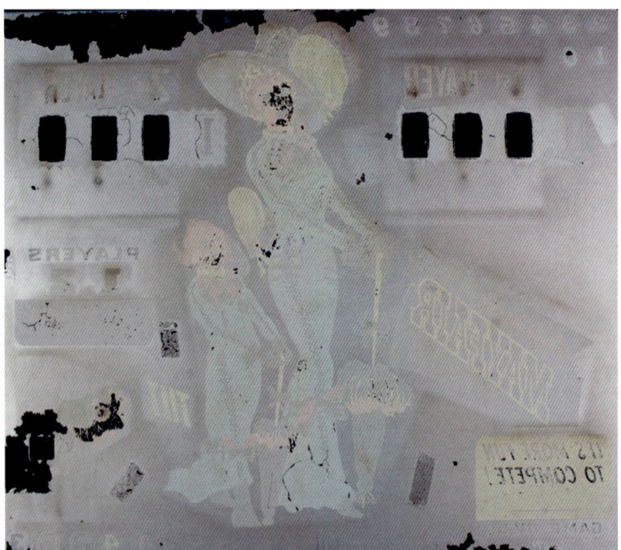

A sure way to determine the condition of a backglass is to remove it from the machine or get behind it and examine the back of the glass. Any missing or scraped or bubbled paint, and any attempts to patch paint damage, will be obvious. On this poor Gottlieb **Mayfair** backglass, the small black marks are bits of missing paint, and the large black marks (other than the score windows) are big chunks of missing paint.

PHYSICAL CONDITION: PLAYFIELD

Most older machines will have some paint missing from the playfield after years of that pinball rolling all over it, particularly around pop bumpers and kick-out holes. Paint is often missing where flippers, set too low, have rubbed on the playfield (an easy fix covered in the "Repairs: Flippers" chapter). The more paint that is missing, the less attractive and less valuable the machine.

Old playfields sometimes have faded paint, usually caused by exposure to direct sunlight over a period of time. There is no fix, so you will have to decide if the machine is attractive as is.

Old playfields are often dirty. Some will clean up nicely and sparkle again and some will not. It's hard to tell, until you start cleaning, how well a playfield will clean up.

Avoid any machine that has physical damage to the playfield: gouges, any signs of water damage, or worn down areas near pop bumpers, flippers, and kickout holes. This kind of damage is difficult to repair.

Replacing a worn or damaged playfield with one in better condition is not an option for most people. The replacement would have to come from another machine being stripped for parts, or from someone's stash of never-assembled "new old stock" parts—both unlikely to locate. And then you'd have hours of work removing, reassembling, and rewiring all the components.

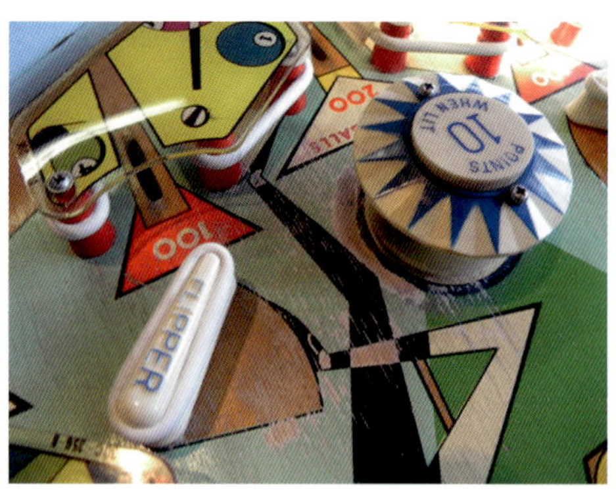

PHYSICAL CONDITION: CABINET

A cabinet that still has the original paint in decent condition is more attractive and more valuable than a cabinet with faded and peeling paint, or one that has been gouged, or ornamented with the names of long-ago high school lovers or expletives deleted, or one that has been repainted.

Check the physical condition of the cabinet and backbox. Most pinball cabinets are plywood, but some cabinets are partially particle board that is often crumbling and chipped at the edges.

Avoid a machine if the plywood is rotted, if the laminations are peeling, or if the seams are coming apart, unless you are good at cabinet repair. If you see any signs of water damage take it as a serious warning. A machine that has been in a flood may be damaged beyond repair.

I've heard warnings, particularly from the southeastern United States, about termite infestation in pinball cabinets. Termites can live inside the wood, eating it from the inside out. If your finger can push in the wood, or you feel a hollowness to it, you may have an unrepairable termite hotel.

PHYSICAL CONDITION: PLASTICS AND BUMPER CAPS

Examine the playfield plastic shields (light shields), the decorated flat pieces of plastic that cover lamps and hide mechanisms on the playfield—commonly called the "plastics." Each game has a different set of plastics, and no two games are alike. If the plastics are missing, broken, or badly warped you may be searching for years for replacements.

Broken or missing bumper caps are easy to replace. Many games used identical or similar caps, and new reproduction bumper caps are available for most machines.

For more information see "Playfield Plastics" in the "Components and Features" chapter.

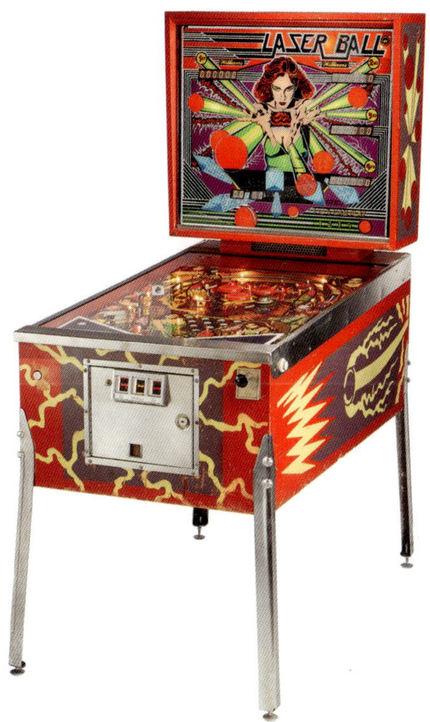

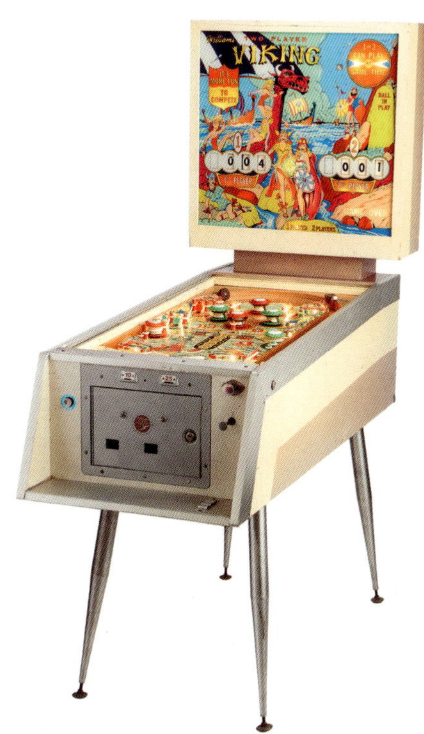

Two dramatically different cabinets found on only a few pinball machines. Williams **Viking**, 1960, featured their "Styling of the '60s" tubular legs, a shelf for drinks, and a cigarette holder. Williams **Laser Ball**, 1979, is a "widebody" machine, several inches wider—and a lot heavier to move—than typical pinball machines. It helped to have long arms to work the flippers.

PHYSICAL CONDITION: PARTS

If there are broken or missing targets or flippers, these are still being manufactured and easily replaceable. Balls, rubber rings, and most electrical components are available.

The plastic bumper skirts on the bottom of pop bumpers are often chipped. Replacement skirts are available, but budget a half hour of work to disassemble and reassemble a pop bumper. Pop bumper repair is covered in the "Repairs: All Machines" chapter.

PLAYFIELD GLASS

Is the playfield glass (the large piece of glass covering the playfield) in good condition or is it all scratched up? Can you tell if the glass is tempered, which it should be, or is it a replacement made of window glass, which it should not be?

If the playfield glass is missing, check extra carefully that nothing on the playfield is also missing, particularly bumper caps that on some machines are held in place only by friction and may be laying on the floor somewhere.

Playfield glass is inexpensive to replace if you are near a pinball supplier and can pick up a glass. Shipping is likely to cost more than the glass itself.

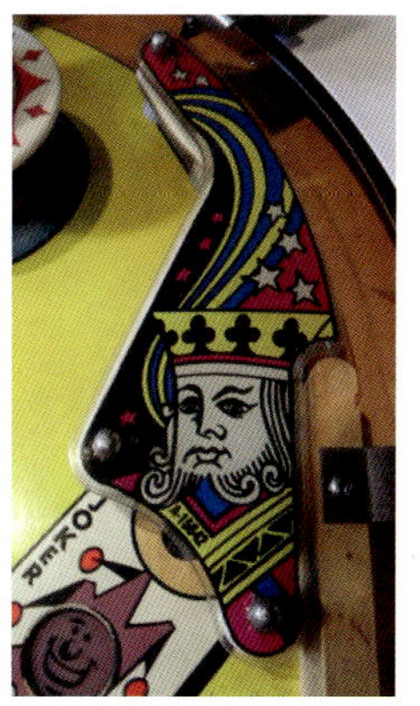

Playfield plastic from Gottlieb **Royal Flush.**

DISPLAYS (ELECTRONIC MACHINES)

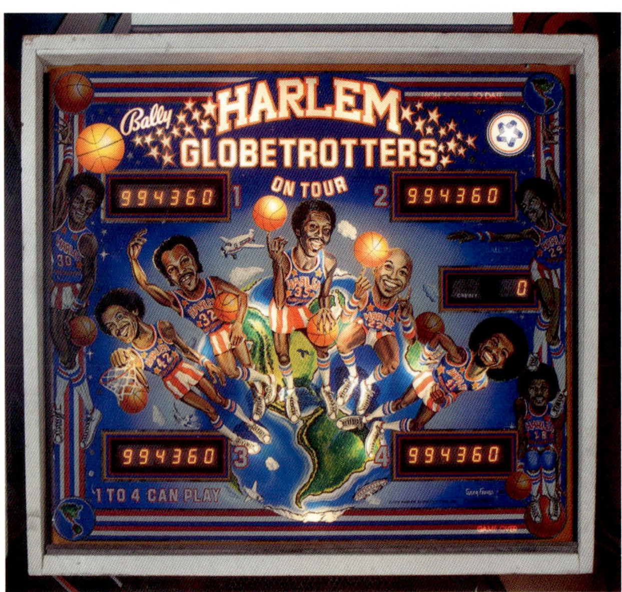

Bally **Harlem Globetrotters On Tour**, 1979.

Are the digital displays working correctly? Until the late 1980s, electronic machines had five or six separate displays: one for each player and additional displays for ball in play and match. Later electronic machines combined everything into one large display.

If a display is not working, or only partially working (parts of words or numbers missing), or displaying unintelligible gibberish, the problem may be something as simple as a bad connection, but more likely either a bad display or a circuit board in need of repair. Replacement displays and board repairs can get expensive. See "Digital Displays" in the "Repairs: Electronic Machines" chapter.

COIN MECHANISMS

Look inside the coin door. Are the coin mechanisms still in the machine? Most pinball machines have two coin slots, and each slot held a removable coin mechanism mounted on the back of the coin door. If you want coin operation, you will need one coin mechanism. The second mechanism was an extra, in case one jammed. Older electro-mechanical machines accepted dimes and quarters, and often had three coin mechanisms. If you don't care about coin operation the machine can easily be set for free play, and you don't need the coin mechanisms.

Don't confuse the coin mechanism with the mounting hardware that holds the mechanism. The hardware is bolted to the coin door. The mechanism is a brass or plastic unit that snaps in and out of the hardware.

You can learn more about coin mechanisms and converting a machine to free play in the "Components and Features" chapter.

WHAT ELSE IS INSIDE THE COIN DOOR?

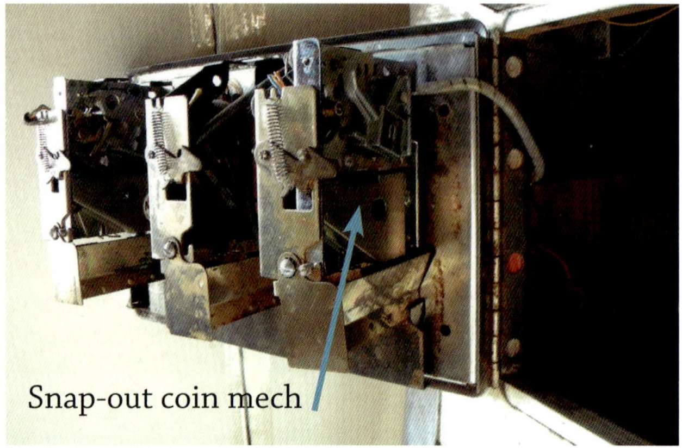

Look inside the machine. All new pinball machines came with a removable coin box that sat just inside the coin door. Is the coin box still there? Is there anything in the coin box or lying on the floor of the cabinet? You may find keys, paperwork, and extra lamps and fuses. You may also find screws, clips, old coils, and other pinball parts—parts that probably came off the machine, parts that maybe still should be on the machine. (You may find $20 in quarters!)

Have a look on the back of the coin door. Most pinball machines have a key hook mounted on the door. You may find keys hanging there.

BACK DOOR (ELECTRO-MECHANICAL MACHINES)

Electro-mechanical machines have a removable access door behind the machine. The door protects the machine from dirt, animals, and curious people. The door muffles the noise of the moving score reels and relays. The door (a panel actually, not hinged) was usually made of sheet metal, though some manufacturers used particle board.

The back door is secured by a lock, requiring a key that often turned up missing. You might find the key inside the main cabinet coin door, hanging from a hook on the door, or in the coin box, or on the floor of the cabinet. You might find that the lock has already been jimmied, requiring only a screwdriver blade to turn it. And you may just have to drill the lock out and replace it. See "Missing Keys" in the "Things to Know" chapter.

Sometimes the back door is missing. Replacement doors are not easy to find. Back doors came in many sizes and shapes. A sheet metal shop may be able to fabricate a door for you, or you can make one out of plywood.

Electronic machines do not have back doors. The inside of the backbox is accessed by removing the backglass (see "Getting Behind the Backglass" in the "Things to Know" chapter).

WIRING SCHEMATICS AND INSTRUCTION MANUALS

All new pinball machines came with a schematic. Schematics are wiring diagrams showing the electrical components of the machine, identifying wires by color, and giving fuse ratings. It is very difficult to trace wiring—to trace a non-working part of the machine to whatever is keeping it from working—without the schematic. There are similarities in schematics for different pinball machines, but each machine's schematic is unique.

Beginning in the mid-1960s, Williams and Bally starting printing instruction manuals for individual games. Gottlieb started printing instruction manuals around 1970. For electro-mechanical machines, the schematics and manuals were separate items. For electronic machines schematics and manuals were bound together.

The instruction manuals—also called owner's manuals, game manuals, service manuals, and for Chicago Coin machines, parts manuals—included playing instructions, adjustments, playfield diagrams, and parts lists. Some instruction manuals include the start-up sequence (which switches activate which relays to start a game), extremely helpful information when troubleshooting a machine that will not start. Manuals for electronic machines include testing and diagnostic procedures. In addition to the instruction manuals, and especially before manuals were created, operating instructions were often stapled to the inside of the cabinet, such as score motor settings on electro-mechanical machines and game adjustments on early electronic machines.

It is a real plus when the machine you are buying comes with the original "paperwork": the schematic, instruction manual, and the score and instructions cards (the two cards at the bottom of the playfield near the flipper buttons). If the owner of the machine doesn't know where the paperwork is, look inside the coin door. On electro-mechanical machines people often taped schematics to the inside of the back door.

Schematics and instruction manuals are essential documents to enable you to set up, adjust, and service your pinball machine. You will find a host of helpful and easy-to-understand information in the instruction manual. Anyone repairing your machine will need the schematic.

If the instruction manual or the schematic is missing you can purchase copies, and sometimes originals, for most pinball machines from pinball suppliers for $15 to $30. If the score cards are missing, replacements are not easy to locate, but you might find a copy to reproduce on the International Pinball Database website (IPDB.org).

When the paperwork is missing from a machine, you may be able to use that as a negotiating point when buying the machine.

OPEN AND EXAMINE THE MACHINE

Now that you've gotten this far, and if the owner of the machine will let you, remove the playfield glass (see "Opening Up the Machine" in the "Things to Know" chapter), raise up the playfield—remove the balls first so they don't come crashing down—and brace the playfield up with the pivoting wood or metal bar mounted on the right side of the cabinet.

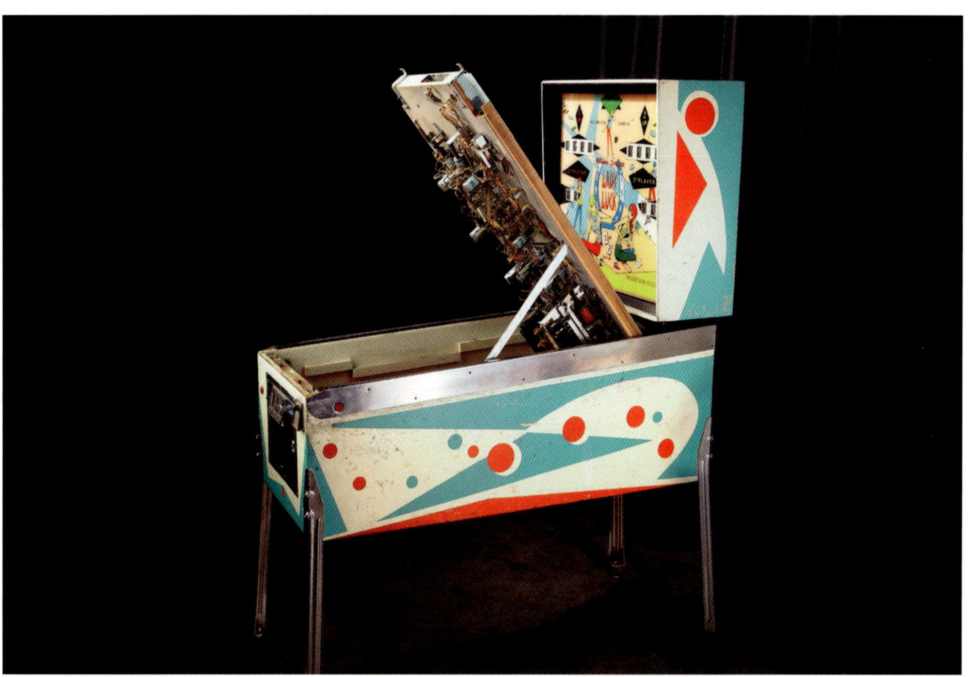

Williams **Lady Luck**, 1968, with the playfield propped up for repair. Over a span of fifty years (1937-1987) six different pinball machines were named "Lady Luck." Only one other name, "Circus," was given to six different machines, from 1948 to 1980.

Some new playfields stay upright without a support bar.

How clean is the inside of the machine? Is there any water damage? Old mouse nests? Any other obvious damage? Give it the sniff test. Stick your nose inside and if it smells like it's been wet, moldy, or mildewy, don't buy it.

Be wary of machines that have rusted or corroded components, and machines that are very dirty or greasy. Rusted coil plungers and dirty, sticky switches will require a lot of time to clean so the machine will play well. There are plenty of clean, well-cared-for machines out there. There is no reason to buy a poorly maintained machine unless maybe it's a give-away price.

Are there cut wires? Take a quick glance at all of the components (at least the components that are easy to see), looking for any with the wires cut off—the wires are most likely just lying there unattached. Look for any obviously missing components: wires leading to nowhere, an open space where there are screw holes in the wood, or possibly a lighter color to the wood than the surrounding area. Cut wires and missing components often indicate that the machine had a problem that someone "solved" by removing or disconnecting a part or a circuit. The problem is most likely still there.

Are there signs of amateur patch job rewiring? Wires of different colors connected together? Lone, unbundled wires that are obviously not original? Black electrical tape wrapped around wires? Again, someone tried to solve a problem and may or may not have succeeded. More than once I have discovered a rewiring job where the wires were reversed, causing even more problems.

Check the fuses. Be wary of blown fuses and any missing fuses. Why didn't the owner just replace the fuse? Maybe because as soon as the fuse was replaced it blew again. Blown fuses—look for a melted or burned element visible inside the glass—are usually caused by a short circuit (a "short"). The short circuit could have been caused by someone carelessly touching two wires with a screwdriver, not usually causing any damage. But a short circuit in the machine's wiring or in a component might be difficult and time consuming to trace. A short circuit is something every pinball owner should know about. See "Shorts /Short Circuit" in the "Repairs: Basics" chapter.

Look inside the backbox for any obvious problems. (Don't know how to open the backbox? See "Getting behind the Backglass" in the "Things to Know" chapter). Look for cut or broken wires, dangling or missing components, and other obvious signs of trouble or neglect.

CIRCUIT BOARDS (ELECTRONIC MACHINES)

Most of the electronics, most of the circuit boards, and most of the fuses in electronic pinball machines are in the backbox. Even if you know nothing about circuit boards or solid state technology, open the backbox and look inside.

Look at the connectors—the white or colored plastic plugs that attach wires to the circuit boards. Be wary of connectors that show burn marks or discoloration from overheating. Many electronic problems can be traced to these connections.

Look for circuit boards that don't appear to be original, perhaps a different color or different markings. There may be nothing wrong if a circuit board has been replaced, but people often substitute circuit boards that they took from other games made by the same manufacturer. The replacements look identical, they take the same connectors, and they even work—sort of. But replacement boards from other games have different programming and do not work exactly as the original circuit boards. I've had machines in the shop for repair that I could not figure out what was wrong, only to finally discover that one of the boards came from a different game.

CHECK THE BATTERIES (ELECTRONIC MACHINES)

Most important on electronic machines: check the batteries, their mounting bracket, and the circuit board the batteries are mounted on for signs of battery corrosion. Battery corrosion can destroy circuit boards. Many people who own electronic machines don't even know there are batteries in the machine, and those batteries could be years old and lethal (to the machine).

A corroded circuit board can give you lots of trouble. Corrosion can continue to grow and eat away at a circuit board even after the leaking battery is removed. Corroded circuit boards can sometimes be repaired, but the fix is usually temporary. The corrosion will eventually creep along the wiring and destroy the board. Sound like an old 1950s horror movie? For your pinball machine it is.

If there is only a small amount of corrosion on the battery holder, corrosion that has not worked its way onto the board and the wiring on the board, the board is probably okay. The battery holder can be replaced inexpensively.

Sometimes you can't see the corrosion unless you remove the circuit board from the game so you can examine both sides of the board. So again, be warned when buying an electronic pinball machine: Corrosion = Trouble.

For more information about batteries see "Batteries" in the "Components and Features" chapter. Electro-mechanical machines and most home models and cocktail table machines do not use batteries.

TEST THE MACHINE

With the playfield glass still off, lower the playfield, start a new game and, one by one, hand operate every switch, target, sling shot, and bumper—everything that the ball can activate—to see if it is working. If an individual target or rollover or bumper isn't working, the problem is most likely an out-of-adjustment switch and the repair is most likely

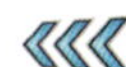

minor—if it's an electro-mechanical machine. A non-working switch on an electronic machine may indicate a minor switch adjustment, but it may indicate a burned transistor, diode, or IC.

If playfield components such as pop bumpers or flippers are not working on electro-mechanical machines, they are often easy and inexpensive to fix, and the fact that they are not working can bring down the cost of the machine significantly. On electronic machines, non-working flippers and pop bumpers may also be an easy mechanical repair, but may be a symptom of electronic failure on one of the circuit boards, requiring a knowledge of electronic circuits to fix.

A DIFFERENT APPROACH: JUST GO FOR IT

If this long chapter of things to check out—and worry about—makes you hesitant to buy a machine, consider instead doing what most everyone else does: Just find a machine you like, try it out, and buy it. Very few people actually take the time to thoroughly investigate a pinball machine (or anything else for that matter) they plan to buy. No studying, no questions, no hesitation. Sometimes that's just the way to go.

And if the machine does turn out to have problems, chances are you can solve them yourself with this manual.

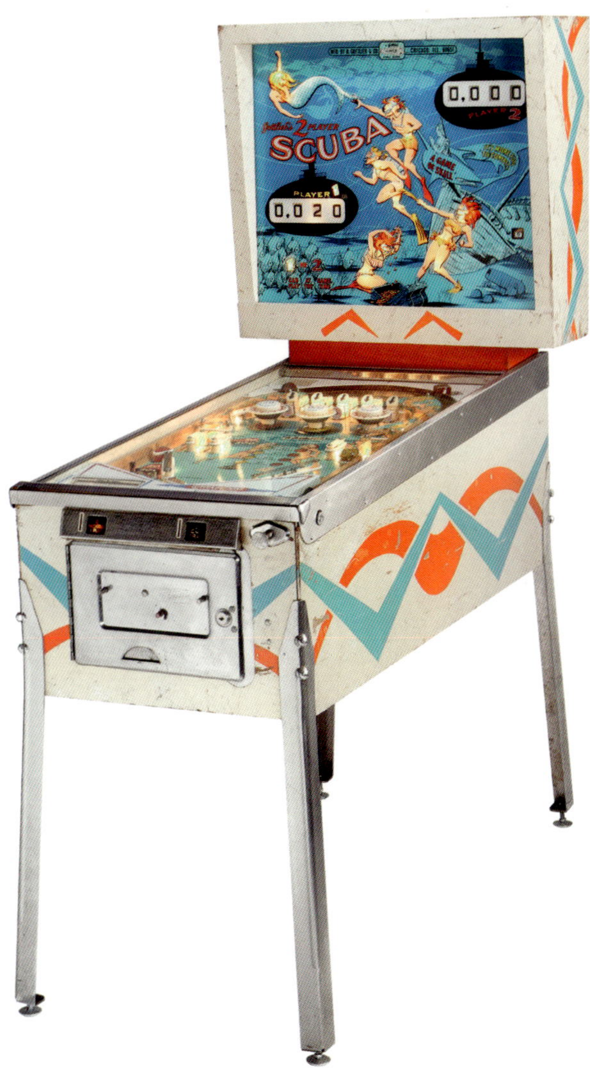

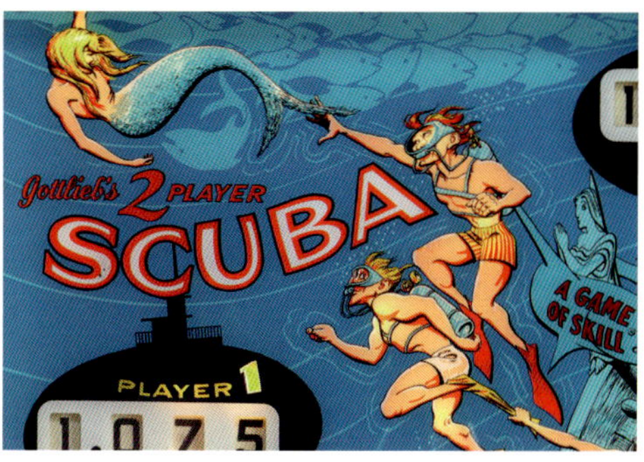

Gottlieb **Scuba**, 1970. The Gottlieb company liked mermaids enough to produce three games featuring their charms: Mermaid in 1951, North Star in 1964, and Scuba in 1970. Williams also created a mermaid pinball, Fish Tales, in 1992.

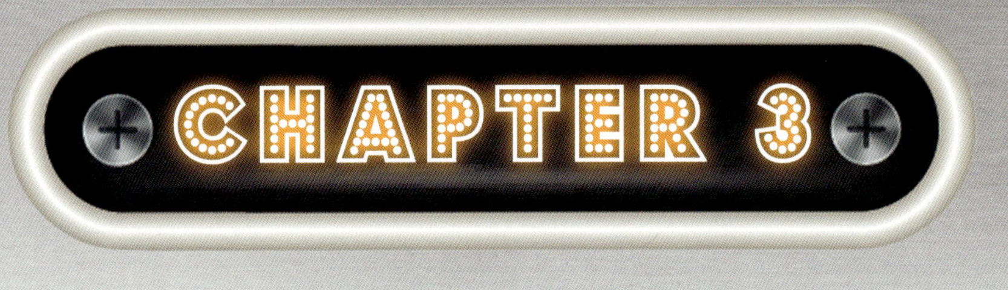

THE MANUFACTURERS

> If you put a box with a quarter slot on a bar counter, somebody will put a quarter in to see what it will do. But it's the second quarter, and the third quarter, that defines its success.
>
> —David Gottlieb, founder, Gottlieb Pinball Co.

Over the years many different manufacturers made pinball machines. The three companies that made the most games and stayed in business the longest were Gottlieb, Bally, and Williams. David Gottlieb founded the Gottlieb Company in 1927. Raymond Moloney founded Bally in 1932. Harry Williams founded Williams in 1943. Another long-time pinball manufacturer was Chicago Coin (Chicago Dynamics), founded by Samuel Gensburg in 1931. All four companies were based in Chicago.

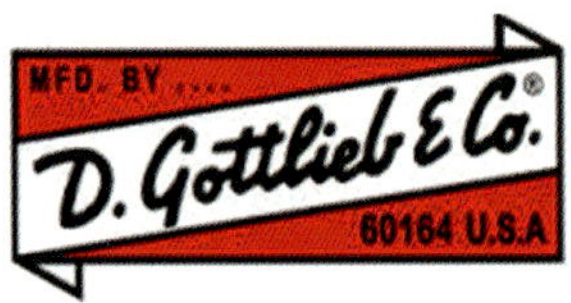

In the early days of pinball, the 1930s through the 1950s, companies from the dust bin of history created and manufactured often unique machines: United, Jennings, Daval, Mills, A.B.T., Western, Success, Stoner, Keeney, Genco, Exhibit Supply, Pacific. These are true collector games today, but you're likely to be on your own trying to repair or find parts for them.

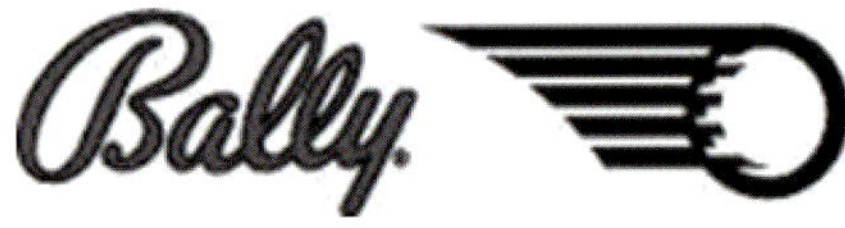

In the 1960s and 1970s, Midway made a few machines and later merged with Bally. A Spanish company, Sonic, made electro-mechanical machines in the 1970s, many of which have migrated to the US, often using Williams parts. In the electronic era, Micro Games of Phoenix, Arizona, made a few machines in the mid-1970s, and is remembered primarily as the manufacturer of the first electronic machine, Spirit of 76. Allied Leisure made cocktail table size pinball machines in the late 1970s. An Italian company, Zaccaria, made machines in the 1970s and 1980s. Atari, Game Plan, and Data East made machines in the 1980s. Capcom and Sega made pinball machines in the 1990s.

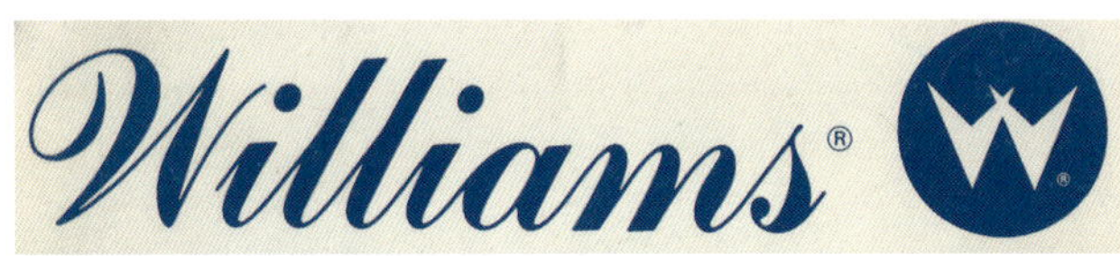

All of these pinball manufacturers are long gone. Chicago Coin sold out to Stern in 1976 (more on Stern later). Gottlieb was purchased by Premier in the 1980s and then went out of business in 1996. Williams bought out Bally in 1988, but they both quit making pinballs by 1999. The other manufacturers lasted only a few years.

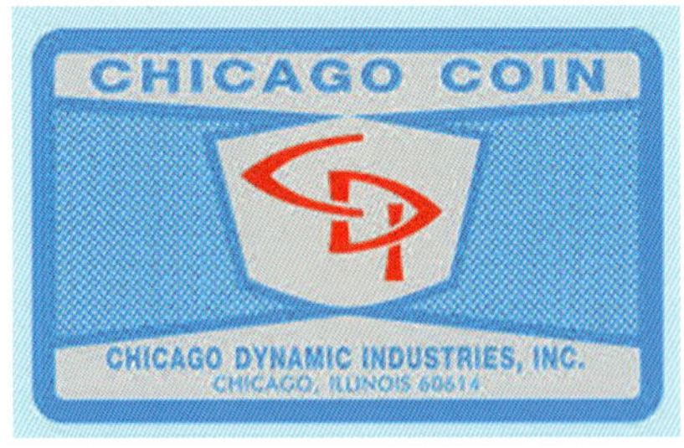

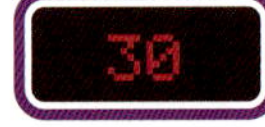

Each pinball manufacturer had its own way of designing and building a machine, its own parts, its own style of artwork, and even, in the electro-mechanical days, its own sound to the chimes. Parts and schematics for Gottlieb, Williams, Bally, and Chicago Coin are usually available from most pinball parts suppliers. Parts and schematics from some of the other manufacturers are still available, but some may be more difficult to locate.

Which brings us to the only historic pinball manufacturer that survived into the twenty-first century: Stern Pinball.

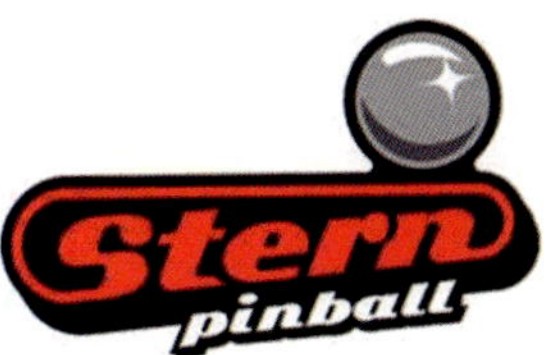

No one knows how to assemble pinball machines like Gary Stern. Watching the Stern factory in action is like watching the electronic revolution. Pinball machines are complicated and labor intensive to manufacture. Each pinball machine assembly line is hundreds of feet long, with dozens of workers spaced along the line assembling thousands of parts, many of which were handmade at other areas of the factory.

—George Gerstman, "Clear and Convincing Evidence"

Stern was actually two different companies owned and operated by father (the first Stern company) and son (the second). The first Stern company started in 1976, when Sam Stern bought the old Chicago Coin company and manufactured pinball machines until 1984, when he closed the doors. In 1999, Sam's son Gary Stern purchased the Sega Pinball company and renamed it Stern, and they've been going ever since. And in keeping with the great history of pinball, they are still based in Chicago. *Thank you, Gary Stern.*

New Stern machines, and the machines being manufactured by a handful of new companies, are technological and mechanical marvels, awesome behemoths controlled by software as complex as any computer network. The Lure of the Silver Ball lives on.

We don't say we're the last pinball company. We say we are the only pinball company. Within about 100 feet is where just about all the pinball machines in the world are made. Mechanical action pinball. That's what my father made. That's what we make. We've done this forever.

—Gary Stern, Stern Pinball Co.

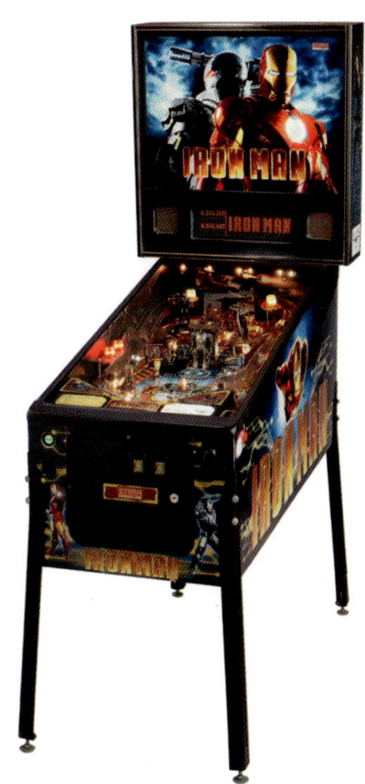

Stern **Iron Man**, 2010. Based on the Marvel Comics movie Iron Man. Almost all Stern machines are licensed themes: celebrities, movies, TV shows, and video games.

CHAPTER 4

DISASSEMBLING AND TRANSPORTING YOUR PINBALL MACHINE

Most people will have to partly disassemble a pinball machine to get it home. These machines are very heavy. Some weigh more than 200 pounds. You can injure your back if you are not careful. If you must lift a machine, get help. Machines can be broken down to fit into most vans and SUVs. The main cabinet can be transported flat, or upright on its back (there are rubber feet on the back), or if necessary on its side.

STEP ONE: REMOVING OR FOLDING DOWN THE BACKBOX (THE HEAD)

The upright part of the cabinet, showing the scores, ball in play, and other information, is called the backbox, or the head. On electro-mechanical machines and some early electronic machines, the backbox is bolted to the main cabinet and can be removed for transportation. On most electronic machines the backbox is hinged and folds down for transportation.

Before doing anything with the backbox, check that the backglass (the painted glass mounted in the front of the backbox) is firmly in place. If the backglass is loose or vibrates, use small pieces of corrugated cardboard as shims. You can remove the backglass entirely to transport the machine, but sometimes that's just asking for trouble, as the glass is much more exposed and vulnerable to getting broken.

BOLT-ON BACKBOX (ELECTRO-MECHANICAL MACHINES)

Depending on the weight of the pinball machine and how much headroom you have in your vehicle, backboxes can be transported still attached to the cabinet, or can easily be disconnected, unbolted, and removed.

To get to the backbox bolts and the wiring, remove the back door to the backbox. If the door is locked and you don't have a key you will have to drill out the old lock. See "Missing Keys" in the "Things to Know" chapter.

Before unbolting the backbox, disconnect the wiring. The wiring between the cabinet and the backbox is connected by two, three, or four plug-in wiring harnesses, also called connectors or "Jones plugs" (named after Howard P. Jones, who invented and first manufactured the plugs). The connectors are usually different lengths, each with a different number of wires, so they can't get mixed up when you reconnect them. If the connectors are easily confused, color code or number them before you unplug them so you can get the right ones back where they belong.

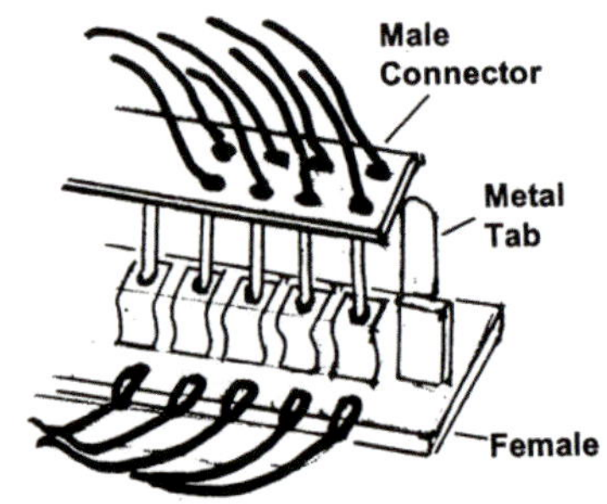

Carefully unplug each connector from its socket. The wiring connectors are made of bakelite, a hard, reddish-brown, fiber-like material about 1/16 in. thick. Bakelite gets brittle with age and can break. Grab both ends of the connector and ease, or slowly wiggle, the connector off. Don't pull on just one end of the connector because you could break the

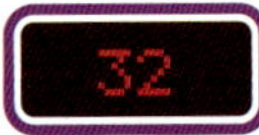

connector in half — which I just did. I was disconnecting an old Williams machine to bring it in for repairs and the plug broke right in half. I was much more careful removing the second plug, and it broke right in half! If the bakelite plugs do snap in half, chances are good that the wiring and the metal prongs are still intact. You just now have two plugs instead of one to reattach.

Chicago Coin: The last machines made by Chicago Coin in 1976 had plastic snap-together connectors, not the standard connectors described.

After the wiring is disconnected, remove the four bolts. Hold on to the backbox when removing the last bolt; the thing can topple right over!

BOLT-ON BACKBOX (1970S ELECTRONIC MACHINES)

Most electronic machines have fold-down backboxes (covered next), but some machines from the 1970s had bolt-on backboxes like electro-mechanical machines. You can tell if the backbox is bolted or hinged by looking at the back of the machine where the backbox sits on the main cabinet. Fold-down backboxes will have latches that open, bolt-on backboxes will not.

To get to the backbox bolts and wiring, remove the backglass and open the hinged door behind the backglass. If you don't know how to remove the backglass see "Getting Behind the Backglass" in the "Things to Know" chapter.

There are a dozen or more plastic wiring connectors between the cabinet and the backbox that have to be disconnected before you can unbolt and remove the backbox. Be sure each connector is labeled so you can get it back where it belongs—correct socket and correct direction—when reconnecting. These connectors usually have a "key," which is a blocked, unused space to prevent the connector from being connected to the wrong socket. If you get a connector in the wrong socket or connected backwards you can damage your machine.

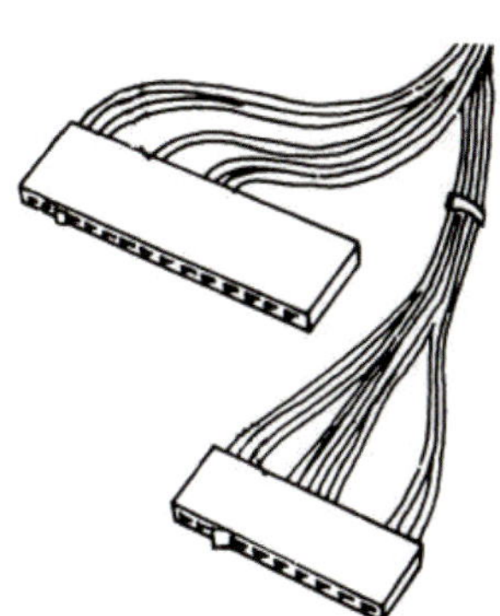

These connectors, by the way, are the weakest link in electronic machines. Poor connections are common. Non-operation and intermittent operation can often be traced to dirty, broken, burned, or loose connectors. Poor connections are covered in the "Repairs: Electronic Machines" chapter.

After the wiring is disconnected remove the bolts. Hold on to the backbox when removing the last bolt to prevent the backbox from falling over.

If all this unplugging of wires sounds like something you'd like to avoid—and, to tell the truth, it is—see if there is enough slack in the wires that you can unbolt the backbox and lay it down without disconnecting the wires.

Put protective cushioning—corrugated cardboard, a towel, foam rubber, or something similar—between the face of the backbox and the cabinet to protect the surfaces. Be careful not to let the backbox slide down and pull on, and possibly yank out, wires. I recommend strapping the backbox in place to prevent movement.

FOLD-DOWN BACKBOX (ELECTRONIC MACHINES)

Backboxes that are hinged should always be folded down for transportation, to eliminate the possibility that the latches, or whatever else is holding the backbox upright, do not open or come apart and allow the box to fall over. Before folding down the backbox, put protective cushioning between the backbox and the cabinet to protect the surfaces.

Some hinged backboxes are bolted to the cabinet; you have to remove the bolts before you can fold down the backbox. To access the bolts, remove the backglass and open the door behind the backglass to get to the inside of the backbox. If you don't know how to remove the backglass, see "Getting Behind the Backglass" in the "Things to Know" chapter. Be careful when removing the bolts that the backbox doesn't fall over.

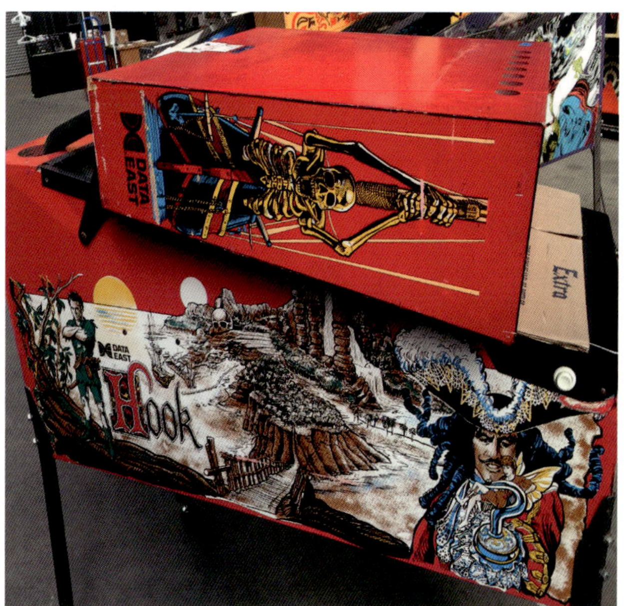

STEP TWO: REMOVE BALLS, COIN BOX, LOOSE OBJECTS

Before you move a machine, remove all the balls. Otherwise they can go crashing all over the place and could break a plastic part. Reach into the outhole between the flippers with your finger, or the hole near the ball shooter where the ball comes back into play, to kick out the balls. You can also use a long-handled magnet.

Open the coin door and remove the coin box if it's still there and any loose objects, particularly if the cabinet will be transported on its back or side. If the coin door is locked and you don't have a key you will have to drill out the old lock. See "Missing Keys" in the "Things to Know" chapter.

STEP THREE (ONLY IF ABSOLUTELY NECESSARY): REMOVE THE PLAYFIELD

If the machine is too heavy to carry, you can remove the playfield glass and disconnect and remove the playfield itself from the cabinet. This is not recommended, as it is too easy to damage the components mounted on and under the playfield. But once, when we had to carry a machine up a flight of stairs, the only way we could carry it was to remove everything that would remove, including the playfield.

The playfield connects to the cabinet with plugs very similar to those connecting the backbox.

STEP FOUR: REMOVE THE LEGS

Transport your machine with the legs off to prevent wobbling and tipping over, especially in a pickup truck. The legs come off easily, two bolts per leg. The leg bolts screw into threaded brackets inside the cabinet. The threads on the brackets are sometimes stripped on old machines and the leg bolts will instead be secured with nuts. Look or feel inside the cabinet to determine if there are nuts on the leg bolts. Remove the nuts before removing the bolts.

You don't have to remove all four legs at once and have people injuring their backs holding up the machine. Here's what you do:

1. Move the machine with the legs still on as close to the vehicle as possible, angling the machine so one rear leg is next to the vehicle.
2. Have someone hold up the corner of the machine closest to the vehicle and remove the rear leg (just one rear leg) closest to the vehicle.
3. Slide that leg-less corner of the machine into the vehicle just far enough so the vehicle is supporting the corner of the machine.
4. Remove the second rear leg from the machine and slide the entire back of the machine into the vehicle far enough so the weight of the machine is off the front legs. If the vehicle floor is lower than the bottom of the pinball machine cabinet (with the front legs on) you will either have to hold up or prop up the cabinet to remove the legs.
5. Remove the two front legs and slide the machine all the way into the vehicle.

STEP FIVE: GETTING THE MACHINE IN YOUR VEHICLE

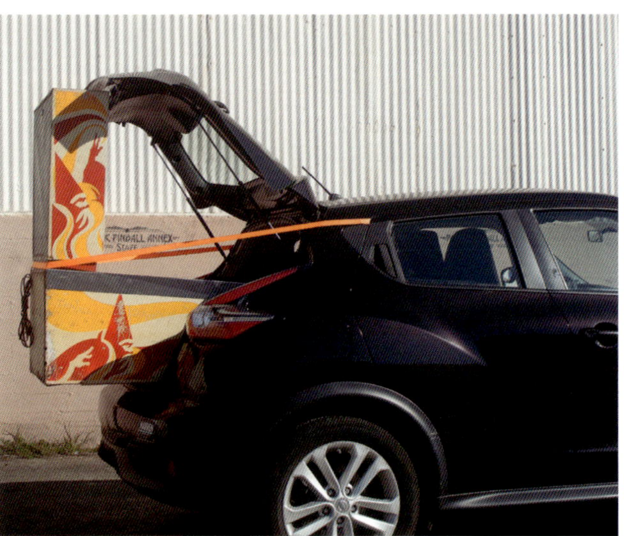

Your pinball machine is never going to fit in a Volkswagen bug, but I've seen machines hauled home in all kinds of vehicles. The main cabinet is about 4½ feet long and will fit in just about any SUV, pickup, van, or station wagon. Most hatchbacks can transport a pinball machine, though often the machine will be sticking out the back a little. I've even seen someone fit a machine in the back seat of a car.

Bring a large piece of cardboard to slide the machine on. It really helps, and it protects the upholstery on the back of fold-down seats. Some people use a few pieces of PVC pipe about 3 feet long to act as rollers, but be sure that the machine won't roll around in transit.

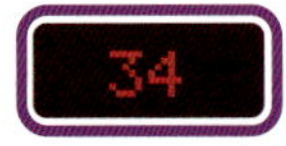

STEP SIX: TIE EVERYTHING DOWN

Pinball machines are heavy, but they can still slide around, and they can—and do—fall out of vehicles. Just about everyone in the pinball world can tell you a story about someone's pinball machine being destroyed when it hit the ground during transit.

Be especially careful with the backbox if it has been unbolted from the cabinet. Cushion and protect the backglass. Strap or secure the backbox firmly so it doesn't get to shaking and tilting, and falling right onto something that might break the glass.

STEP SEVEN: MAKE SURE YOU HAVE EVERYTHING

Check to be sure you didn't forget the legs, the back door, bolts, keys, any loose parts, and the paperwork.

If you are buying the machine from a pinball shop or from someone who has several pinball machines and probably has plenty of spare parts, check that the leg levelers (the removable screw-in feet) and the balls are in good condition. If any look rusted or damaged ask for new ones. If any of the rubber rings look old, or worn, or cracked, ask for new ones. If any lamps appear not to be working ask for some replacements. Of course, you really shouldn't have to be checking up on these parts if you're buying from a reliable seller, but even the best shops can get careless once in a while.

WHEN YOU GET THE MACHINE HOME

Before you put the legs back on and haul your machine into the house, tilt the main cabinet onto its back and see what rolls out from hiding. Particularly on old electro-mechanical machines, you're almost guaranteed to find parts, nuts, washers, labels, and maybe fuses (maybe quarters?). If you have some sort of blower, blow out the cabinet and the backbox; get rid of years of dust and cobwebs.

DON'T HURT YOUR BACK

I'll tell you over and over: If you have to lift a machine, get help. Don't hurt your back.

> Your own pinball. Now you can solo your favorite machine before breakfast or after midnight. You can steal a game while supper simmers, during the seventh-inning stretch, or while waiting in line for the family shower. Use your pinball's lusty sounds to drown out the neighbor's leaf blower, Junior's drum practice, or the lonesome silence of an empty apartment. Your game will be the hit of your next party, the liveliest conversation piece in the den, the softest colored lighting in the bedroom. A center for family competition, a small world of private escape.
>
> —Jim Tolbert, *Tilt*

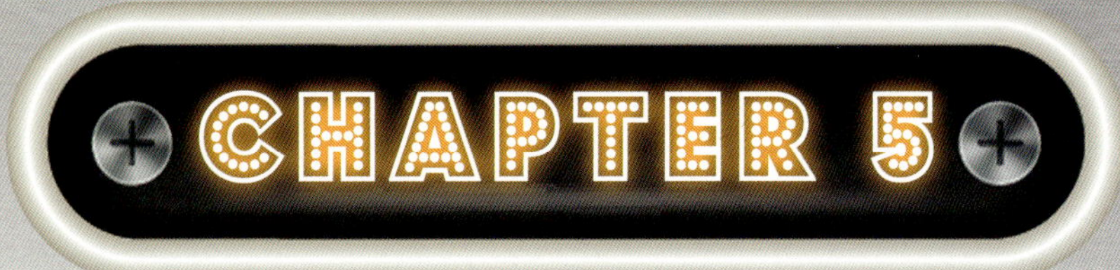

SETTING UP THE MACHINE

Don't let excitement and eagerness to start playing your new pinball machine overtake the important steps in setting up the machine. Take your time and follow these steps. You'll be glad you did.

LOCATION: SUNLIGHT, TEMPERATURE, HUMIDITY, AND CLEARANCE

A pinball machine, like all fine machinery, needs protection from the elements.

Sunlight: Bright sunlight will fade the paint on the backglass, playfield, and cabinet. If your machine is exposed to direct sunlight, cover it when not in use.

Temperature: Extreme temperatures will damage the painted glass and plastic parts, and could damage electronic components.

Do not turn on a pinball machine if it is very cold. Let it warm up to room temperature first. And if the room temperature itself is getting hot—85°F or more—I suggest turning off the machine, particularly if it is an electronic machine. Old electro-mechanical machines seem to be impervious to dry, atmospheric heat, but electronic machines can be damaged by heat if the machines are turned on.

Humidity: Humidity can loosen cabinet joints and peel backglass paint, can corrode electrical parts, and can damage solid state electronics. If your machine is in a humid location, cover it when not in use to keep moisture out. Run down to the local thrift store and buy a blanket or quilt, something that will cover the head and playfield. Don't use plastic wrapping or a tarp because moisture could build up and get trapped under the plastic.

Clearance: You will occasionally need to remove the playfield glass to replace a broken rubber ring or a burned lamp, or tighten a loose post, or maybe to retrieve a ball that got stuck. You'll need four feet of clearance in front of the machine to remove the glass. Try not to box your machine in so tightly that you can't work on it.

LEG LEVELERS

Before you reattach the legs, check the feet. Each leg should have a screw-in foot, called a leg leveler (also called a leg adjuster) that is used to level the machine. Leg levelers are often rusted and won't budge. Use WD-40 and lots of patience to free them. Do not force a rusted leg leveler. You could actually twist the leg itself out of shape trying to free the leveler. I've done it. If the leveler won't unscrew, let the WD-40 soak in overnight; it often helps.

Once you get the leveler off, if it can't be cleaned and if it is difficult to screw in, new leg levelers are inexpensive and available from any pinball parts business.

Many people set the front leg levelers as low as possible and set the back leg levelers as high as possible to get maximum tilt to the playfield for fast play. But leave at least ¼ inch of the leveler visible above the screw hole. Otherwise the leveler might bend or break. If the leg levelers don't already have a nut on them put

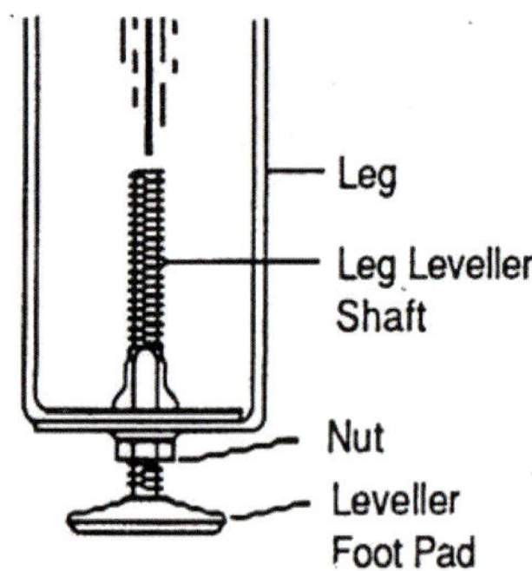

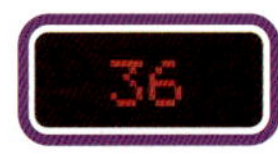

one on (standard ⅜ inch nut). The nuts go underneath the flat part of the leg. Screw the nut onto the leveler before screwing the leveler onto the leg of the machine. Don't tighten the nuts yet. Once the machine is completely set up you'll be leveling the playfield (see following). Then the nuts get tightened.

Your machine's instruction manual may specify the slant that the manufacturer recommended. The angle varied with manufacturers and different games, but typically it was 3½% to 5%. But you may prefer a steeper slant.

Leg levelers come in 2 inch and 3 inch lengths (2 inches or 3 inches of thread). Electro-mechanical pinball machines and most older electronic machines used 2 inch levelers on all four legs. I sometimes replace the back levelers with 3 inch levelers to get that extra slant to the playfield. You could also put a piece of wood—1" × 4" or 2" × 4"—under the back legs.

If you get too much of a slant, you may find that the flippers don't have enough power to overcome the steep angle. If the ball can't make it all the way up the playfield when hit by the flipper, consider lowering the back legs a little. (You can also install a more powerful flipper, which is covered later in this manual.)

ATTACH LEGS

Before you reattach the legs, clean them up a bit. Most legs on electro-mechanical and early electronic machines are chrome, though some Bally legs were painted a dull grey. Old legs are often rusted. Surface rust can be removed, or at least reduced, using Brillo or SOS pads (steel wool impregnated with soap). Rub the dry pads on the legs and then wipe off with a damp cloth. It's a bit of a mess, but it often brings nice results.

Examine the leg bolts that came with the machine. Make sure the threads are not stripped. If you are unsure, try threading a standard ⅜" nut onto the bolt. The nut should screw on easily. If your bolts are damaged, or if you are missing leg bolts, a standard ⅜" bolt 2½" long should work fine. But you can buy nice looking pinball leg bolts from any pinball supplier. You can also make your old bolts look nicer by shining the heads using a wire wheel or a wire brush.

When you reattach the legs, if the bolts give resistance, a tiny squirt of WD-40 on the tip of the bolt's threads might help it get started. Make sure the legs are on tight. Wobbly legs make for a wobbly game. If the bolts will not tighten securely, put washers and nuts on the bolts inside the cabinet. Most leg bolts take standard ⅜" nuts.

Try not to drag the machine around by the legs. The legs are made of steel, but they can bend. If you have a bent leg, straighten it out in a vise or replace it.

Different machines and different manufacturers used different lengths of legs. If you need a replacement leg, know the machine's name and manufacturer and the length of the leg before ordering a new one. New legs are being manufactured, but only in a few lengths. Most pinball repair shops have an assortment of used legs in the back room.

On most machines all four legs are the same length. A few machines that had deep cabinets used different length legs on the front and back.

If your machine is on a rug, putting a piece of ¼" plywood or some other sturdy, flat material about 3" square under each leg leveler will help keep the machine level and protect your rug.

BACKBOX (BOLT ON)

If the backbox was unbolted and removed, before reattaching the backbox, make sure the line cord and plug are outside the cabinet and set in the little notch between the cabinet and backbox—not pinched or jammed. I can't tell you how many times I've reattached a backbox completely forgetting to get the line cord out and then having to unbolt it again.

Set the backbox on the cabinet and align the bolt holes in the backbox with the holes in the cabinet. Backboxes bolt on with four bolts and four large washers. If any of the bolts or washers are missing, get replacements. Most backbox bolts are standard ⅜" threads with hex heads, but different machines used different length bolts. Take one of the old bolts to the hardware store to match it.

BACKBOX (FOLD DOWN)

If the backbox is folded down tilt it back up, reattach the bolts if there are holes for bolts, and relatch the latch if there is a latch. If bolts are missing, get replacements. Don't rely solely on latches to hold up the backbox. If the latch is broken or missing, get a replacement latch from a hardware store.

Check the line cord. If the line cord comes out between the backbox and the cabinet, make sure the line cord and plug are outside the cabinet. On many electronic machines

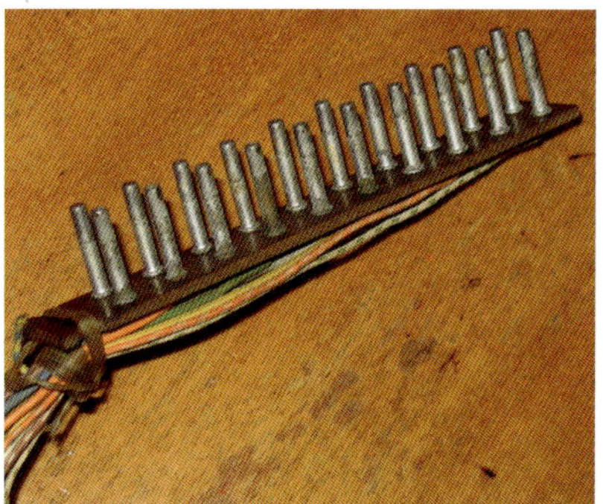

Wiring harness ("Jones plug") connects backbox to cabinet and playfield, unplugs for disassembly. Clean pins with a wire brush before reconnecting.

the line cord comes out of a hole in the cabinet, not between the cabinet and the backbox. The newest machines have universal power cords that unplug.

REATTACH WIRING (ELECTRO-MECHANICAL MACHINES)

Before reconnecting the wiring harnesses check for corrosion or dirt on the connector plugs (the Jones plugs). Sometimes a game that was working fine before you disassembled it will suddenly not work when reassembled. Corroded or dirty connectors are often the cause. Even if you don't see any problems, clean the pins with a wire brush, steel wool, or similar abrasive.

When reconnecting the wiring, each connector plug is usually a different length, with a different number of wires, so you can't get them mixed up when you reattach them. The male connectors go in one direction: up or down, or right or left. You don't want to reverse them. The female socket (the one attached to the backbox) has a metal tab at one end, making it nearly impossible to reverse the male connector when plugging it back in.

But if you've done the impossible, you'll know it as soon as you try to start the game. Nothing will work right. Don't panic, you haven't damaged the machine. If you are unsure which connector goes where, compare the wire colors. The wires on the male connector should match up in color with the wires on the female socket. (You *do* know which plug is male and which is female? The male connector has the prongs that stick out and are inserted into the female socket.)

REATTACH WIRING (ELECTRONIC MACHINES)

If the wiring to the backbox was disconnected, have a look at the plastic connectors before reconnecting them. Be exceedingly careful that you reconnect the plugs where they belong. A wrong connector in a wrong slot or the wrong direction could cause serious damage to your machine.

CHECK FUSES

Check all the fuses. Pull each fuse and make sure it is the proper ampere rating. Make sure slow-blow fuses are used only where they are called for. Never use a fuse with a higher ampere rating than the manufacturer specifies. Most fuse holders have the ampere rating labeled on them or next to them. If they don't, you'll have to look at the schematic or the instruction book to determine the correct rating.

See "Fuses" in the "Components and Features" chapter. The wrong fuse can lead to damage to your pinball machine.

CHECK AND RELOAD BALLS

Make sure you have all the balls. Until about 1965, all pinball machines had five balls. Most electro-mechanical machines made after 1965 had only one ball. Newer electronic machines might have as many as ten balls. See "How Many Balls In A Pinball Machine" in the "Components and Features" chapter.

Make sure the balls are shiny, with no pitting, dents, or rust. If the balls are imperfect, do not play the game until you get new balls. Damaged balls will damage your playfield. Put the balls into the outhole between the flippers. See "The Ball" in the "Components and Features" chapter.

LEVEL THE PLAYFIELD

The playfield should be level side to side. If the machine is tilted to one side, even slightly, the ball will roll to that side and will affect the play.

Lay a long level across the cabinet at the back of the machine next to the backbox, and then at the front of the machine. If you don't have a long level, place a short level on the playfield glass. The leg levelers can be adjusted up and down to level the game. If the levelers are mounted correctly there will be a nut threaded onto the leveler under the leg. Loosen the nut. Put a small wrench on the leveler and rotate to adjust the height. Then tighten the nut. You should be able to do this without lifting the machine off the ground.

CHECK THE TILT

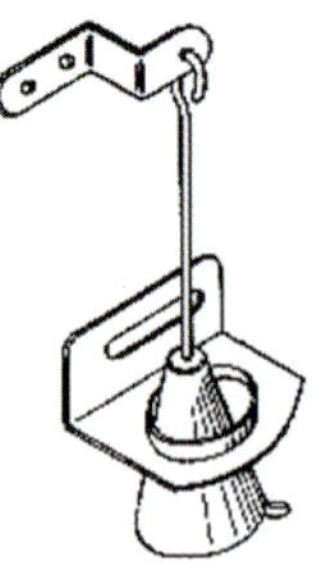

After you level the machine, check that the swinging tilt pendulum is centered in its metal ring (the strike plate). The tilt is just inside the coin door on the left side of the cabinet. If the tilt rod is not centered in the ring, the tilt rod will be closer to one side of the ring. The slightest shake will likely tilt the machine.

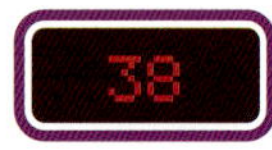

The metal ring is attached to the cabinet with two screws. The ring has slotted screw holes, allowing some movement when the screws are loosened. Loosen the screws and adjust the ring so that the pendulum is centered within it. This is an easy but important adjustment to make your game more fun to play. See "Tilt" in the "Components and Features" chapter.

EXAMINE THE INSIDE OF THE CABINET

If you haven't already done so, check the bottom of the cabinet for loose parts. Use a magnet. If the cabinet has been stored upright on its back end, all those loose parts will have rolled and slid down there. Every one of those parts goes somewhere on the machine, so be on the lookout for mounting hardware that is missing screws or nuts.

If you have an electro-mechanical machine, you will probably find paper labels that identify different relays and rotating units lying on the cabinet floor. If you can match the labels to the relays or units, tape or glue them back in place. If you have no idea where the labels came from, save them in an envelope in case a repair person might be able to identify them if the machine ever goes in for repair. But don't guess. Don't stick a label back on a relay unless you know it is the right relay. That's just what a repair person needs, a misidentified relay! See "Unmarked Relays: Missing Labels" in the "Repairs: Electro-mechanical Machines" chapter.

GAME ADJUSTMENTS

Pinball machines have custom adjustments you can make, such as high score settings, number of balls per game, match on or off, add-a-ball, and conservative/ liberal settings (more or less challenging). Most of these adjustments are described in this manual.

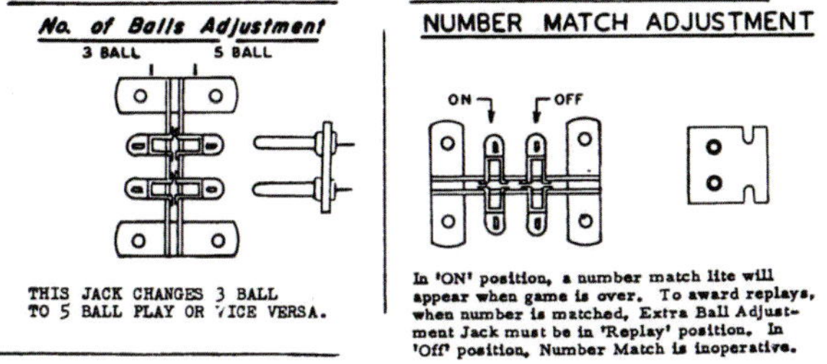

Typical adjustment instructions for electro-mechanical machines.

Electro-mechanical machines: Game adjustments are accomplished by plugging wiring connectors into different sockets. You will find the connectors inside the cabinet, usually near the coin door, and inside the backbox. The adjustments are explained in your machine's instruction manual, and often on labels next to the connectors inside the game.

Electronic machines: On electronic machines from the 1970s, the game adjustments are made on tiny slide switches (called dip switches) on the CPU board in the backbox. There is usually a card stapled inside the backbox showing the switch settings, or you can find them in your machine's instruction manual. If you don't know how to open the backbox see "Getting Behind The Backglass" in the "Things to Know" chapter.

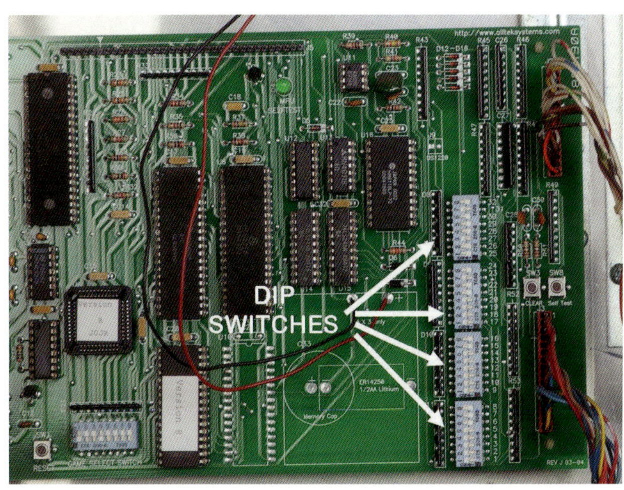

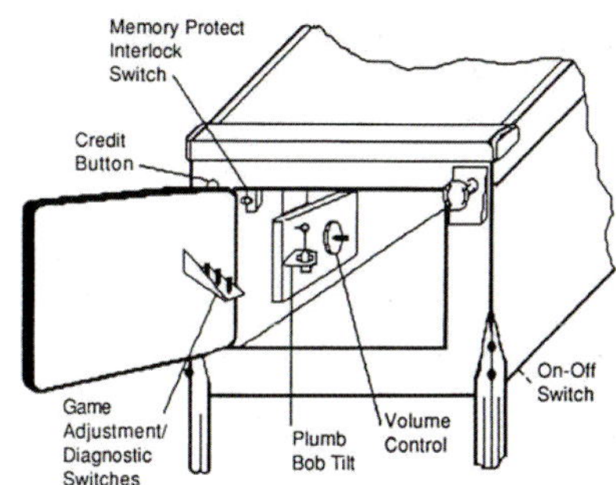

On machines manufactured after the 1970s, game adjustments are made using a set of push buttons mounted on or next to the coin door. Open the coin door and you will see two or three buttons mounted together. The game manual includes instructions on how to use these push buttons to make any adjustments you want. If you do not make any adjustments the machine automatically gives you the factory default setting, or whatever settings the previous owner made.

SURGE AND ELECTRICAL PROTECTION (ELECTRONIC MACHINES)

A lightning strike can cause extensive damage to electronic pinball machines. It is a good idea to plug your pinball machine into a surge protector, like the ones people use to protect computers. Whenever there is an electrical storm coming, or if you will be away from home for an extended period, unplug the machine from the wall.

Some people leave their machines turned on all the time in game over mode, just to admire the lights, particularly at night. I suggest that you turn off your machine when not in use. It's just an added level of protection, particularly for digital displays and electronic circuits.

Electro-mechanical machines are not as sensitive as electronic machines to being left on all the time and are not likely to be damaged by surges, but it never hurts to be cautious.

> The game itself needs expert explaining. What it is is a head-on confrontation to acquire signs, symbols, messages, and instructions that are satisfying and useful.
>
> —Playwright William Saroyan, describing a pinball machine, 1939

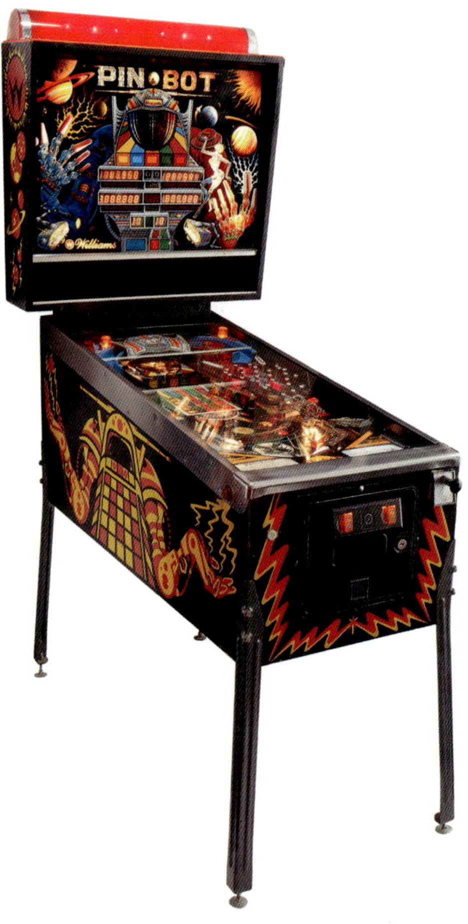

Williams **Pinbot**, 1986. One of Williams' most popular games, over 12,000 were built.

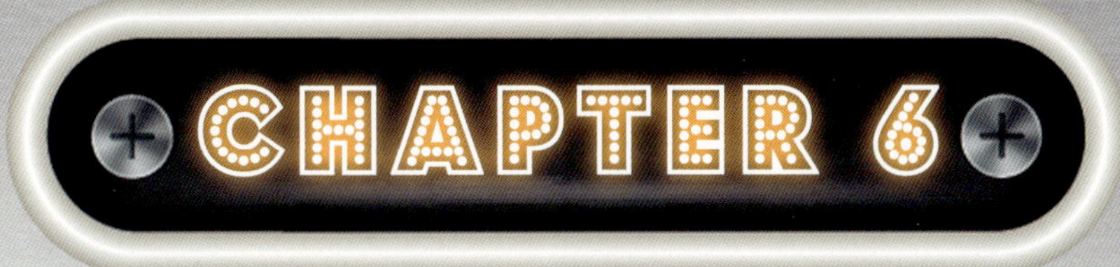

THINGS TO KNOW ABOUT PINBALL MACHINES

Easy to learn, hard to master.

—Michael Gottlieb, Gottlieb Pinball Co., grandson of David Gottlieb

WHAT YEAR WAS YOUR PINBALL MACHINE MANUFACTURED?

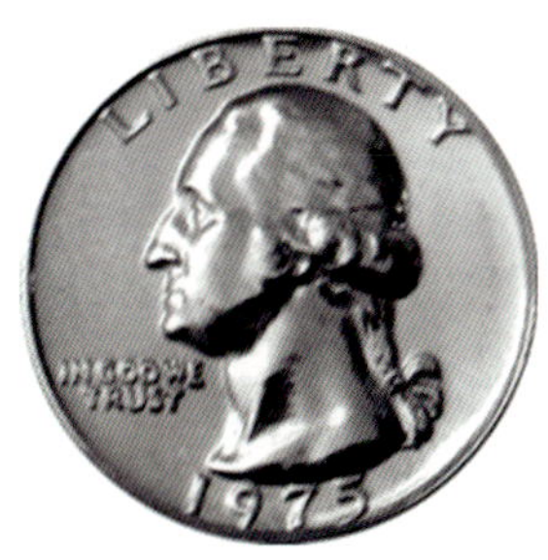

Flipper pinball machines were first manufactured in 1947, and new machines have been made every year since. People often want to know what year their machine was made. There are several references in this manual to adjustments and repairs that vary depending on what year a machine was made.

Some machines have a year shown on the bottom of the playfield, just above or just below the flippers. Some machines have a year printed somewhere on the backglass. Some schematics and game manuals have the year of manufacture.

Pinball price guides, which you can purchase from pinball parts dealers, include a complete listing of machines and the years manufactured. You can log onto the Internet Pinball Database (ipdb.org), type in the name of your machine, and find the date of manufacture, and a lot of other information about the machine.

PINBALL, QUARTERS, AND GEORGE WASHINGTON

The third Monday in February is Presidents Day, in honor of George Washington and Abraham Lincoln, who were both born in February, seventy-seven years apart. Back when a pinball game cost a dime to play, kids got two school holidays in February: one on February 12 for Abe and one on February 22 for George. If kids thought they got cheated when the cost of a pinball game went from a dime to a quarter, they *knew* they got cheated when their two days off of school got cut down to one.

Abe Lincoln wound up on the penny, good for checking tire tread but not much else. But George wound up on the quarter, still good for one play on many pinball machines.

The George Washington quarter was first minted in 1932. What year did your pinball machine get here? One way to "celebrate" the creation of your machine is to locate a quarter from that year, drill a hole in it, and put it on the key ring. How cool is it to have a 1980 Black Knight with a 1980 quarter on the key ring? Or a 1976 Spirit of 76 with one of the Bicentennial 1776–1976 quarters on the key ring?

You don't need to go to a coin dealer or spend any money to find a quarter with your machine's date on it. Coins from as far back as 1965 are still in circulation and very common, except for 1975, when no quarters were minted. Just start checking the dates on every quarter you get in change. It probably won't take more than a couple weeks to turn up the date you're looking for. Quarters from 1964 and earlier had silver in them and are no longer in circulation—and are worth a lot more than 25¢.

Is it illegal to drill a hole in a quarter? According to the US Treasury, no it is not illegal. It is illegal to "fraudulently deface a coin" and to "represent it to be other than the altered coin that it is." You can't, for example, scratch out the word "quarter" and write "dime" on the coin; that's illegal. But a quarter with a hole in it is not only legal, it still can buy a quarter's worth of whatever a quarter will buy these days.

TURNING A MACHINE ON AND OFF

I once got a service call from the owner of a bar, saying his new pinball machine wasn't working: nothing would come on, no lights, no nothing. I suspected that a main fuse blew, though I couldn't guess why. I loaded my tool box, meters, and extra fuses and drove over. I went to the machine, turned it on, and it worked fine. The owner of the bar watched me, and was shocked and embarrassed to realize that the machine had an on-off switch mounted on the underside of the cabinet.

Most pinball machines have a master switch—either a toggle switch or a push-button switch that turns all power to the machine on and off. It is almost always under the cabinet, in front near the right hand side of the machine. If you have an electronic "home" model, the on-off switch is behind the machine. On cocktail table machines the on-off switch is often underneath the machine. Machines manufactured before the mid-1960s usually did not have an on-off switch. Plugging them in turned on the machine and unplugging them turned them off.

The on-off switch is not the same as the start button you push to start a game. The on-off switch and the start button are two different things. The machine must be turned on before the start button will work (See "Starting a Game").

It is a good idea to check the on-off switch. Make sure it is fastened securely to the cabinet, and that the wires are secure and not frayed. There is 120 volts to the on-off switch, so **unplug the machine before touching any wires**.

When you turn a machine off in the middle of a game, most machines will reset to Game Over when you turn the machine back on. Some machines will pick up right where they left off in the middle of the game.

Electronic machines: Do not turn electronic machines on and off repeatedly, to protect electronic parts from repeated surges of current. If you turn an electronic machine off, wait a few seconds before turning the machine back on.

Do not let electronic machines sit for a long time in the middle of a game, ball ready to shoot. The computer circuits sometimes go a little crazy, wanting someone to shoot the ball and get on with the game. The machine will actually start activating playfield components, suspecting that a ball might have gotten stuck somewhere on the playfield. If you are not going to finish a game you've started, turn the machine off. Electro-mechanical machines do not have this problem.

Once a game is over it is okay to leave a pinball machine on, all day if you want, in the game over mode. A pinball machine, sitting with its 30 or 40 miniature lamps glowing, draws about 125 watts of power, the equivalent of about two 60 watt incandescent light bulbs.

STARTING A GAME

Pinball machines have a start button to start a new game. The start button is on the coin door, or on the front of the cabinet. Once the machine is turned on you drop in your quarter or quarters and then press the start button.

If the machine is set for free play, or already has credits on it (games showing in the credit window on the backglass), don't add money, just push the start button.

On multi-player machines, you press the start button once for one player. If adding a second player, wait until the machine resets (before shooting the ball), then press the start button again. Repeat for each player.

On some machines that take coins, just putting in your quarter(s) will automatically start a game. On other machines you put in the money and then push the start button. On yet other machines, when you put in the coins the machine will rack up one or more credits (games) on the credit wheel or digital display, and then you push the start button. Confusing? Just put in your quarter and see what happens.

If you press the start button in the middle of a game, most machines will automatically start a new game if there are credits already on the machine or if the machine is set for free play. If there are no credits or free play, pushing the start button in the middle of a game, or after a game is over for that matter, will do nothing.

ONE PLAY 25¢

The coin information above is about putting quarters in the pinball machine. One play 25¢. Except that many electro-mechanical machines from the 1970s offered two or even three plays for a quarter. That's why when you drop a quarter in the slot the machine might step up the credit unit one or two steps, showing the number of games remaining. These machines had adjustable settings—connectors mounted on the floor of the cabinet inside the coin door—so that the operator could choose how many games players would get for their quarter.

Pinball machines from the 1960s took dimes, 10¢ a play. Many of the 1960s machines also took quarters, with a quarter giving you three plays. These machines had a 10¢ coin slot and a 25¢ coin slot. And machines from the 1940s and 1950s took nickels, 5¢ a play. A roll of nickels could keep a kid playing pinball—and playing hooky from school—all day.

By the 1980s, 25¢ to play a game of pinball was just a distant memory. The 1980s machines were requiring two quarters to start a game. By the 1990s inflation really rolled in. Pinball machines from the 1990s had dollar bill acceptors: a buck a game, and only three balls, too! And then, last summer at a lake resort, I saw a pinball machine that took credit cards.

GETTING THE LIGHTS TO LIGHT

Some pinball machines light up as soon as you turn on the master switch. On some machines you have to push one of the flipper buttons—usually the left button—to get the lights to come on. On some machines from the 1960s and earlier, the lights will not come on until you start a game. Once the game ended the lights would stay on until you turned off the machine.

NOISES (ELECTRO-MECHANICAL MACHINES)

When you turn on an electro-mechanical machine, you may hear a thump or similar short sound. Some relays are activated when the on-off switch is turned on and you might hear them move the mechanisms attached to them. Some relays hum or buzz all the time. They don't usually cause problems, but they can be very irritating.

You may hear a short whirring sound, which is the score motor resetting. That sound should stop in a couple seconds. If not turn off the machine; something is wrong.

You may hear a constant low hum, as some transformers have a ground hum. Many old machines did not have a ground wire to quiet the hum. Many machines with ground wires, indicated by a middle prong on the wall plug, have the ground wire disconnected or the wall plug prong broken off, thus causing the hum. See "Line Cords and Plugs" in the "Repairs: Basics" chapter.

One noise your pinball machine should not be making when you turn on the machine or start a game is a loud buzzing, or a constant mechanical clicking or whirring sound. That usually means something is stuck, and activated when it should not be. Turn off the machine.

All of the above noise problems are covered in the "Repairs: Electro-mechanical Machines" chapter.

NOISES (ELECTRONIC MACHINES)

When you turn on an electronic machine you will hear a thump or similar short sound from the speaker. This is normal.

You may hear a constant hum. The hum could be a bad ground connection or a broken ground prong on the wall plug. See "Line Cords and Plugs" in the "Repairs: Basics" chapter. The hum could be caused by a capacitor in the circuit, which requires a knowledge of solid state electronics to trace. If the hum does not bother you or distract from the game there usually is no problem with just letting it hum away.

If you hear a loud buzzing, that usually means a solenoid coil from a flipper or pop bumper or kickout hole is stuck, and activated when it should not be. Turn off the machine and look up your problem in the "Repairs: All Machines" chapter.

MISSING KEYS

If the pinball machine is locked and the keys are missing—and you're no good at lock picking—the cheapest and fastest remedy is to replace the locks. Having replacement keys made is a lot more expensive than buying a new lock.

Try shoving a screwdriver in the key slot and see if you can force the lock to open. If that doesn't work, drill out the lock. A drill bit right into the center of the key slot will usually do the job. The lock may just fall apart, or you should be able to insert a screwdriver in the drilled out hole and turn it to free the locking mechanism.

Start with the coin door, because once you get the coin door open you may find keys to the back door (on electro-mechanical machines) or to the backbox (on electronic machines) inside the machine. There is a hook on the inside of the coin door for keys, or the keys may be lying on the floor of the cabinet.

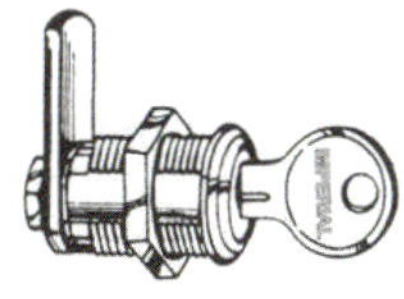

Replacement locks are available from pinball parts suppliers and many hardware stores. Most pinball machine locks are a uniform diameter—13⁄16" to fit through a 3⁄4" hole—but different machines used different lengths of locks. Pinball machines also used different types of cams (the metal arm that rotates up or down to latch the door): straight or angled with varying lengths. Finding the right replacement lock can sometimes be a chore.

OPENING UP A PINBALL MACHINE

To get inside a pinball machine, underneath the playfield glass, first open the coin door (Key missing? See "Missing Keys").

1. Remove the lockdown bar: Just inside the cabinet, at the top of the coin door opening, is a lever that pivots to the right or to the left. That lever opens a latch so you can remove the lockdown bar (hold down bar) that goes across the front of the pinball machine. The bar then lifts off, exposing the front edge of the playfield glass.

On machines from around 1960 and earlier the lockdown bar was bolted down, using two threaded posts attached to

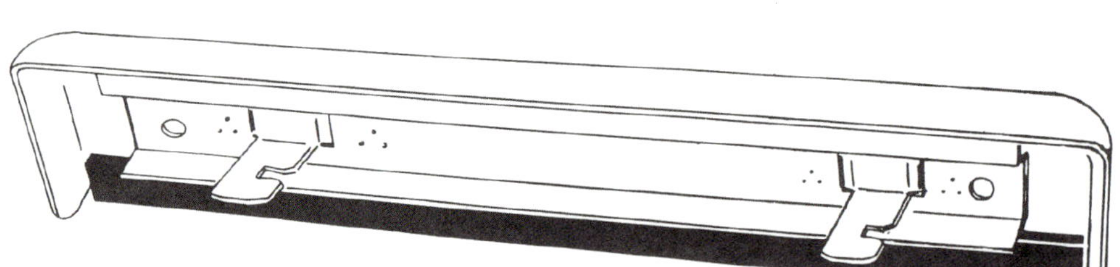

the underside of the lockdown bar. On Bally electronic "home" pinball machines the bar is held in place by friction; you pull it up and off.

If the machine has not been opened for several years, you may find the latch quite difficult to open, stuck shut by rust or corrosion or dirt. It can get frustrating trying to get the latch to budge, but don't take your frustration out on the machine. Try some WD-40 and some gentle persuasion with whatever tools seem to fit the job. "Talk to it," as my good friend Bob O'Neill used to say when we were trying to get two corners of a woodshed we were building to go together. Don't hit the sucker with a hammer—not too hard anyway. You'll bend the metal for sure, and then you'll really be frustrated.

Lockdown bars on many Gottlieb electro-mechanical machines had two adjustable posts (sometimes called "buttons") that the latch in the machine hooked to. The posts could be screwed up or down to better connect with the latch. These posts almost never need adjustment. If you are having a problem getting the lockdown bar to latch, look elsewhere before adjusting, and probably messing up, the post height.

2. Remove the Playfield Glass: Slide the playfield glass out of the machine and park it somewhere safe, somewhere it won't fall and you won't walk into it.

The glass may be very difficult to remove if it has been in the machine a long time. Dirt and grime will work into the slots holding the glass and darn near weld it in place. Try nudging the glass side to side to break the seal. You have maybe ¼" clearance side to side—not much, but enough to finally free it.

Once you get the glass out, scrape off the grit and grunge along the edge of the glass so you won't have to repeat that job the next time you open the machine. This is a good time to measure the glass and write the dimensions down in case the glass ever breaks and needs replacing.

On some Bally electro-mechanical machines from the 1970s, there is no removable lockdown bar and the glass does not slide out. The glass is mounted in its own metal frame and hinged. It lifts up on its hinges and is held up in place by a metal rod that swings up to hook to the frame. If necessary, the glass and its frame can be removed completely from the machine.

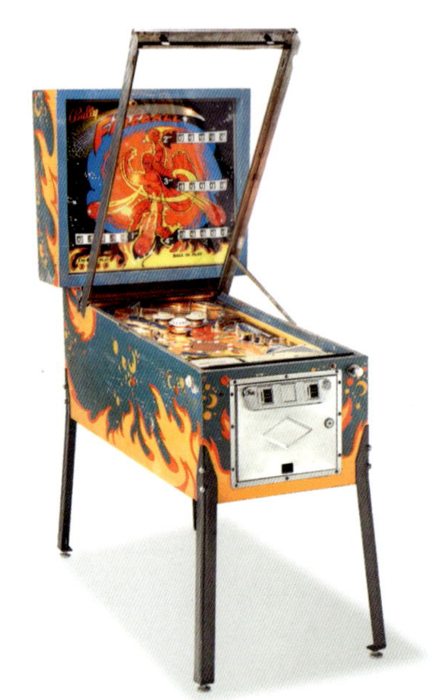

The playfield glass for Bally's **Fireball** was mounted in a hinged, removable frame.

3. Remove the Balls: With the glass out of the way, remove the ball (or balls) from the machine before lifting the playfield. Otherwise the balls will come crashing down when you lift the playfield and could break a plastic part. Stick your finger into the outhole between the flippers or in the hole near the ball shooter where the ball comes back into play. You can also use a long-handled magnet.

4. Lift the Playfield: With glass and balls removed, lift the playfield starting at the front edge. You may have to pull in the shooter tip a little to get it to clear the playfield as you lift it. You may need to pull the playfield forward about an inch so the top of the playfield at the far back of the machine, often covered with a decorative plastic or metal shield, doesn't grab or rub as you lift the playfield. Don't force anything if the playfield doesn't come up effortlessly. Stop and look for what's hanging it up.

On some electro-mechanical machines, the playfield may be latched in place. There is a long bar mounted to the underside of the playfield, about a foot back from the coin door, that you pull forward to unlatch the playfield. On some machines from the 1950s, the playfield is screwed down along the left and right sides, held by three or four screws on each side. On some old machines, the playfield is held in place by the same latch that releases the lockdown bar.

There is a hinged prop bar mounted on the right side of the cabinet that swings up and holds the playfield up in the air so you can work on it. If propping the playfield up does not give you enough working room you can slide the playfield forward onto the wooden or metal rails until you can lean it back against the backbox.

Electronic machines: Newer electronic machines have hinged playfields. Some have large handles mounted on the bottom of the playfield. You can use the handles to lift the playfield. You can slide the playfield forward so the handles will rest on the front panel of the cabinet.

GETTING BEHIND THE BACKGLASS (ELECTRO-MECHANICAL MACHINES)

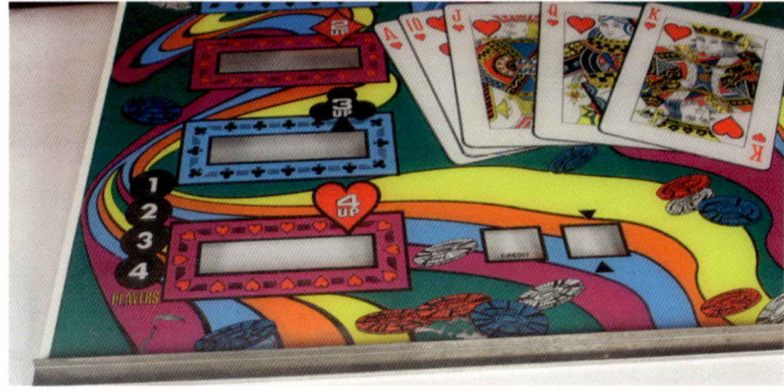

Illumination lamps, score reels, step-up units, and relays are mounted on a panel behind the backglass. Open and remove the back door of the machine to get to the backglass. If the back door is locked and there is no key then drill out the lock (see "Missing Keys").

On Williams and Bally machines there are two sliding latches at the top of the backbox that slide toward you (toward the back of the machine) that free up the backglass. The glass then lifts up and out from the front. There is a metal lip at the bottom of the backglass to aid in removing the glass. If the backglass hasn't been removed in a long time it may be difficult to remove. The metal lip gets stuck in its slot, "glued" in place by dirt and grime. Just keep nudging it carefully until you free it.

On Gottlieb machines, the backglass does not lift out from the front. The panel behind the backglass (the board with all the mechanisms mounted to it) is held in place with a latch that allows the panel to pivot on hinges down from the top, just enough so you can reach between the panel and the backglass and access the illumination lamps—hopefully. Often there isn't enough room to get your fingers in there. You will have to carefully remove the backglass, lifting it up and out the back of the machine, which will require delicate maneuvering.

Sometimes you'll find a backboard panel that has no latch, but is held in place with four or six screws. You have to remove the screws and then lean back the board to get to the backglass. We did not truly appreciate the manufacturers who decided to save a few bucks by mounting the board with screws instead of hinging it.

When you remove the backglass, put the backglass in a safe location. That glass is most likely irreplaceable.

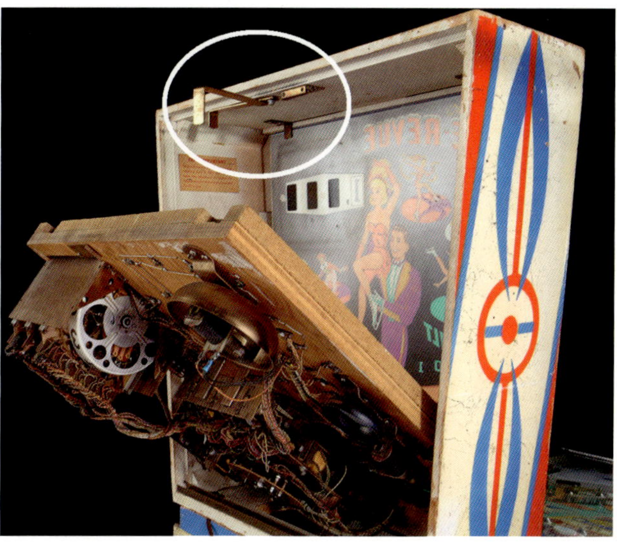

The latch (*circled*) pivots to let the hinged panel fold back, allowing access to the lights and the backglass. Be careful when opening the latch: The panel and the backglass can fall over!

GETTING BEHIND THE BACKGLASS (ELECTRONIC MACHINES)

To get inside the backbox of an electronic machine you have to remove the backglass. The backglass is usually locked in place by a latch—actually a horizontal metal bar inside the backbox. On the side or the top of the backbox is a lock that will rotate the bar, freeing the backglass. (No key? See "Missing Keys.") The backglass then lifts up and out of the backbox. There is a metal or plastic lip at the bottom of the backglass to aid in removing the glass. On newer electronic machines the backglass is actually a flexible translite, a thin, decorated sheet of plastic mounted behind a clear pane of glass.

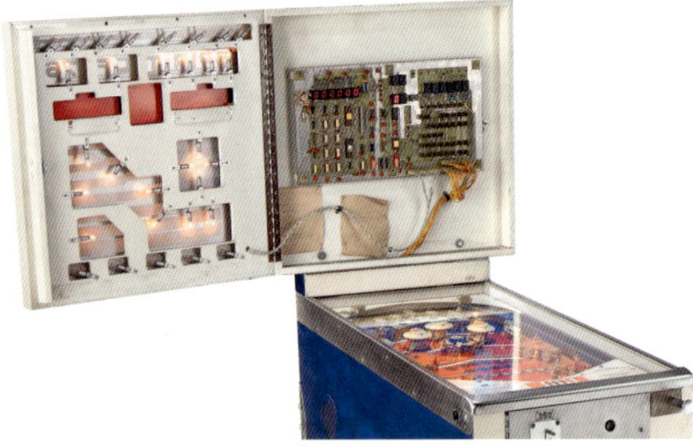

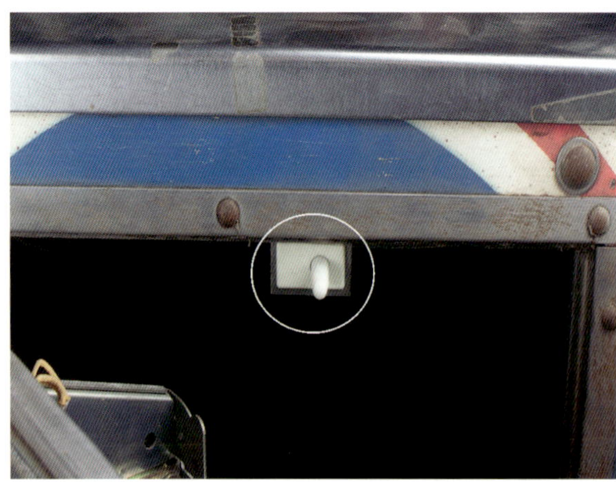

Behind the backglass is a hinged door that holds the lamps and score displays. There should be clearance between the door and the backglass. A door will occasionally warp or not close all the way and press against the backglass. Any pressure on the backglass could break the glass. I've seen a backglass that was scratched by the edge of a display module sticking too far out because the door didn't close all the way. Fix the door. Behind the hinged door are the circuit boards for the machine.

COIN DOOR SHUTOFF SWITCH

Starting in the 1970s, some states and some cities added an electrical code provision for pinball machines, requiring an automatic shutoff of power to the machine when the coin door was open. To meet this requirement the manufacturers added a spring-loaded "interlock switch" mounted inside the cabinet near the top of the coin door, with the switch button pressing against the door. Whenever the coin door opened, the switch button sprung out and cut the power. Repair technicians working on these machines had ways to override the switch if necessary.

Over the years those machines were resold and moved and shipped all over. I think I've had about ten of them come through my shop. So if you open the coin door of your pinball machine and suddenly all the power dies that's probably why.

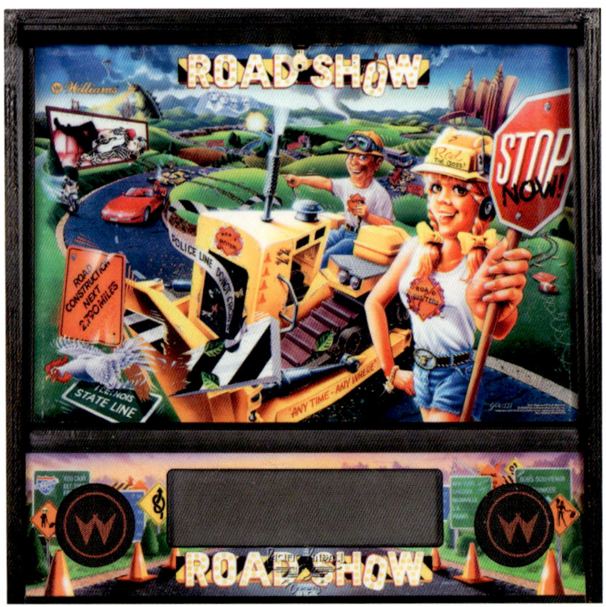

Williams **Road Show**, 1994. Official name is Red & Ted's Road Show, and features two large plastic heads on the playfield (Red and Ted) that can swallow the ball. Donated to the PPM by John Lee. Photography by RobPerica.com

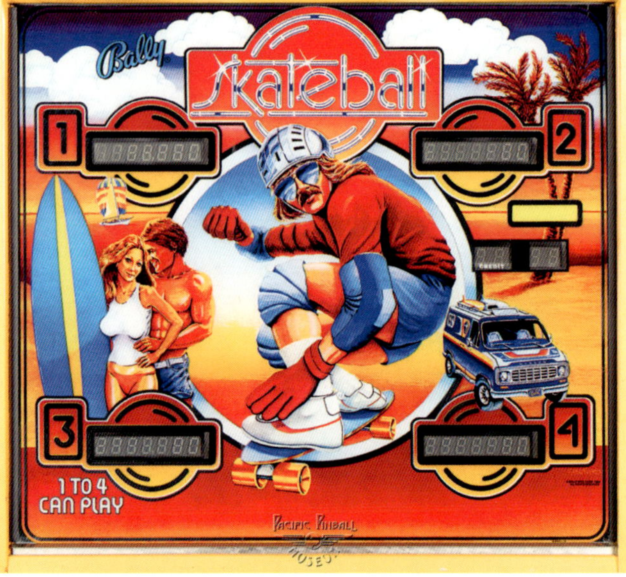

Bally **Skateball**, 1980. The shooter lane ran only half way up the playfield where it curved and sent the ball into action. Donated to the PPM by Tom and Deborah Rader. Photography by RobPerica.com

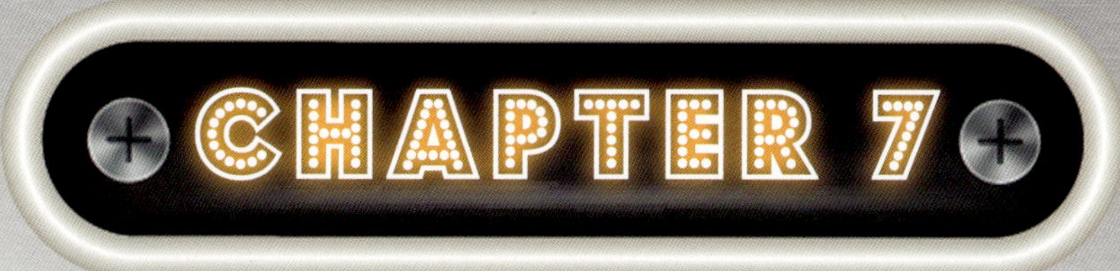

COMPONENTS AND FEATURES

The ball is wild.

—Harry Williams, founder, Williams Pinball

THE BALL

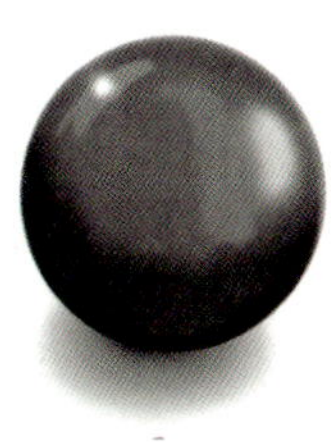

The silver ball, the reason for pinball, the center of the known universe. Lightning caught in a bottle. Only 1 1/16" diameter, only a few dollars to replace.

A ball that is pitted or rusted is like a rolling piece of sandpaper, grating harshly and destructively on what was a beautiful playfield. Your playfield is not replaceable. Check the ball.

Balls can get dented, particularly in multi-ball games where balls often slam together. A dented ball can damage the playfield and definitely will not play well. If you see any visible dents in a ball, replace it.

Standard pinballs are made of carbon steel. If not damaged or rusted they last forever. Pinballs do sometimes lose their shine and get a dull appearance, but if the surface is still smooth the balls are still quite usable. Some pinball people recommend polishing pinballs with jeweler's polish: the shinier the pinball, the smoother the roll. I've never polished a pinball, but if you do, make sure you've wiped the excess polish off so it doesn't get on the playfield.

All pinball suppliers sell new pinballs. If you live near a store that sells bearings, a 1 1/16" grade #25 chrome steel bearing should work fine (except read the warning below about chrome balls). Grade #25 is considered high quality.

CHROME BALLS

Some very shiny pinballs are chrome. They look mighty nice, but I don't think they play any better than regular pinballs, and they cost more than regular pinballs.

The problem with chrome pinballs is that if they are subjected to a magnet they can become magnetized. Regular steel pinballs cannot get magnetized. Magnetized pinballs can get "stuck" to other balls and to metal parts, particularly ball troughs under the playfield (the channel where the ball rolls from the outhole to the shooter). Still, the ball would have to be subjected to a strong magnet to get magnetized. Most pinball machines do not have magnets to worry about. But if you have a pinball machine with ball-saver or ball-grabber electromagnets you should avoid chrome balls.

Chrome balls are shinier—that's their only attraction—but it's difficult to recognize which type of ball is which unless you are looking at both types at the same time.

If you are not having problems with your balls I wouldn't worry about what kind of balls you have. But if the balls start "sticking" to each other, or getting stuck in the outhole, held in place by magnetism, you'll know they're chrome and need to be replaced. You could also take each ball out and see if it acts like a magnet: if it picks up small metal items. If it does, it's chrome and needs to be replaced.

When you buy balls from a pinball dealer, they'll send the regular balls and not the chrome balls unless you specify chrome.

HOW MANY BALLS IN A PINBALL MACHINE?

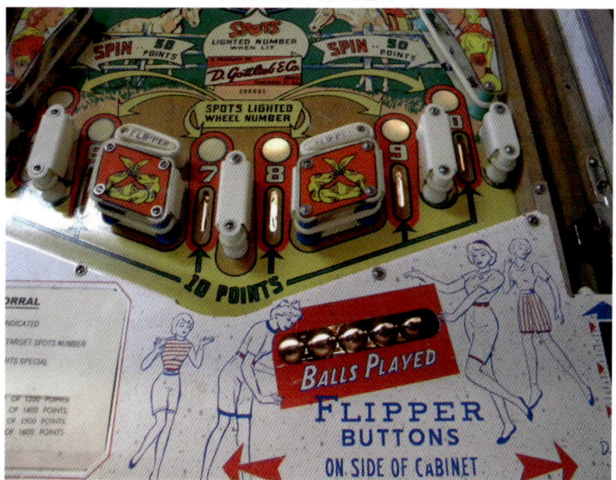

1961 Gottlieb **Corral** with the five balls sitting under the apron, ready to roll down (*out of sight*) when you start a new game. Notice how nearly useless the flippers are. Disabling the tilt and giving the machine a good shove when needed made the game more fun to play.

Back when flipper pinball was first invented and right up to the early 1960s, pinball machines had five balls. When you put your coin in the slot and started a game you could see—and hear—the five balls slide down the trough and into position. If you ever play an old pinball machine with five balls, you might get an appreciation of the anticipation created by the sight and sound of those five balls all rolling together, getting ready for the game.

The five balls rolled under the playfield, awaiting your first move. You pushed in the manual ball lifter and one ball would pop onto the runway, ready to be shot into action. After you lost that ball, after it dropped off the playfield, you could see it sitting in the trough for the rest of the game, inaccessible. You knew how many balls you had left by looking at the balls already used.

Kids often would pop out two or three balls in succession and create their own multi-ball games. If you were fast on the draw you could get all five balls playing at once—and end the game very quickly! Although newer pinball machines are often more fun to play than those old machines, something truly was lost when the pinball manufacturers invented the automatic ball return and eliminated the need for five balls.

When pinball went from the push-up ball lifter to the solenoid powered ball return only one ball was needed. The machine would kick out the same ball over and over. A light on the backglass or at the bottom of the playfield would tell you what ball you were playing. This not only saved the manufacturers the cost of providing five balls, it enabled the machines to be set for five-ball games or three-ball games. That one ball could kick out five times every game, or if the operator was greedy, merely moving a plug would convert the game to three-ball play.

Eventually by the mid-1980s all games went to three-ball play, although multi-ball was added as an extra attraction: The ability, once again, to have more than one ball in play at the same time.

When I bought my first pinball machine some years ago, the first thing I asked the guy who sold it to me was, "Where are the rest of the balls?" He thought it was a pretty funny question.

In the early 1960s, Gottlieb made a few pinball machines that had manual push-up ball lifters but only had one ball, not five. When the ball landed in the outhole it rolled down to the ball lifter, waiting to be lifted. Like modern machines, these machines shot that same ball five times, and the backglass had "Ball in Play" lights. What is interesting is that Gottlieb was making these manual lift machines at the same time they were making solenoid powered ball returns on other machines. I guess they had a large inventory of manual lift mechanisms to use up.

CAPTIVE BALLS

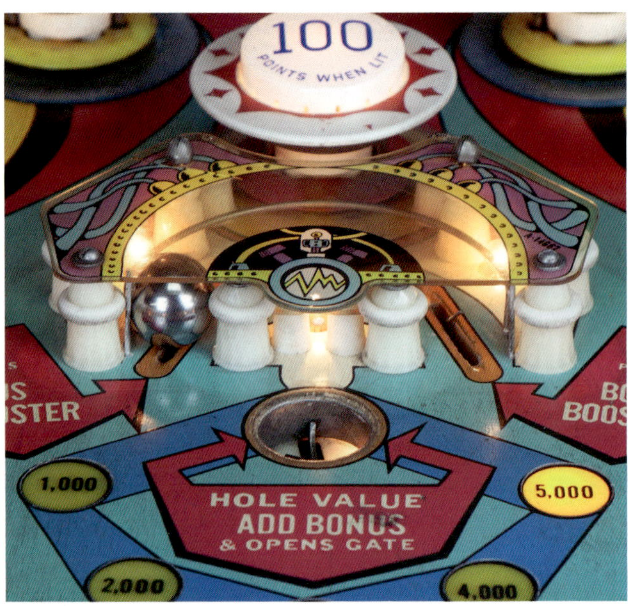

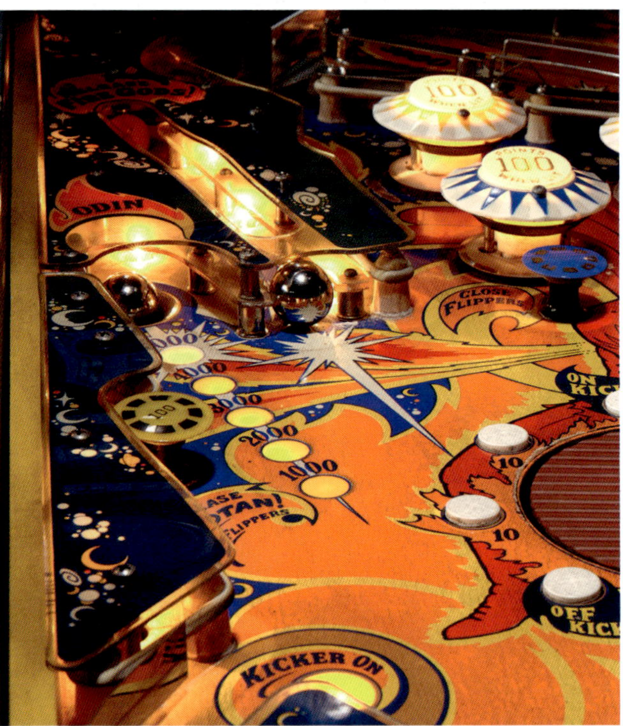

Top: Gottlieb **Duotron**, 1974, with a captive ball trapped in a semi-circle. At each end is a rollover switch that scores points and raises the bonus when the ball in play shoves the captive ball across the loop.

Bottom: Bally **Fireball**, 1972. The captive "messenger ball" in the middle is permanently trapped in the chute. When a ball in play hits the captive ball, the captive ball flies up the lane, hits the target, and then drops back down.

Some pinball machines have what's called a captive ball feature: a ball permanently enclosed in a short lane or in a half-circle horseshoe. When the ball on the playfield hit the captive ball, it would shove the captive ball up the lane or around the horseshoe to score points. Captive balls in short lanes are also called "messenger balls." Captive balls are the same as regular pinballs: they're interchangeable.

TILT BALL

Most pinball machines have two tilt switches: the swinging pendulum, and a rolling ball on a track just about the pendulum. The tilt ball is a smaller diameter than regular pinballs (see "Tilt").

RUBBER RINGS

It's amazing how many pinball machines have worn out rubber rings. Old rubber rings get hard and brittle, lose their elasticity, shed ring "dust," and make a game much less fun to play. New rubber rings can make an old game come alive. Rubber rings are inexpensive and are available from any pinball supplier.

Rings come in sizes specified by the inside diameter of the ring. Sometimes the ring itself will have the diameter molded onto it, and sometimes you have to measure the old ring (inside diameter). Very often the game manual will give the ring diameters. However, some ring diameters specified in game manuals are no longer manufactured. You'll have to use a slightly smaller or larger ring. I prefer using a smaller ring if it will fit. Taut rings give the ball a better bounce.

Until about 1995, all pinball machines came with white rings. Starting about 1995, some pinball machines came with black rings. White rings have more "bounce" and flexibility than black rings, which are made from a harder rubber. Unless your machine originally had black rings, use white rings. And there is nothing wrong with using white rings even if the machine originally came with black rings.

If you want an entire set of rings, many pinball suppliers will provide a full set of rings simply by knowing what game you have. Talk about dedication to their customers: Suppliers will look up your machine in their files, find out what rings are needed, and assemble a set of rings just for you, all for about $15.

On most electro-mechanical machines and early electronic machines, changing out a set of rubber rings will only

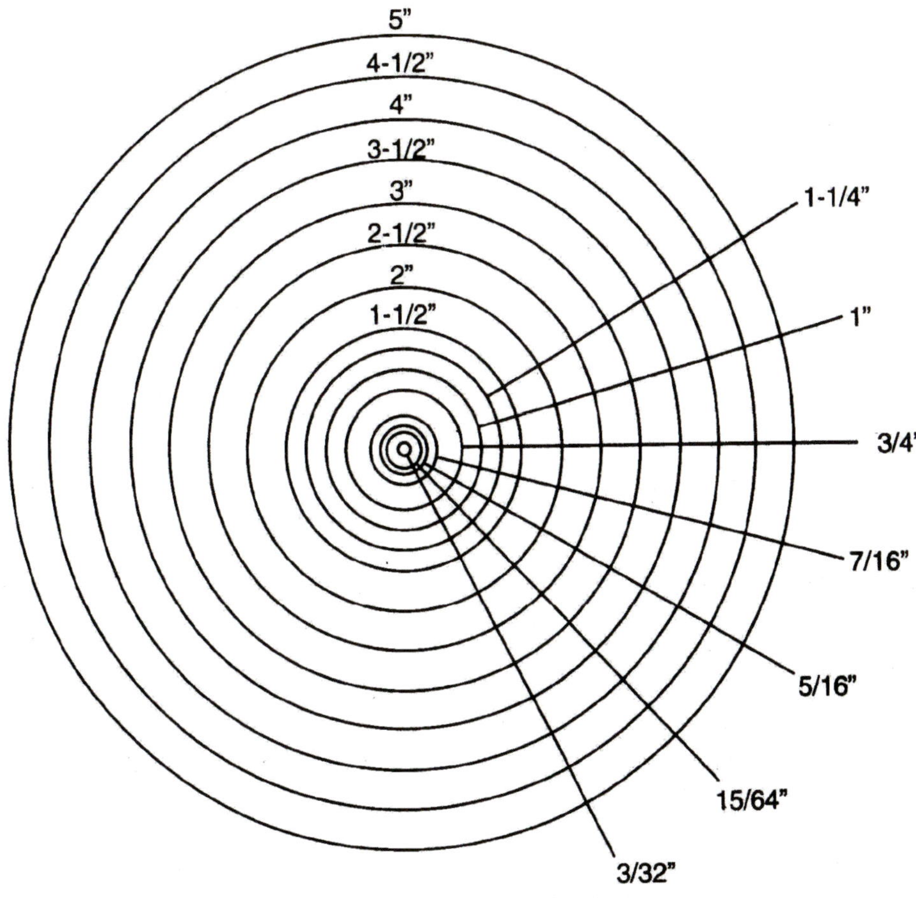

The ring sizes shown are the inside diameters of the rings. If your old ring size is no longer available, buy a ring the next size smaller. Better to have a tight, bouncy ring than a loose, flabby ring.

take about fifteen minutes. Wash your hands before putting on new rubber rings. Don't smudge those brand new white rings. If they do get smudged, rubbing alcohol on a tissue will usually clean them right up.

Save old rubber rings. Old rings can be used as backups if a new ring breaks. Small rings can be used to repair chimes. See "Chimes" in the "Repairs: All Machines" chapter.

Newer machines: Some newer machines with multi-levels of plastics and ramps sometimes require that you disassemble part of the playfield just to get to some of the rings. It can be quite a job. Make a diagram or take photos of all of the parts you remove—what goes where. If there are lamps buried down there with the rubber rings, check for burned out or loose bulbs while everything is already disassembled.

FLIPPER RINGS

Most pinball machines made after the 1960s have 3" flippers and use a standard ½" wide flipper ring available in yellow, red, and black. Some Gottlieb machines had narrower flipper rings, ⅜" instead of ½". Some machines with three and four flippers also have smaller 2" flippers that use mini-flipper rings. A few pinball machines used a curved flipper, called a banana flipper, that required a molded flipper rubber that fit over the flipper like a protective cover.

Pinball machines from the 1940s through the 1960s, and a few machines made in the 1970s, had 2" flippers that used a mini-flipper ring or a regular 1" rubber ring, and sometimes two 1" rings, one above the other.

REBOUND RUBBER

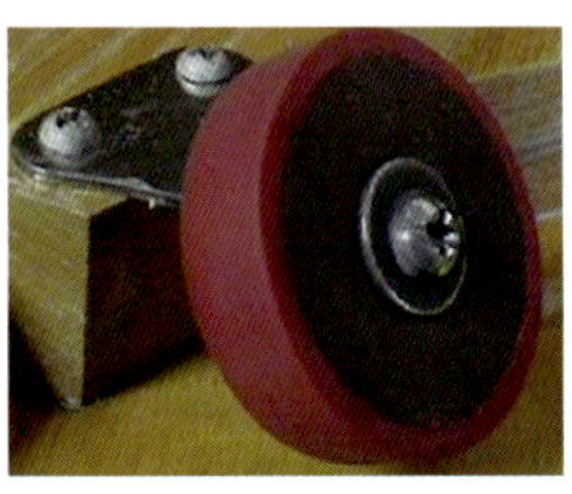

Many pinball machines have what's called a rebound rubber at the top left side of the playfield, a 1½" diameter rubber wheel that the ball hits coming around the top arch. Some pinball machines do not have a rebound rubber, but instead have a metal stop (a "gate").

Over the years the rebound rubber gets hard and loses its elasticity. The ball does not rebound as well as it should. You can buy a replacement rebound rubber from a pinball supplier. The original rubber is probably attached with a rivet, requiring drilling out the old rivet and riveting or using a bolt to attach the new rubber.

Instead of replacing the rebound rubber, you can put a mini flipper ring around the old rebound rubber to increase the elasticity if there is enough clearance to fit the ring. This may or may not work, and may or may not look good, but for about 75¢, it's worth a try.

SCORE REELS (ELECTRO-MECHANICAL MACHINES)

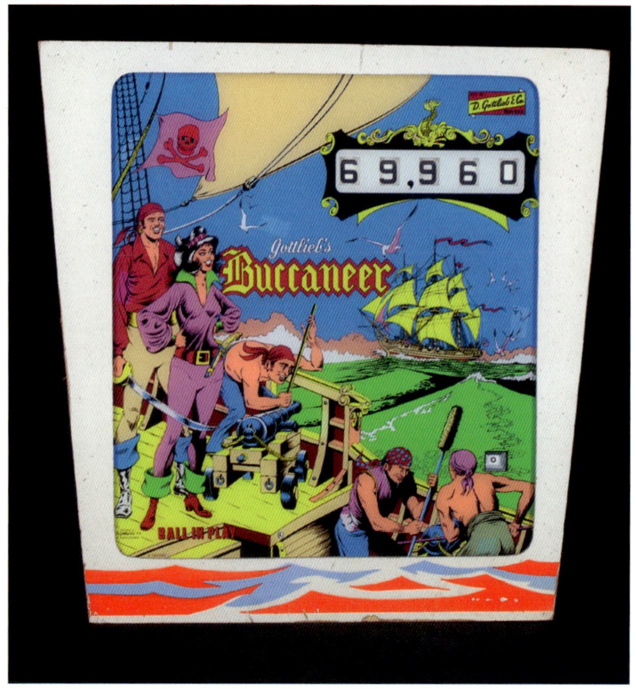

Gottlieb **Buccaneer**, 1976. When machines added the 10,000 score reel they dropped the 1's reel, replacing it with a dummy reel permanently set at 0.

Electro-mechanical scoring is accomplished by score reels, with the numbers 0 through 9 on each reel. Three or four score reels are mounted next to each other in the backbox to score 10s, 100s, and 1,000s. Machines from the early 1960s also had a 1s score reel, scoring 1 to 9. Later machines dropped the 1s reel, replaced with a dummy reel permanently set at 0, and added a 10,000s score reel. If the score went over the maximum score the reels could register, the game would "turn over" and start at zero again, but continue to rack up points.

A few machines have a 100,000 score reel. Some machines have a 100,000 light. Some pinballs had a light that says "Over The Top" that flashes when you turn the game over. Some Bally games also have a buzzer that rings when you hit 100,000.

Score reels often get dirty and gummed up, get sluggish and have difficulty turning, or stop working altogether. A main reason a game will not start when you push the start button is stuck score reels that don't reset. See "Score Reels" in the "Repairs: Electro-mechanical Machines" chapter.

THE MATCH

The match is a number that appears somewhere on the backglass of the pinball machine when the game is over. On electro-mechanical machines it is a painted number that lights up. On electronic machines it shows on one of the displays. If the last two digits on your score (or the last single digit on machines from the 1960s and earlier) are the same as the lit match number on the backglass, you get a free game. The match is a much-loved part of pinball—the great equalizer. No matter how bad or good your score, you still have one chance in ten of matching and getting a free game.

Electro-mechanical machines: The match unit is a rotating disk with ten contacts, each contact corresponding to a different number: 10, 20, 30, etc. One or more of the bumpers or switches on the playfield activated the match unit. Every time the ball would hit the bumper, it would advance the match unit one step. So all through the game the match unit was rotating. It was pure random chance where the match unit would stop when the game was over. Most machines have a plug inside the backbox or on the floor of the cabinet that can disable the match. If the match isn't working see "Match Unit" in the "Repairs: Electro-mechanical Machines" chapter.

Electronic machines: The match number is controlled by a computer chip. Unlike electro-mechanical machines that always gave the player one chance in ten (there were ten steps on the rotating match unit), electronic machines allow you to set the percentage of free games anywhere from zero on up. Most electronic machines also have a match on-off adjustment setting.

TOTAL PLAY METER (GAME COUNTER)

Pinball route operators—the businesses that owned the pinball machines put into bars and bowling alleys and other locations—usually split the money earned with the location's owner. So route operators needed a way to check on how many games were played since the last time the operator came by to pick up the quarters; a way, as they say, to keep honest locations honest. That's why every pinball machine has some sort of record of total plays.

Electro-mechanical machines: Electro-mechanical machines have a game counter called the "total play meter" that added one game to the count every time a quarter dropped in the slot. The total play meter could not be reset; it just kept adding numbers—total plays forever. Route operators would record the total plays on the meter each time the operator came to the location and then count the quarters to be sure the money and the numbers agreed.

Most electro-mechanical machines still have their total play meters. And even after years and years, many of those meters are still working. How many thousands of times did your pinball machine get played over its entire lifetime? Take a look at the number on the meter, then start a game and see if the meter registered another game or not. If the meter is still working, you'll know. If the meter is not working, at least you'll know that there are at least as many games as the meter shows.

Electronic machines: Electronic machines have bookkeeping totals built into the programming of the machine, accessible only by the owner of the machine.

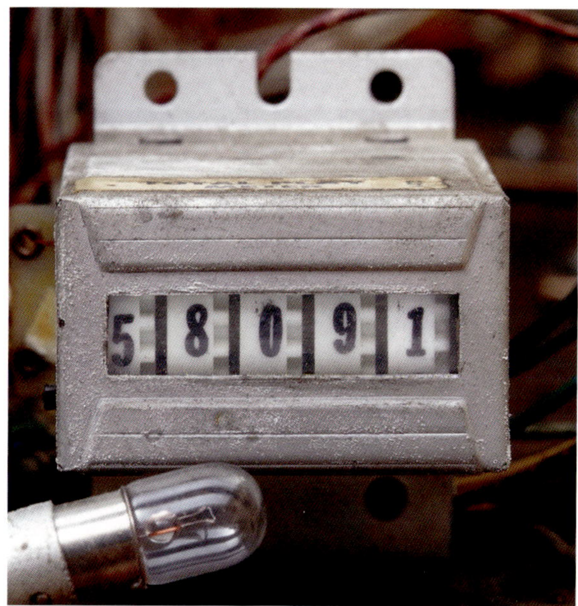

HIGH SCORES AND FREE GAMES

The goal of pinball since pinball was invented: Get to the high score and win a free game. (Well actually, the goal of a lot of early pinball was get to the high score and win $$. The money payoff wasn't legal, but it sure was popular.)

Winning a free game loses a lot of its appeal once you own your own machine and can play all the free games you want. But it is still fun to try to get to the high score and to hear that loud "crack" when the free game registers. A coil shoots its plunger up to bang against a metal stop, making the sound every pinball player loves to hear. That coil is called the knocker. If your knocker isn't working, or if it sounds like a timid tap instead of a good solid wallop, see "Knocker Not Knocking" in the "Repairs: All Machines" chapter.

HIGH SCORE SETTINGS

Every pinball machine has an adjustment where you decide what high score, or scores, win free games. If you find the free game score is too low, if almost everyone who plays the game wins, you can easily raise it.

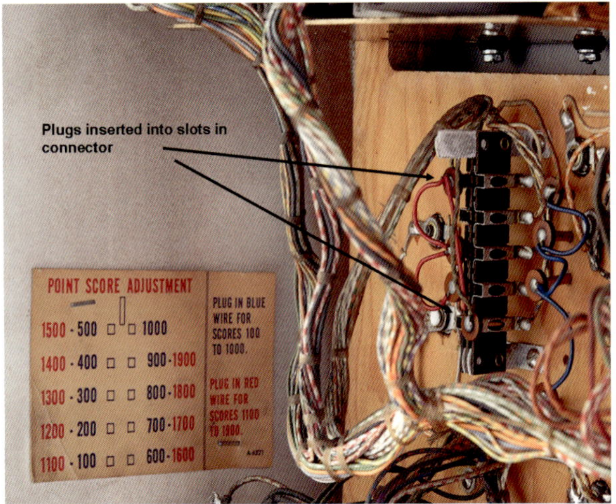

High Score adjustment (to win free games). Metal plug-in connectors, with color coded wires, plug into slots to match the high score shown on the Point Score Adjustment card. Multiple plugs can be plugged into different slots to have more than one high score.

Electro-mechanical machines: In the backbox are several plugs with color-coded wires that plug into different sockets. Different plugs in different sockets determine the score needed to get the first and subsequent free games. There is usually an instruction card stapled to the backbox giving you your choices.

Electronic machines: There is a setting in the game adjustment mode. See “Game Adjustments” in the “Setting Up” chapter.

MULTIPLE WAYS OF SCORING

Some pinball machines have two different scoring setups: the regular points score and a second way to score. Both sets of scoring happened simultaneously, independent of each other.

Gottlieb’s Pro Football had a “High Score” for the regular scoring and a “Point Score” for making plays and touchdowns. Bally’s Twin Win, a race car theme, had “Points” for the regular scoring and “Laps” for scoring how many times two competing race cars (actually lights on the playfield) completed a lap. Each “car” had its own lap counter. Twin Win is a two-player game, so players could compete for both high score and the most laps. In both Pro Football and Twin Win the regular high score determined if you would win a free replay.

Williams’ Klondike had, in addition to the regular scoring, three slot machine reels on the playfield that spun when you hit certain targets and awarded free games if you lined up different combinations. Both the slot reels and the regular high scores could win free replays.

Other multiple scoring machines were manufactured over the years, but not many. They were more complicated to design and more expensive to manufacture. But they definitely added an extra dimension of fun to playing a game.

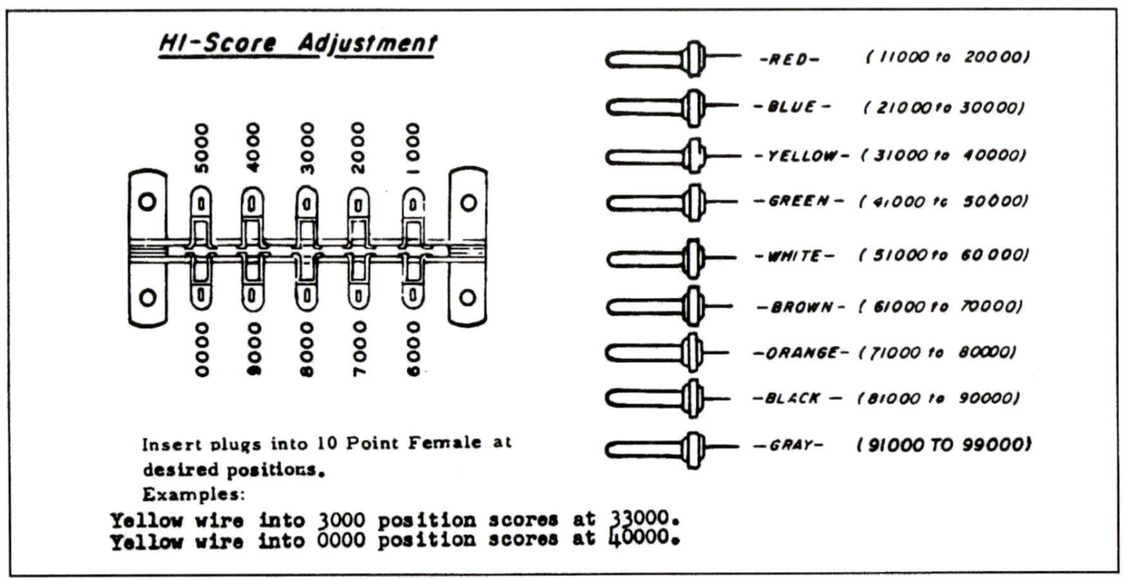

HI-Score Adjustment

5000 4000 3000 2000 1000

0000 9000 8000 7000 6000

-RED- (11000 to 20000)
-BLUE- (21000 to 30000)
-YELLOW- (31000 to 40000)
-GREEN- (41000 to 50000)
-WHITE- (51000 to 60000)
-BROWN- (61000 to 70000)
-ORANGE- (71000 to 80000)
-BLACK- (81000 to 90000)
-GRAY- (91000 TO 99000)

Insert plugs into 10 Point Female at desired positions.
Examples:
Yellow wire into 3000 position scores at 33000.
Yellow wire into 0000 position scores at 40000.

High score adjustment instructions for an electro-mechanical machine. Electronic machine high scores are set using the adjustment buttons on or near the coin door.

"SCORE TO BEAT"

Around 1960, the Gottlieb Company came up with an idea to put a "Score to Beat" window (or a "Previous High Score" window) in some of their backglasses. There was a removable card behind the window that showed a score that players would try for. Gottlieb provided a collection of score cards with different "scores to beat."

However, when a player hit the Score to Beat, the machine didn't do anything. No lights flashed. You didn't win a free game. And the Score to Beat didn't change, as route operators never bothered to change the score card. I guess the Gottlieb Company realized quickly that the Score to Beat feature was not such a great idea. In fact, it worked against the machine making money. When players hit the Score to Beat and found out, the next time they played, that they were trying for the same score to beat, the player often lost interest in that machine.

Today, some people who own pinball machines with a Score to Beat (or Previous High Score) window make their own cards that show the highest score made on their machine, making it more fun to keep trying to beat the latest "score to beat."

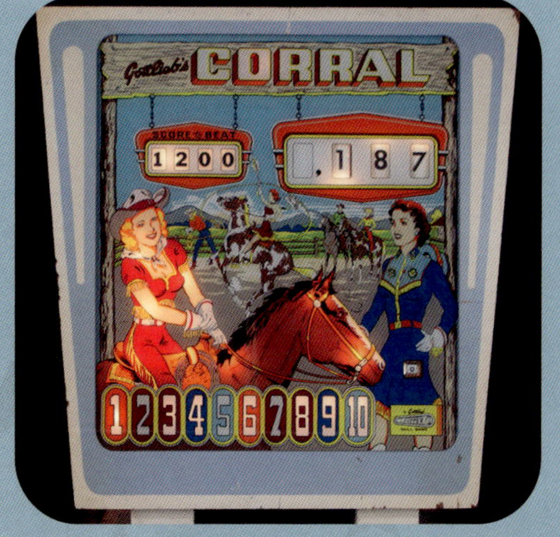

Gottlieb **Corral**, 1961. The 1200 in the Score To Beat window is a card inserted behind the backglass. Note in the regular score window that the thousand window is not a score reel, but a light that lit a 1 if someone got to 1,000.

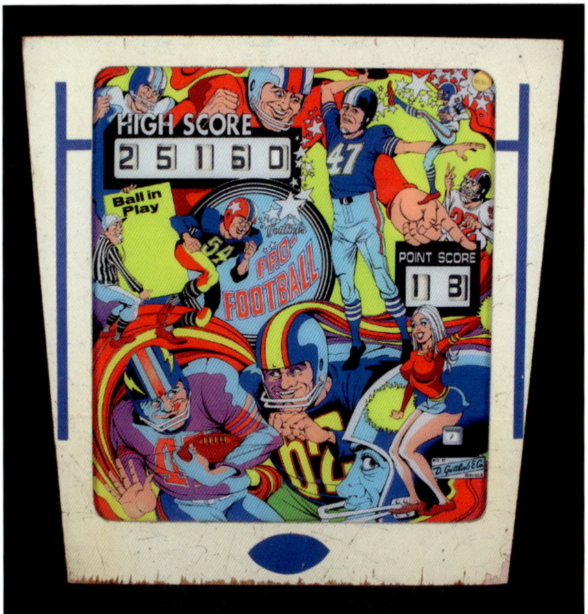

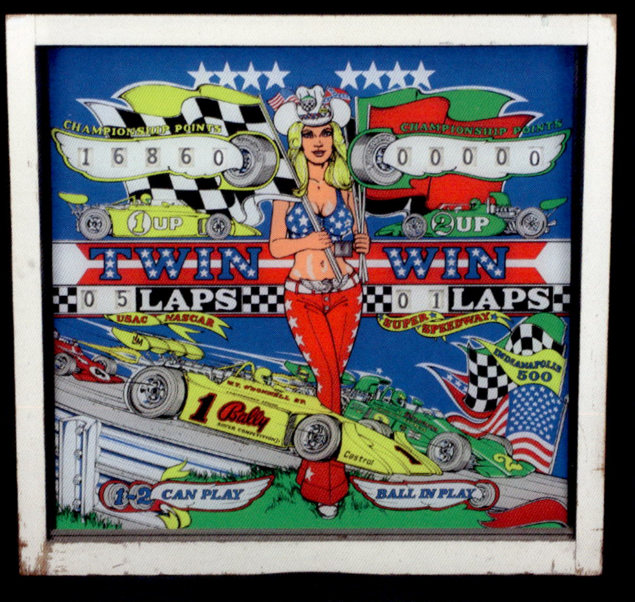

In Gottlieb **Pro Football**, 1973, you scored points moving the football up and down the playfield, in addition to the regular scoring. In Bally **Twin Win**, 1974, two players competed with their race cars (actually lights moving around a race track), scoring points for making laps, in addition to the regular scoring. Interesting that both machines have solenioid powered shooters, launching the ball from between the flippers when you push a shooter button.

NO SCORE: TRY AGAIN

Most pinball machines are designed to give you a friendly "Try Again" if you don't score anything on the first ball. The machine will not advance the ball count to the second ball until you score something. Some machines offer the same "Try Again" on every ball, not just the first.

Electronic machines: Some newer electronic machines give you another try even if you do score, if the score is very low or if the ball drains too quickly. In the popular Capcom game Breakshot, if you score poorly a female voice laughs at you and says, "Come on, you can do better than that," and you get another try. And get to be embarrassed a little if anyone is listening.

FUSES

Fuses protect your pinball machine should a coil, or wiring, or a circuit short out. See "Short/Short Circuit" in the "Repairs: Basics" chapter.

Do not—DO NOT—bypass fuses. Do not put a jumper wire across them, do not wrap aluminum foil around them, and do not use a nail instead of a fuse. You can destroy your machine. You can burn your house down.

Before you plug in a machine you just brought home, check all the fuses. Almost every time I get a machine in the shop I find at least one incorrect fuse.

Electro-mechanical machines usually have up to six or seven fuses. Electronic machines have as many as twenty fuses. Some fuses are on the floor or side of the cabinet, often grouped together. Some fuses are mounted on the underside of the playfield, sometimes in odd places. The main power fuse is often by itself, near the transformer. In electronic machines, most of the fuses are in the backbox.

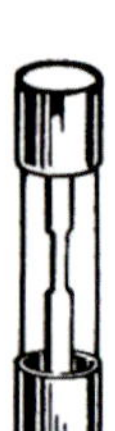

Fuses are rated by amperes and volts, and they are either regular (fast blow) or time delay (slow blow). The volt and ampere ratings are stamped on the metal ends of the fuses. A letter code on the fuse will tell you if it is fast or slow blow (see Letter Code).

The correct fuse ratings for your pinball machine are usually shown on a card or a label on or near the fuse holders. The schematic will show the fuses and their ratings, and the instruction manual may have fuse information.

AMPERES

Amperes ("amps") are a measure of current flow. Do not use a fuse with an ampere rating higher than the rating specified by the manufacturer. If the machine calls for a five amp fuse don't substitute a 10 amp fuse. If you use a larger amperage fuse than specified, you can damage components in your machine, and you could overheat the circuit enough to start a fire.

If you use a lower amperage fuse than the rating specified—for example, putting a 10 amp fuse in a circuit that specifies 15 amps—you will not damage the machine, but you may blow the fuse.

12 ampere fuses: Some electro-mechanical machines from the 1970s specified a 12 amp fuse in the lamp circuit. 12 ampere fuses are no longer readily available. However, if most of the lamps in your machine are #47 lamps instead of the original #44 lamps (covered under "Lamps"), the lamp circuit is drawing less power. You can safely substitute a 10 amp fuse. It should not blow.

VOLTS

The general rule is that you can use a high voltage fuse in a low voltage circuit, but you should not use a low voltage fuse in a high voltage circuit.

Voltage ratings indicate how well a blown fuse protects a circuit. When a fuse "blows," the fuse does not really "blow" as in "blow up." The metal element in the fuse separates (it actually melts), stopping current from flowing.

If a blown fuse is in a high voltage circuit, sometimes the voltage is powerful enough to jump the space between the melted ends of the metal element—called arcing—sending the current right through the fuse as if the fuse was still okay. *Not good.* A high voltage fuse is designed to keep that from happening.

So a fuse rated at 120 volts will work just fine in a 30 volt circuit. But a 32 volt fuse in a 120 volt circuit might cause arcing, and cause serious problems.

Now don't confuse volts and amperes. They are different things. A high voltage fuse is okay in a low voltage circuit. A high amperage fuse is NOT okay in a low amperage circuit.

In pinball machines, most of the circuits are between 6 and 50 volts and should work fine with 32 volt fuses, the most common fuse being manufactured today. It is difficult to find regular fuses rated at more than 32 volts.

The 120 volt circuit coming into your machine (the line cord from your home electrical outlet) is usually protected by a slow-blow fuse (explained next). That fuse is high voltage, 120 volt or 250 volt, and low amperage (½ amp, 1 amp, or 5 amp), still readily available.

FAST BLOW, SLOW BLOW

Most fuses are fast-blow fuses. You won't see the term "fast blow," but regular fuses are all fast blow. If there is a surge of current or an increase in current above the fuse rating the fuse will blow, right away—fast.

Slow-blow fuses are delayed action fuses that can withstand a momentary surge of high current. You will usually find slow blow fuses on 120 volt circuits, because when you first turn on the machine the line voltage coming from your house wiring sometimes surges briefly.

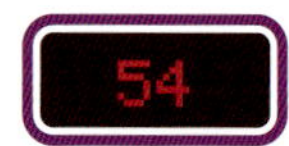

If the pinball machine calls for a regular fast-blow fuse—it will just specify "fuse"—don't substitute a slow-blow fuse. Even a momentary surge of current could damage electronic components.

TYPES OF FUSE

Most pinball machines use a standard size glass fuse, 1¼" long and ¼" diameter (sometimes referred to as a #3 size) available from hardware stores, electronics stores, and pinball suppliers. Glass fuses have a visible metal wire or metal strip so you can see if the metal has melted and separated.

Fuses are also available as ceramic fuses, which are opaque white or off-white. You cannot see the metal wire, so you can't tell by looking if ceramic fuses are blown. Ceramic fuses are used in appliances where they are subject to a lot of heat, such as a stove. Ceramic fuses can be used in pinball machines, but since you can't see inside of them, and they cost quite a bit more than glass fuses, few people use them.

LETTER CODE

Most fuses have a letter code, or combined number and letter code, stamped on one of the metal end pieces of the fuse. You can substitute different letter codes as long as the fuse type—regular (fast blow) or slow blow—is correct. Don't confuse a letter code number with an ampere rating. For example, a 3AG fuse refers to a #3 size fuse, not 3 amperes. The fuse will also have the ampere rating stamped on one of the metal end pieces.

Common Fuse Codes:
3AG: regular (fast blow)
3AGSB: slow blow
ABC: regular ceramic
AGC: regular glass
AGX: regular, designed for electronics
ANB and ASB: slow blow
GMA: small 5 mm × 20 mm regular. Not used in pinball machines.
GDC: small 5 mm × 20 mm slow blow. Not used in pinball machines.
MDA: slow blow ceramic
MDL: slow blow glass
MDQ: slow blow. Not used in pinballs.
MKB: regular fuse, low resistance. Not used in pinball machines.
MSL: slow blow.
SFE: regular glass.

CHECKING FUSES AND FUSE HOLDERS

Look at each fuse in your machine to see if it looks burned. But be aware that fuses may be defective with no visible signs. Test each fuse with a multi-meter or test light. Remove one side of the fuse or remove the entire fuse from the fuse holder. If you test a fuse mounted in its holder you may get a false reading. **Unplug the machine from the wall before touching a fuse**. Some fuses are wired directly to the 120 volt circuit. If you have a blown fuse see "Troubleshooting Blown Fuses" in the "Repairs: All Machines" chapter.

Examine fuse holders for broken, loose, or corroded connections. Poor contact between fuse and holder will cause all kinds of problems. If all the lights on the machine are out, the fuse holder is sometimes the culprit. Clean and adjust or replace the fuse holder.

LAMPS

There are dozens of different lamps used in pinball machines: different shapes, different bases, different voltages, different levels of brightness, and different applications.

HOT AND BRIGHT LIGHTS

For many years all pinball machine lamps ran on a 6 volt circuit that was completely separate from the circuits that powered the operating components in the machine. Lamps rated at 6.3 volts were commonly used. Modern pinball machines with strobe lights often have higher voltage lamp circuits.

Even in the low 6 volt circuits, some lamps burn much hotter and draw much more current than others. Reducing heat and high current draw can extend the life of your pinball machine. Backglasses, bumper caps, and playfield plastics can be damaged if lamps are too hot.

Lamps with blackened or silver-looking tops—caused by carbon buildup inside the lamp—should be replaced. The black and silver retain heat, and the bulbs burn much hotter than clear bulbs.

Some people remove some of the lamps from behind the backglass, such as those that are purely for decorative illumination (as opposed to the lamps that serve a purpose, such as lighting up scores, ball in play, etc.), to extend the life of the backglass paint. Usually this is unnecessary and unattractive, but if your old backglass is starting to peel or bubble, fewer lights may help the remaining paint to survive.

I mention this elsewhere in the manual, but don't turn on a pinball machine in a cold room. The dramatic temperature change behind the backglass when the lamps come on can peel paint.

The newer pinball machines with translites are not as susceptible to temperature change as older painted backglasses.

#44 VERSUS #47 LAMPS

Most electro-mechanical and early electronic machines use #44 or #47 lamps. #44s and #47s look identical and are interchangeable. Both have a bayonet socket. The lamp number is printed or stamped on the base of the lamp.

#44 lamps are brighter, draw more current, and burn hotter than #47s. The heat generated by #44 lamps can damage backglass paint over time. You may add years of life to your backglass by switching #44 lamps to #47s.

#44 lamps can cause damage on the playfield. #44s under the colored, decorative plastic shields (the "plastics") can warp the shields over time, especially if left on for hours at a time. #44 lamps are hot enough to scorch bumper caps. Switching to #47 lamps can save the plastics and the bumper caps.

Sometimes #44 lamps are a better choice than #47s. Some playfield and backglass lights are part of the scoring, indicate bonuses or extra balls, or serve some other important function. It is important that these lights shine brightly. You may want to use the brighter #44 lamp to get better illumination.

#47 lamps draw less current than #44s. In electronic machines, switching to #47s will put less stress on power supplies and possibly extend their life. High-current lamps are a common cause of burned wire connectors (the plastic plugs that attach to the circuit boards) in electronic machines.

#47 lamps cost about 20¢ each. It's not much of an investment to switch from #44s to #47s.

Many automobile light bulbs look like #47 bulbs and will fit in pinball light sockets, but automobile bulbs are 12 volts. The lamps will be dim, if they light at all, in a pinball machine's 6 volt circuit. Don't waste your money on them.

FLASHERS

Flashers are lamps that blink on and off constantly. Every pinball machine should have two or three flashers behind the backglass, just for the great effect. Flashers use the same bayonet base as the #44 and #47 lamps but have larger, fatter globes, often requiring a wider space to fit the bulb. Quite often the manufacturer made a wider opening for the flashers where they were originally installed. If not, you can easily file or route out the opening in the backbox wood just enough to allow the fatter bulb to fit.

Some electro-mechanical machines use flashers as part of a relay circuit. The flashers are mounted on relays inside the cabinet. See "Relays with Lamps" in the "Repairs: Electro-mechanical Machines" chapter.

#455 is a commonly used flasher lamp.

#555 AND #259 LAMPS

These two lamps are almost identical and are interchangeable. They have a glass wedge base that snaps into the socket. Many newer electronic machines came with #555 lamps. The #555 burns a little hotter than the #259. Switching from #555 to #259 lamps might help with overheating of connectors and the stress on power supplies.

LED LAMPS

LEDs (light emitting diodes) generate no damaging heat, draw less current than incandescent lamps, shine brightly, and last just about forever. LEDs are more expensive than regular incandescent lamps, though the cost of LEDs has been dropping steadily in the last few years. Some people love bright LED lights, but some people find LEDs way too bright.

LEDs are straight replacements. They are manufactured with bayonet sockets and wedge sockets, so they will fit whatever type of lamp socket your machine has.

Not all LEDs are the same. Be sure to get LEDs made specifically to work in your pinball machine. LEDs in pinball machines need to handle AC and DC current, need to have correct polarity, and may need built in diodes and bridge rectifiers. Most pinball suppliers sell LEDs. Some suppliers will put together a complete set just for your machine.

LEDS in electronic machines: New electronic machines are coming from the factory with LEDs. In older electronic machines—those that originally came with incandescent lamps—LEDs may or may not work the way the original incandescent lamps work. LEDs are not just a brighter version of traditional incandescent lamps. LEDs will operate on much lower voltages than incandescent lamps. Pinball machine switching circuits and voltages were designed for the characteristics of incandescent lamps. You may find that LEDs that should turn off and on stay on all of the time, they may flicker, or they may exhibit all kinds of strange behavior. You may need a different driver board.

Most pinball suppliers are familiar with converting to LEDs and can tell you what your machine needs. And if you're just experimenting on your own, the good news is that I have yet to hear of LEDs damaging anyone's pinball circuits. So if you like LED's brightness and can afford their higher cost, I guess it doesn't hurt any to give them a try.

ILLUMINATION ADJUSTMENTS

Some pinball machines have an illumination adjustment, allowing for brighter or dimmer settings. Look in the instruction manual or inside the coin door.

LAMP CHART

This chart is a list of lamps that you might find in a pinball machine. These are incandescent lamps, not LEDS.

Lamp number: This is a standard universal number used by all manufacturers.

Color: Lamps are clear glass. #44, 47, 555, and 906 lamps are also available in colors.

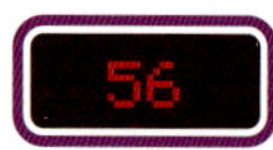

Letter type:

A, B, and C. All had the same bayonet base used in most electro-mechanical and early electronic pinball machines. The different shapes and sizes of the glass bulbs (A, B, and C) will fit some pinball machines and not others, depending on how the lamp sockets were mounted on the machine

D. Larger bayonet base. Some of the new strobe lights use the D bayonet base.

E and F. Wide and narrow wedge (snap in) base used in newer electronic machines.

G and H. Screw-in base (G) and two-pin (H) used on some very old machines.

I. Long wire leads used on some newer electronic machines.

Note that the letter designations are just to help read the chart. The letters are not universal, and are not used to identify lamps. Lamp numbers identify the lamps.

Special thanks to Pinball Resource and to Marco Specialties for their extensive lists of lamps.

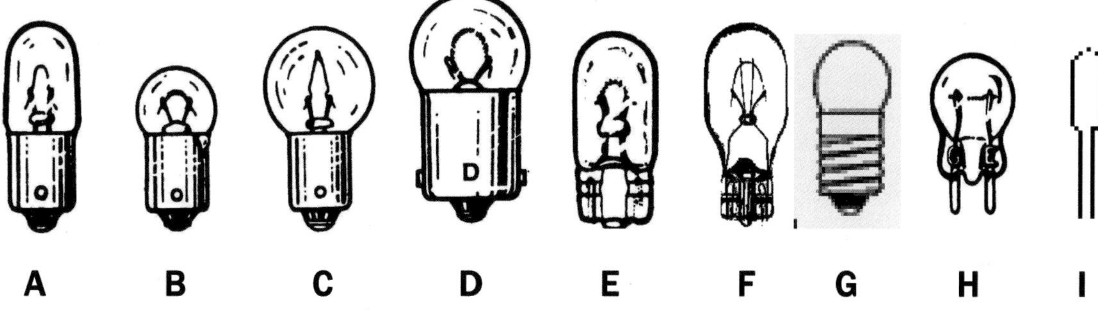

12. Type H. Two-pin. Some machines from the 1950s.
37. Type E. Wedge. 14 volts.
40. Type G. Screw. 6 volts. Rarely used after the 1950s.
44. Type A. Bayonet. 6.3 volts. Original most EMs, early electronics. Interchangeable with #47.
46. Type G. Screw. Rarely used after the 1950s.
47. Type A. Bayonet. 6.3 volt. Most EMs, early electronics. Interchangeable #44.
50. Type G. Screw. 6.3 volts. Rarely used after the 1950s
51. Type B. Bayonet. 7.5 volts.
52. Type G. Screw. 14.4 volts. Rarely used after the 1950s.
53. Type B. Bayonet. 14.4 volts.
55. Type B. Bayonet. 7 volts. Can handle vibration. Some Bally and Williams games.
57. Type C. Bayonet. 14 volts.
63. Type D. Large bayonet. 7 volts.
67. Type D. Large bayonet. 13.5 volts. Used in many electronics.
73. Type E. Wedge. 14 volts.
74. Type E. Wedge. 14 volts.
79. Type E. Wedge. 6 volts.
81. Type D. Bayonet. 6.5 volts.
85. Type E. Wedge. 28 volts.
86. Type E. Wedge. 6.3 volts. Clock, chaser circuits.
88. Type C. Bayonet. 6.8 volts.
89. Type D. Large bayonet. 13 volts. Flasher electronics.
90. Type B. Bayonet. 13 volts.
147. Type E. Wedge. 7 volts. Midway home.
158. Type E. Wedge. 14 volts.
159. Type E. Wedge. 6.3 volts. Interchangeable #259, 555.
161. Type E. Wedge, 14 volts Coin doors for redemption machines.
168. Type E. Wedge. 14 volts.
192. Type E. Wedge. 14 volts.
193. Type E. Wedge. 14 volts.
194. Type E. Wedge. 14 volts.
222. Type G. Screw. 2.25 volts.
257. Type B. Bayonet. 14 volts.
259. Type E. Wedge. 6.3 volts. Interchangeable #444, #555. Burns cooler.
286. Type E. Wedge.
303. Type D. Large bayonet. 28 volts.
313. Type A Bay. 28 volts. Black Hole, Haunted House.
330. No illustration. 14 volts.
367. No illustration. 10 volts.
387. No illustration. 28 volts.
444. Type E.. Wedge. 6.3. volts. Interchange #259, #555.
455. Type C. Bayonet. 6.5 volts. Flasher. Common in EMs, early electronics.
555. Type E. Wedge. 6.3 volts. Most newer electronic. Also available in colors. Interchangeable with #444 and #259, which burn cooler.
585. Type F. Wedge. 28 volts.
656. Type E. Wedge. 28 volts.
657. Type E. Wedge. 28 volts.
658. Type F. Wedge. 14 volts.
680. Type I. Wire leads. 5 volts.
755. Type A. Bayonet. 6.3 volts. Interchangeable #44, #47, but dimmer, cooler.
756. Type A. Bayonet. 14 volts.
757. Type A. Bayonet. 28 volts.
904. Type F. Wedge. 13.5 volts.
906. Type F. Wedge. 13 volts. Flasher in many electronic. Also available in colors.
912. Type F. Wedge. 12.8 volts. Flasher used in many electronic machines.
921. Type F Wedge. 12 volts.
1073. Type B. Bayonet. 12.8 volts. Substitute for #1156. Williams Getaway beacon.
1129. Type A base (different globe). 6.4 volt.
1156. Type B. Bayonet. 12.8 volts. Substitute for #1073. Williams Getaway beacon.
1251. Type D. Large bayonet. 28 volts.
1449. Type G. Screw. 14 volts.
1458. Type B. Bayonet. Replace with #1464.
1464. Type B. Bayonet. 22 volts. Used as a replacement for #1458. Found in bingos.
1683. Type D. Large bayonet. 28 volts. Rotating beacon F-14 Tomcat, High Speed.
1813. Type A. Bayonet. 14.4 volts.
1815. Type A. Bayonet. 14 volts.
1819. Type A. Bayonet. 28 volts.
1829. Type A. Bayonet. 28 volts.
1847. Type A. Bayonet. 6.3 volts. Same as #47, but with a longer life.
1864. Type A. Bayonet. 28 volts.
1873. Type A. Bayonet. 28 volts.
1891. Type A. Bayonet. 14 volts.
1892. Type A. Bayonet. 14.4 volts.
1895. Type C. Bayonet. 14 volts.

TILT

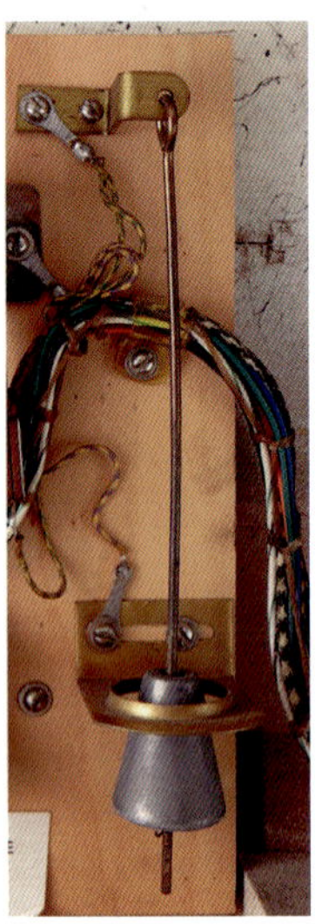

Shake the machine too hard and it will tilt. The game will shut down. The tilt is a swinging pendulum mounted inside a metal ring. The pendulum and the ring are really a switch. One wire is connected to the pendulum, the other wire is connected to the ring. If you shake the machine too hard, the pendulum touches the metal ring, the switch closes, and the machine tilts. On most machines you forfeit the ball in play. On older machines (until about 1960) if you tilt, you forfeit the entire game. A few electro-mechanical machines have a plug that gives you the choice of whether the tilt will cost you only one ball or the entire game, but most EMs have no options.

ADJUSTING THE TILT

The tilt can be adjusted for delicate or rough action. Most players like to have the tilt working, but not too delicately. You want to be able to shake the machine some, it's part of the fun of playing, without tilting the game every time. It may take experimenting to get the tilt "sensitivity" just where you want it.

The tilt is on the left side wall of the cabinet, just inside the coin door. The tilt pendulum is a wire rod with a cone-shaped metal plumb bob fastened to the rod with a set screw or metal clip. You can move the plumb bob up or down on the metal rod to increase or decrease the space between it and the ring (the strike plate). Less space makes it easier to tilt the machine. More space allows more movement, more shaking the machine, before it tilts. Some pinball machines had a second set of screw holes to mount the strike plate higher up, close to the top of the metal rod, to make it harder to tilt the machine. You can further reduce the likelihood of tilting the game by removing the plumb bob and letting the wire rod swing by itself.

If you want, you can disable the tilt completely by disconnecting the wire attached to the bracket holding the wire rod or by removing the wire rod or by shoving some insulating material, such as foam rubber or tissue paper between the swinging pendulum and the metal ring.

ROLLING BALL TILT

Most pinball machines have a second tilt mechanism: a rolling ball sitting on a track. If some clever person tried to lift the front of the machine to manipulate the ball on the playfield, the rolling tilt ball would start to roll down its track until it hit and closed a switch, tilting the game. The rolling ball is usually just above the swinging tilt mechanism.

If the rolling ball is missing from the track, which it often is, and if you want to replace it (if you have beefy friends who like to lift up the front of the machine and hold it up there), the rolling ball was usually a smaller diameter than regular playfield pinballs. Be sure to tell your pinball supplier that you want a tilt ball, not a pinball.

Some very old pinball machines had a mercury switch instead of the rolling ball.

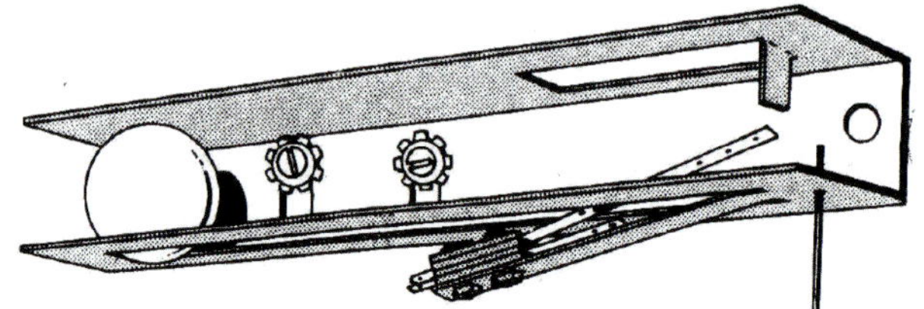

MORE TILT SWITCHES

It is not common, but some pinball machines also have tilt switches mounted under the playfield or on the floor of the cabinet. These are flexible metal blades that close (or in some machines, open) if the machine is bounced up and down or kicked.

This can be confusing because almost all pinball machines also have a separate anti-cheat circuit that uses the same kind of switch. The anti-cheat circuit is separate from the tilt circuit. See "Anti Cheat Switches."

Electronic machines: Many electronic machines have several tilt options: choosing whether the tilt will cost you only one ball or the entire game; choosing to disable the tilt entirely; and choosing one or two tilt "warnings"—often a sound that resembles a groan or a growl—before the game tilts. The tilt warning option can add a lot of fun to a game, enabling you to give the machine a good shove when you need to get the ball out of a problem area without tilting. See "Game Adjustments" in the "Setting Up" chapter.

THE STOOL PIGEON

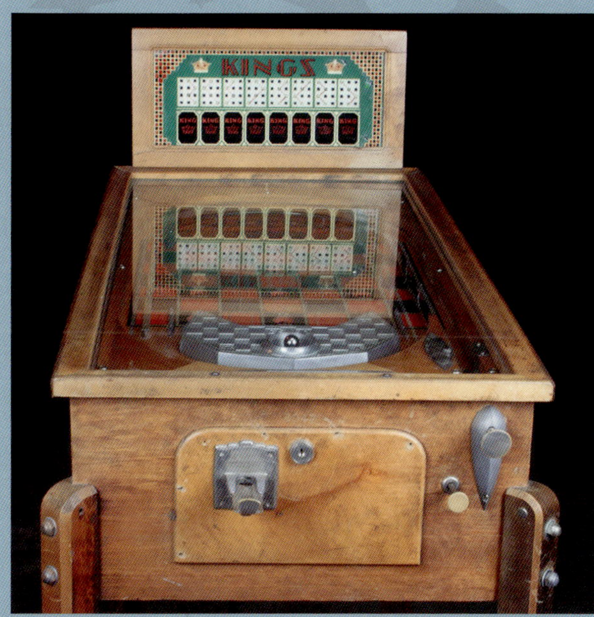

Genco **Kings**, 1935, with the "stool pigeon" tilt. One of the first tilt machines, it is also one of the first machines with backglass lights, which attracted players—and apparently store owners as well: The advertising for the machine boasted: "The merchant checks the score without leaving the counter!" Players paid 5¢ to play 12 balls. Don't know what the winning payout was.

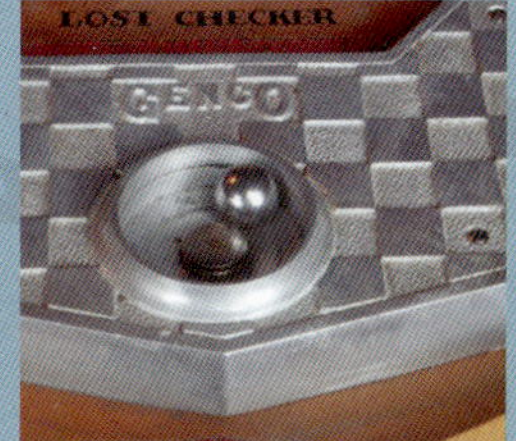

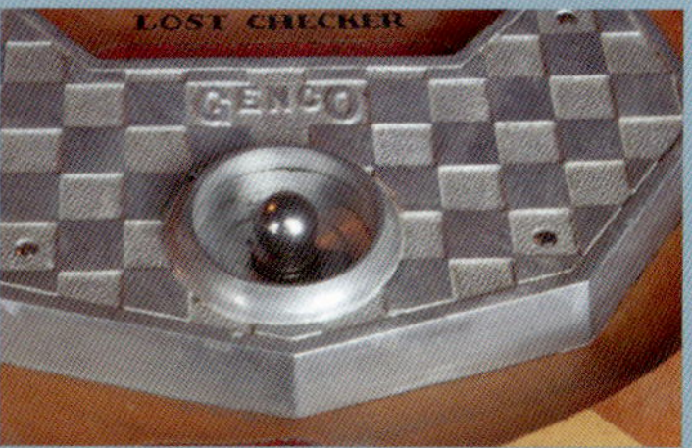

Harry Williams, founder of the Williams Pinball Company, invented the tilt in 1934 after watching someone beat up on one of his machines. Harry's original invention was a steel ball balanced on a pedestal mounted on the playfield. If you shook the machine too hard, the ball fell off the pedestal and touched a metal band that circled the pedestal, causing the machine to tilt and ending the game. When the game reset the pedestal dropped down, the ball rolled back onto the top of the pedestal, and the pedestal was lifted up again, balancing the ball. It was a very cool looking contraption. Harry called the device a "stool pigeon."

Harry took the first machine with his "stool pigeon" out to a location and he stayed and watched. Soon a player tried to shake the machine. The stool pigeon worked. The ball fell off the pedestal and the game ended. "I tilted it!" exclaimed the player. Harry immediately changed the name from Stool Pigeon to Tilt.

By 1935, every pinball manufacturer was adding some version of the stool pigeon to their machines.

ANTI-CHEAT SWITCHES

In addition to the tilt switches, all pinball machines have game protecting anti-cheat switches—also called slam, bounce, and kick-off switches—mounted on the coin door, the floor of the cabinet, and sometimes the underside of the playfield. These anti-cheat switches will close (or open) if someone hits or shoves the machine hard. Activating an anti-cheat switch will end the game.

Stuck anti-cheat switches are the culprit in many pinball problems. If your machine will not start, check the anti-cheat switches to see if they are stuck open when they should be closed, or closed when they should be open. Different machines had different anti-cheat circuits, so you will have to determine the correct open or closed position by either finding the switch in the schematic, or studying the wiring, or just doing a little trial and error: open a closed switch, or close an open switch, and see what happens.

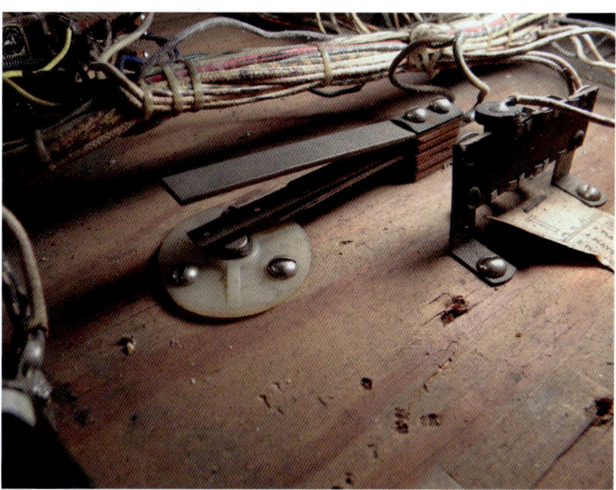

Anti-cheat switch on the floor of the cabinet. Anyone who bangs the bottom of the machine hard enough—or lifts up and drops the machine!—will bounce the switch enough to close it and end the game.

COIN MECHANISMS

Coin mechanisms ("coin mechs")—also called coin acceptors or coin rejectors—snap in and out of pinball machines in a few seconds, no wiring needed. Operators often took the mechanisms out of machines before selling the machines to keep people from putting the machines out on route and competing with them.

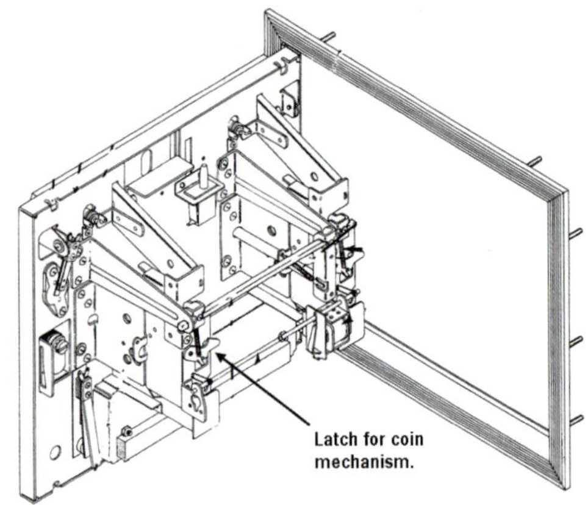

Inside of the coin door with the coin mechanism removed.

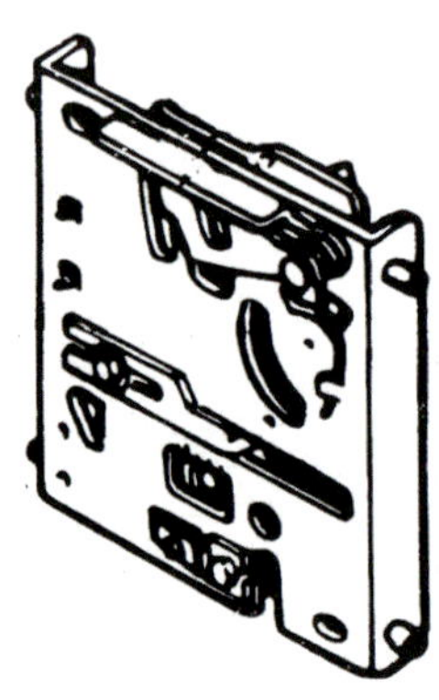

Don't confuse the coin mechanism with the mounting hardware that holds the mechanism. The hardware is bolted to the coin door and is permanent. The mechanism is a brass (or more recently, plastic) unit that snaps in and out of the hardware.

Most pinball machines came with two and sometimes three coin slots and coin mechanisms. On machines made from the 1940s through the 1960s, each coin slot accepted different coins, such as dimes or quarters.

In machines made after the 1960s, both coin slots accepted quarters. A quarter or two quarters in either slot would buy a game. Although there are two coin slots, only one coin slot and one coin mechanism was needed. The manufacturers provided an extra coin slot in case one slot got jammed—someone sticking a washer in it, or chewing gum, or lord knows what. The operators didn't want to lose the income. So if you want a coin-operated pinball machine you only need one coin mechanism, not two.

If you don't care about coin operation, you do not need the coin mechanism to play your pinball machine. Most machines can be easily adapted to free play, bypassing the mechanisms and the coin circuits entirely. This is explained in "Free Play" later in this chapter. But if you want your friends to drop some quarters in the slot and help pay for the machine (or the snacks), or if you want to actually earn some money from the machine, you will need working coin mechanisms.

Different machines used different coin mechanisms. Coin mechanisms for some machines are still available from pinball suppliers, but some types of coin mechanisms are no longer manufactured. If coin operation is important to you, check inside the coin door to see if at least one of the mechanisms is still there.

Old coin mechanisms usually still work, but sometimes they don't. These mechanisms are complicated and at times delicate instruments, sometimes requiring fine adjustments to get them to work properly. I've never been very good at adjusting coin mechanisms. If they don't work I play around with them until they either do work, or I just give up and switch the machine to free play.

COIN CIRCUITS

I should also warn you that the coin circuits in pinball machines—the wiring and the switches—tend to be the most non-operational. Most people have hand operated the little wire switches so many times to get free games that the switches are all bent out of shape. You will often find cut wires and damaged, broken, or missing parts.

On most machines, wiring the machine for free play bypasses the coin circuit so you can ignore any problems. But on some machines, if the coin switches are incorrectly adjusted, the game will not start; sometimes the game will constantly restart. You'll have to grab the bull by the tail and face the situation. See "Coin Switches" in the "Repairs: All Machines" chapter.

FROM A DIME TO A DOLLAR (AND TOKENS)

Over the years the cost to play a game of pinball kept increasing. Some pinball machines were still taking dimes into the 1960s, but most machines had converted to quarters. Many machines were set up for one play for a quarter, but some machines would give you two or even three plays for a quarter. Many machines had an adjustment where you could choose how many plays a quarter would buy. And then, of course, machines started requiring two quarters per play, and then even dollar bills.

People who want to earn more money from a quarter machine (more than 25¢ or 50¢ that is) may be able to convert the machine to take dollar bills. However, most machines made before 1990 cannot be converted to bills.

Older machines can be converted to tokens. You can purchase new coin mechanisms that accept tokens. Then you can buy a quantity of tokens that you'll be able to use over and over and sell the tokens for whatever price you want. Tokens come in different sizes and weights.

Check with your pinball supplier to see what they offer and if your machine can convert to dollar bills or tokens.

When I was a kid there were advertisements in comic books for "slugs"—round metal disks the size of a quarter. You could get a bag of 30 slugs for 50 cents. And just in case you had no idea why anyone would buy a bag of slugs, the ad said, "Not for use in coin-operated machines."

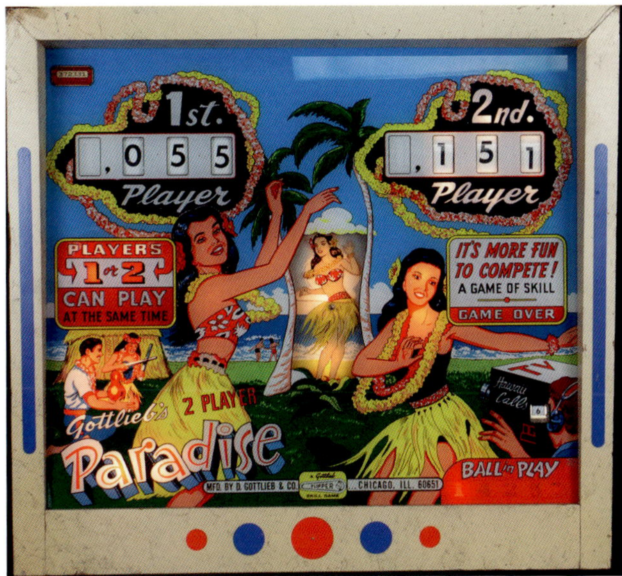

Gottlieb **Paradise**, 1965. The hula dancer in the center does a little wiggle dance during play.

BACKGLASS

A backglass is the painted (silk screened) glass that sits upright in the backbox. Backglasses were on every pinball machine until the 1980s, when the manufacturers switched to translites.

Backglasses have colorful artwork that show the theme of the game—space travel, bowling, pool, card games, buxom ladies, a famous personality or movie—and display the score, ball in play, match number, tilt, and game over. The backglass is the most striking, most prominent, and most visible part of a pinball machine. Many people fall in love with a particular machine, or hate a particular machine, solely from the artwork on the backglass.

Although a backglass is just that, a piece of painted glass, some people consider it the most important piece of the machine. A damaged backglass can make a $500 machine sell for $50. Replacement backglasses may be difficult or next to impossible to locate. A replacement would have to come off another machine, one that was being scrapped for parts, or it would have to be one of the few backglasses so loved by collectors that reproductions are available, at anywhere from $200 to $400 each.

There are businesses that restore old backglasses, working with your original glass and digital images of the original artwork. This, as you can imagine, is an expensive and time consuming process; you have to ship your old glass to them. These restoration businesses can be found on the internet. I personally have no experience with them.

If you are looking for a backglass for a machine you own, an internet search might locate one. If there is a nearby pinball show coming up, there usually are dealers at these shows selling backglasses. Do an internet search for pinball shows in your area.

PROTECT YOUR BACKGLASS

There are protective measures than can help preserve your backglass. Moisture and bright light can damage the paint, causing it to fade, to bubble up, and to fall off. Do not store the machine where humidity or water can get to it. A blanket over the machine will help keep out moisture and light. If the glass is loose, shim it with a piece of foam or cardboard so it cannot rattle. Be extra careful when transporting it. Using cooler light bulbs (#47 lamps instead of #44 lamps) will reduce paint damage. See "Lamps" in this chapter for more information. If you keep your machine in an unheated room or garage don't turn on the machine until the room is up to room temperature. The heat from the lamps can crack and peel paint off a cold backglass.

FLAKING OR PEELING PAINT

If paint is already starting to flake off the backglass it is difficult to know exactly what to do. There are products on the market that claim to stop paint flaking, but none is a sure guarantee of success. Some people have luck with polyurethane spray. I've had good luck with Krylon Crystal Clear acrylic spray, the kind artists use to fix drawings so they won't smear. The Crystal Clear prevented more paint from flaking, paint that hasn't already started to peel, at least in my experience. But sometimes the spray has just the opposite result and removes the flaking paint, leaving an even bigger paintless area. I highly recommend testing the spray in a small spot, maybe near the edge of the glass where it won't be too noticed, to see if the spray helps or makes a bad problem even worse.

I hear from people who have success with a spray called Krylon Triple Thick, but when I tried it, Triple Thick left a filmy, cloudy coating on the clear part of the glass where the score reel numbers show through that needed to be scraped off.

Also be warned that these sprays are toxic and flammable. Use outdoors and do not breathe the stuff.

Touch Up Paint: Unless you are an experienced artist and good at mixing paint colors, repainting or touching up the paint usually makes it look worse, looks like a touch up job, and almost always decreases the resale value. Most people just leave the glass alone. Too often the cure causes more damage.

RUBBER CUSHIONS

Backglasses on most Gottlieb and a few Williams machines made in the 1970s were cushioned with little 1" squares of black rubber placed between the backglass and the light board. Those black cushions often stuck to the backglass, damaging or fading the paint. It's quite common to see little faded squares of paint on old Gottlieb backglasses. If it isn't too late, remove the rubber cushions and replace them with non-damaging cushioning. Place the new cushioning at the edge of the backglass, where they won't damage the most visible part of the paint if the new cushions turn out to be as problematical as the old ones. But be careful: If the old rubber cushions are stuck to the backglass, you are probably better off just leaving them in place. You may damage the paint even more trying to remove them.

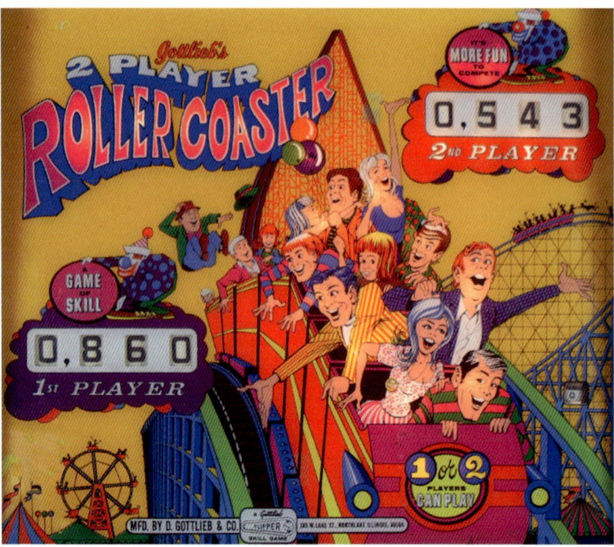

The pinball machine is a wonderful piece of technology with a lot of creativity behind it, but without the artwork, where would it be?

—Gordon Morison, creator of the art for Gottlieb *Roller Coaster*, 1971.

TRANSLITES

In the mid-1980s silk screened backglasses were eliminated, replaced by translites: printed, flexible plastic sheets that were mounted in the backbox behind a piece of clear glass.

Translites hold up quite well as long as they are not exposed to bright sunlight, which will cause them to fade, or mounted too close to hot light bulbs, which can actually burn holes in the translite. If the hinged door behind the translite doesn't latch completely closed or if the door is warped, the lamps mounted on the door may get too close to the translite. I saw one translite that had been badly scratched by the edge of a scoring display that was mounted on a door that did not close all the way.

Replacement translites are easier to locate, less expensive to replace, and much easier to ship than painted backglasses. You can often find translites on the internet selling for about $50.

CHIMES, BELLS, AND DIGITAL SOUND

A pinball machine without sounds is like a day without the sun, like Anthony without Cleopatra, like Elwood without Jake. But sometimes you want to play a game when the kids are going to sleep or when your spouse or housemate wants to read (or when the people in the apartment downstairs are banging a broom handle on the ceiling). You may want to work on the machine without the distraction.

CHIMES

Most electro-mechanical and early electronic machines had three chimes mounted next to each other near the front of the cabinet, usually accessible from the coin door. Each chime was for a different score: 10 points, 100 points, 1,000 points. The common wire to all three chimes is usually black, and it often attaches to one of the chime coils with a small connector that can be unplugged, silencing the chimes.

On some machines the chime unit has a small wiring harness that plugs into a socket on the bottom of the cabinet that easily unplugs. If the chimes cannot be unplugged, you can cut the common wire where it connects to one of the chime coils and rewire it with a connector you can disconnect when you want to.

Some electro-mechanical machines had only one chime, often mounted inside the backbox. Because of the awkward location, this chime is more difficult to temporarily silence. See the suggestions under "Bells," which are usually in difficult-to-access places.

If the chimes don't ring clearly, or don't work at all, see "Chimes" in the "Repairs: All Machines" chapter.

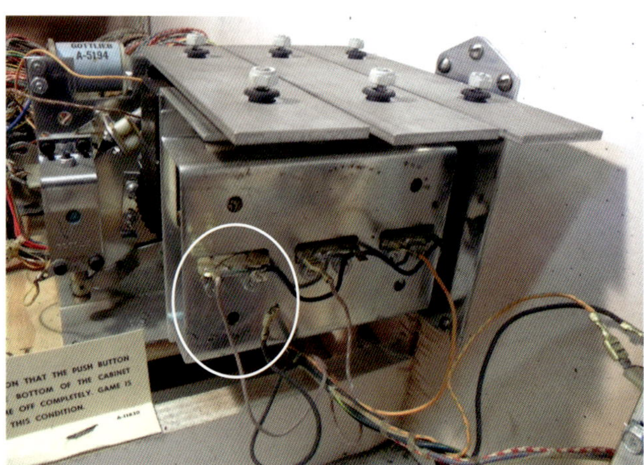

Many chimes have a detachable black wire (circled) that silences the chimes. When unplugging the black wire, use a pair of pliers to hold onto the coil side of the connector so you don't pull out the coil wire along with the black wire.

BELLS

Before chimes were used, pinball machines had one or two bells that rang when you scored points. The bells were in the backbox, accessible through the back door of the machine, and sometimes in the main cabinet.

Some of those old bells are really loud, especially bells that are mounted directly on the wood cabinet. The wood amplifies the sound. You can muffle the sound a little by putting a stick-on piece of felt (available from most hardware stores) on the bell where the striker hits. You might be able to stuff some tissue or cotton in the bell and tape it in place, which occasionally works. You'll probably have to experiment with different methods, and different locations on the bell, to achieve the result you like.

If you want to be able to temporarily silence bells to play a game or work on the machine quietly, it is not as easy to do as with chimes because of the bells' out-of-the-way location. You would have to cut one of the wires going to each bell, add a length of wire to each end of the cut wire, and thread the two wires through the machine to somewhere near the coin door, where you can add an on-off switch. It's tedious, but not difficult.

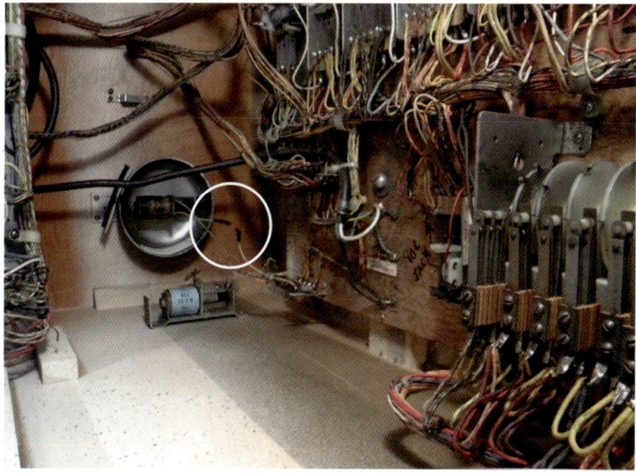

A large, loud bell mounted at the far back of the cabinet. Circled is a snap-apart connector to silence the bell if needed. You can muffle the sound with a piece of stick-on foam or cushioning, but it never seems to stay on very long.

DIGITAL SOUND

Electronic machines with digital sound have a volume control inside the cabinet, on the left side of the cabinet near the front, or on the inside of the coin door. Open the coin door and you can easily reach and adjust the volume. On "home" models the volume control is in the backbox, accessible from the back of the machine. Many electronic machines also have an adjustment that can shut off or alter the sound. See "Game Adjustments" in the "Setting Up" chapter.

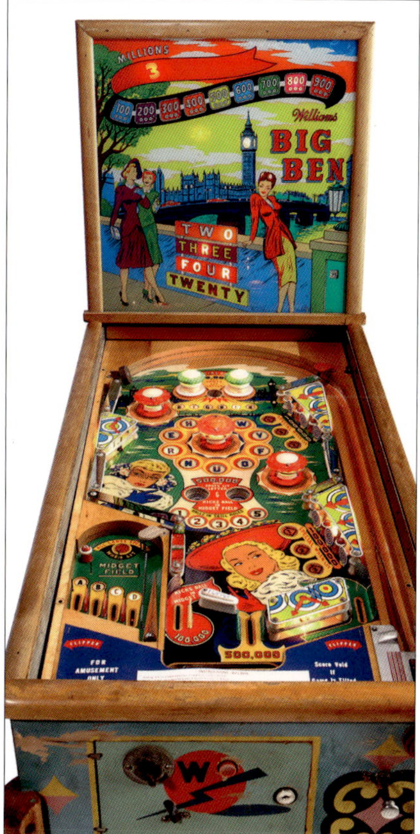

Williams **Big Ben**, 1954. Two games in one: Note the miniature bagatelle pinball game in the lower left of the playfield. Like some woodrails, the back-glass slides out sideways after a piece of wood trim is removed.

PLAYFIELD

The playfield is, well, the playfield, where the ball rolls around, hits targets, lands in holes, and other fun stuff. The playfield is also called a playboard or table. A well cared for playfield will last a lifetime. The playfield should be cleaned and waxed every few months if it is getting a lot of use, or once or twice a year if getting less use. If moisture gets on the playfield, dry it off immediately. Water will ruin the paint and the wood.

Do not clean the playfield with water or water based cleaners. Use a cleaner and wax made especially for pinball machines. My favorite is Mill Wax, a cleaner and wax combined. Mill Wax will leave spots of dry, gray-colored wax if not cleaned off thoroughly, but once cleaned off, Mill Wax holds up well. Other popular brands include Wildcat 125, though I'm told it can damage plastic parts; Novus #2 and Novus #3, though they are only recommended for Mylar and Diamond Plate playfields, and some people report that Novus sometimes leaves white specks of polish dust; and Gemini CP-100, which has been popular for years. You can also use a good furniture cleaner and wax.

Always test any cleaner, even cleaners especially made for pinball machines, in a small, out of the way spot first to be sure the cleaner will not remove paint from the playfield. Use a white or light colored rag and be constantly alert for smeared paint coming off on your rag.

For best cleaning and waxing results, remove as many of the parts mounted on the playfield as you reasonably can. As you remove the parts, make a drawing of the playfield where the plastics went, or take photos, so you can easily put them back where they came from. See "Playfield Plastics."

If the playfield is exposed to bright sunlight the paint can fade, so cover the playfield when not in use. Do not store the machine with the playfield glass off. The glass protects the playfield from dirt and damage. Make sure the balls are in good shape, not pitted and no rust. A pitted or rusty ball is like a rolling piece of sandpaper and can damage a playfield.

Flippers sometimes rub against the playfield, scraping away an arc of paint (and slowing down flipper action). See the chapter "Repairs: Flippers."

If there is missing paint, you probably should just leave it as is. Unless you are an excellent touch up artist, and are able to match the paint colors exactly, chances are you'll make the playfield look worse.

MYLAR ON THE PLAYFIELD

Mylar is a clear plastic sheeting that is glued to surfaces to protect them from wear. Pinball manufacturers started using Mylar on playfields in the early 1980s. Mylar was first used on small areas of the playfield that received the most wear, such as around pop bumpers and kickout holes. Later, large sheets of Mylar were used to cover most of the playfield. By the 1990s Mylar was discontinued in favor of a polyurethane finish.

Mylar held up great, crystal clear and indestructible, for about ten years. Then the Mylar started discoloring, getting a cloudy look, bubbling up in places, and peeling up at the edges.

If the Mylar is scratched or dull looking, the Novus company makes two products: Novus #2 to polish Mylar and Novus #3 to reduce or eliminate scratches in Mylar. I have not used these products, but several pinball suppliers recommend them.

If the Mylar is peeling, you have a real problem. For starters, DO NOT try to peel the Mylar off; you're likely to take the paint with it. In fact, I had a machine in the shop that someone had tried to remove the Mylar from and it pulled a chunk of wood right out of the playfield.

People sometimes use heat guns or hair dryers to loosen the glue and remove the Mylar, then use a glue remover to clean up the mess. Some people are very successful doing this, and some people turn things from bad to worse and wish they'd never started the project. If you have no experience removing Mylar, your pinball playfield is probably not a good place to try this out.

If the Mylar is peeling at an edge and affecting the play of the game, you might consider taking a razor knife and carefully cutting off the loose (raised) Mylar. Be very careful not to cut into the playfield or cut into the Mylar still glued down.

If the Mylar is not peeling, not causing problems other than it doesn't look very nice, and if you aren't sure what you're doing, consider just leaving it alone. Sometimes the best solution to a problem is to ignore the problem.

PLAYFIELD ADJUSTMENTS

Many pinball machines are designed to allow minor adjustments on the playfield to make the game easier or more difficult to play. These adjustments usually involve the plastic posts that are mounted next to outlanes. By moving a post as little as 1/8" to 1/4", you can reduce the opening to the outlane. The ball will be more likely to stay in play instead of drain.

The pinball manufacturers often drilled two or three mounting holes next to each other, concealed under the base of the plastic posts. The machine's instruction manual will have a playfield diagram with the choices marked, usually labeled "LIB" (liberal: a narrower outlane, less likely to drain) or "CON" (conservative: a wider outlane, more likely to drain).

It is your choice where you want to mount the posts: in the "liberal" position or in the "conservative" position. These options can make a significant difference in how much fun a game is or how challenging a game is. It is easy to move posts and try the options, though few people ever do. Almost every machine I've seen had the posts obviously mounted where the factory put them; the optional pre-drilled holes had never been used. If you do move the posts, be sure the posts screw securely to the playfield. If the posts are loose, and if the screws will not tighten, see "Playfields: Loose Components" in the "Repairs: All Machines" chapter.

And sometimes you have to create your own liberal/ conservative options. Every once in a while I encounter a pinball machine that has such wide outlanes the ball drains too often, and the game is simply no fun to play (and the manufacturer did not provide a liberal/ conservative option). Such games sometimes require drastic measures.

I have, on a few occasions, drilled new holes in the playfield and relocated the plastic posts maybe as much as 1/4", closing the outlanes just enough to get my own fair advantage. I did that on a Bally Space Time and turned a machine no one wanted to play into one that got a lot of play. But be warned, I am NOT telling you to drill holes in your playfield. I am suggesting that a pinball machine should be fun to play, period.

PLAYFIELD PLASTICS

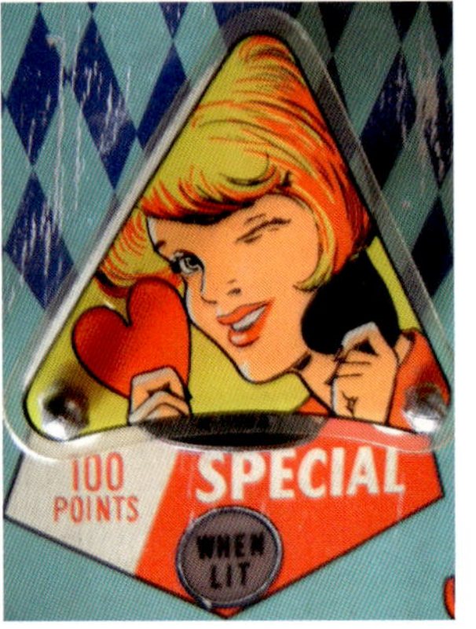

The plastic shields mounted on the playfield, usually just called "plastics," cover lamps and hide mechanisms. Plastics are unique to each game. No two games had the same plastics.

 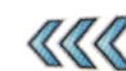

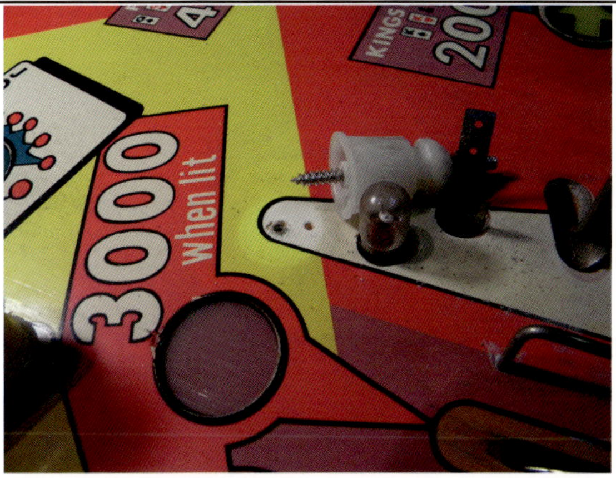

Gottlieb **Royal Flush**, 1976, outlane showing factory-drilled options for placing the playfield post. The large, well-worn hole is the liberal setting (narrower outlane). The smaller hole is the conservative setting (wider outlane), obviously never used.

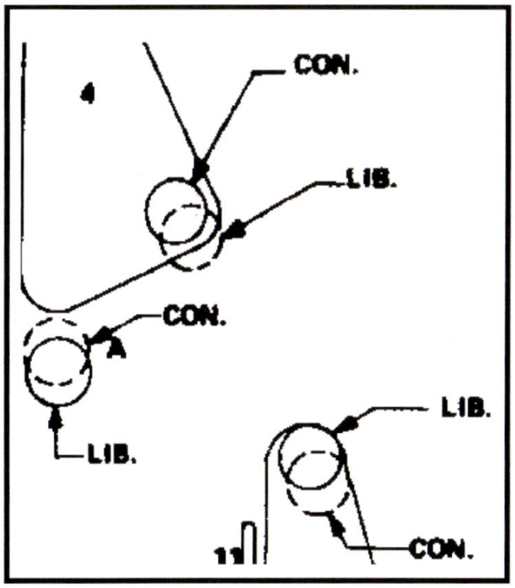

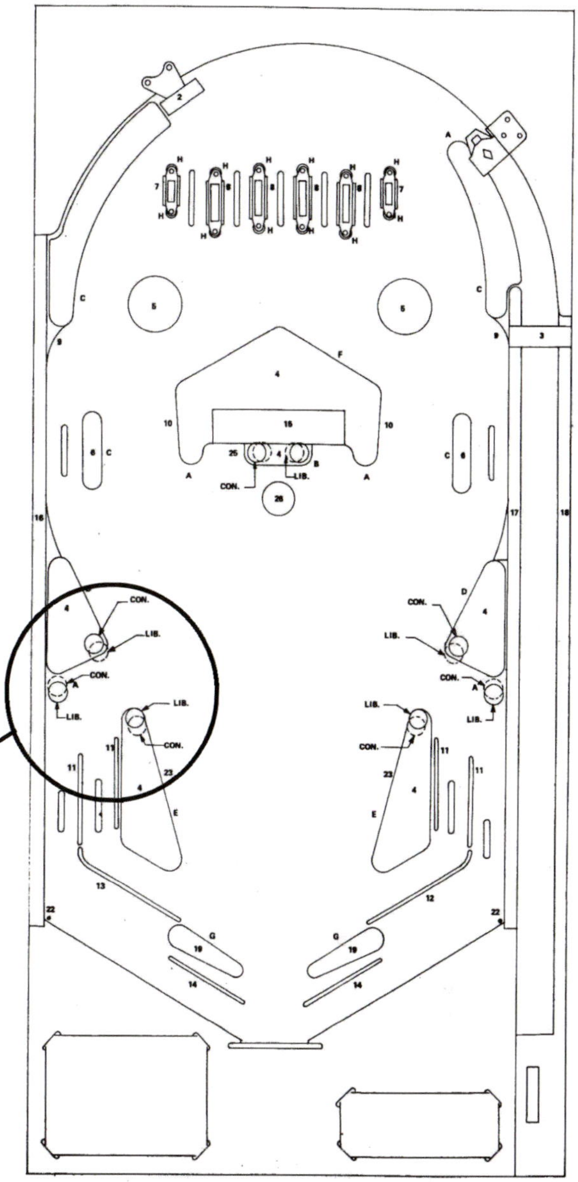

The owner's manual included an illustration of the playfield with the liberal-conservative adjustment options. Two screw holes (for the two settings) were pre-drilled by the manufacturer. The unused screw hole is hidden under the playfield post. If there is no liberal-conservative option, you can create your own by drilling a hole and moving the playfield post, if the design of the playfield and the plastic shield will allow. Make sure that the new hole is close enough to the original hole so that the unused hole will be hidden under the playfield post.

MISSING OR DAMAGED PLASTICS

If any of the plastics are missing or broken or are badly warped, you may be searching a long time for replacements. People sell used plastics on the internet, usually full sets, and usually expensive, but the likelihood you'll find a set for your particular machine is not great. There were hundreds of different shapes, sizes, and designs. Reproduction plastic sets are being manufactured for some of the more popular machines and are available from some pinball suppliers.

If you are able to get to one of the pinball shows held around the country, there are always people selling used plastics. In fact, they'll have boxes full of plastics, and usually for low prices. If you cannot locate plastics for your particular game, you might find plastics from a different game that were the exact same size and shape, but a different design.

You can also make temporary replacements from a piece of Plexiglas if you are good at cutting smooth, even lines.

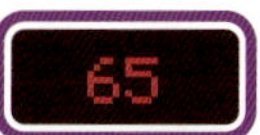

BALL HITS PLASTICS

The edges of the plastics sometimes stick out just enough for the ball to hit them. Some machines have chipped and broken plastics from contact with the ball. Most plastics are mounted high enough that the ball rolls under them without contact. If a plastic is low enough for the ball to hit it, see if you can raise the plastic a little, just enough so the ball clears it, by putting a tiny washer between the plastic and the post it is mounted on.

If that doesn't work, you can put a "fender" washer (one with a small hole so it doesn't move around) under the plastics. The washer should stick out just a fraction of an inch beyond the edge of the plastic, so the ball hits the washer instead of the plastic. Some pinball suppliers sell clear plastic washers—sometimes called "deflector shields"—for this purpose. The plastic washers are less likely to damage your ball than metal washers, and the clear plastic does not detract from the appearance of the plastics as much as a metal washer.

REMOVING THE PLASTICS

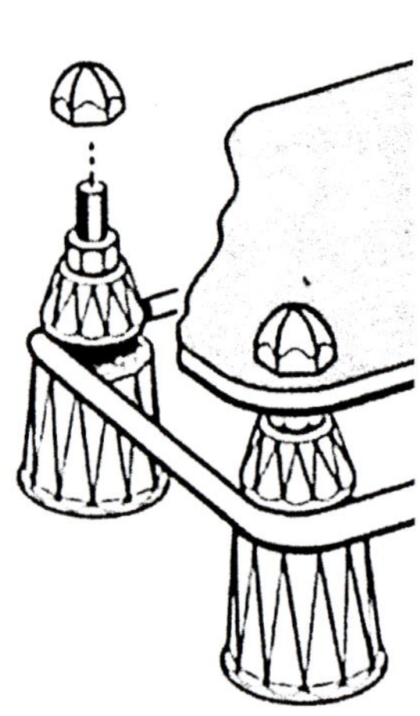

When you need to remove the plastics from the playfield to replace a lamp or a rubber ring or to make it easier to clean the playfield be careful when unscrewing the cap nuts (also called acorn nuts) that hold the plastics in place. The cap nuts are screwed onto playfield posts that are screwed into the playfield. Sometimes the cap nuts are screwed tightly to the posts. When you unscrew the cap nuts the posts also unscrew, something you'd rather avoid. The posts can come loose from too many times being screwed in and out of the playfield. Try using thin, long-nosed pliers to grip the posts underneath the plastics while you unscrew the cap nuts. Better, if you have a "super thin" flat metal 1/4" open-end wrench (car mechanics used them for adjusting distributors) it will fit underneath the plastic and hold the post in place while you unscrew the cap. Another option: Pinball Resource, a parts dealer in New York State, sells a specially designed wrench that originally came with Gottlieb machines for this very purpose.

When replacing the cap nuts, screw them on finger tight. On a few machines cap nuts snap on and off, no screw threads.

If you have a loose playfield post that will not tighten securely it's easy to fix. See "Loose Playfield Components" in the "Repairs: All Machines" chapter.

FREE PLAY

Free play makes pinball a little easier and lots more fun. You no longer need a stack of quarters. You no longer need to stick your finger in the coin door and trip the skinny wire switch over and over until it bends and becomes useless. Setting a pinball machine for free play is quite easy for many machines, a bit more work for some, and downright difficult for a few.

FREE PLAY (ELECTRONIC MACHINES)

Many (but not all) electronic machines have a setting that allows free play. Inside the coin door is a set of push button switches. One of the switch settings is free play. The machine's instruction manual has instructions how to set the free play adjustment. You adjust the setting, and from then on you get free play until you change the setting back.

When the batteries in the machine start to get old and lose voltage, you may discover that your machine does not retain the free play setting. After you install new batteries you have to reset the free play setting again.

If you have a Williams machine from the early to mid-1980s, you may find no mention of a free play setting in the instruction book. Try this: In adjustment mode go to maximum credit adjustment, which is usually Adjustment 18. Set the number at 00. I found that this is the free play setting in many Williams games.

Some early electronic machines, including the Gottlieb System 1 machines, had no free play setting. To get free play on these machines see "Another Way to Get Free Play."

FREE PLAY (ELECTRO-MECHANICAL MACHINES)

On electro-mechanical machines, in the backbox there is a switch mounted on the credit wheel that must be closed to start a game. It is called the "zero credit switch" (also called a "zero switch," or a "zero position switch"). By locating this switch and permanently closing it, or shorting it with a jumper wire, you will have free play. If you are not familiar with "shorting" a component see "Short/Short Circuit" in the "Repairs: Basics" chapter.

Open the backbox and locate the credit wheel. It is a white wheel about 6" in diameter with black numbers printed on it. Those numbers are the credits—how many games you have to play, running from 0 up to 15 or 20 games. (Some credit wheels have a blank space instead of a 0.) Don't confuse the credit wheel with the score reels. They are also white with black numbers, but the score reels are plastic, and there will be three or four reels together. But you know that.

Mounted on the credit wheel is a bank of two or three switches. As the credit wheel rotates, a post mounted on the credit wheel comes around until, when the credits hit zero, the post opens the bank of switches.

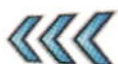

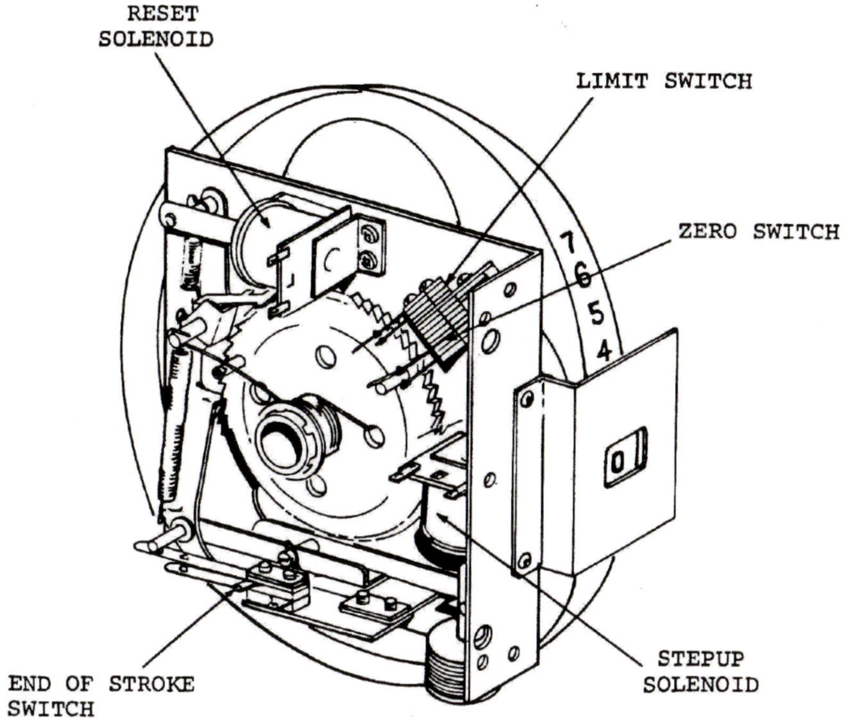

A Bally electro-mechanical credit (replay) unit. On this unit there are only two switches on the switch bank (upper right). The lower of the two switches is the Zero Credit Switch, the one that must be permanently closed to get free plays. Some Bally units had three switches, and most Gottlieb units had three switches. On Gottlieb machines, the middle switch was usually the Zero Credit Switch. On Williams machines, the top switch was usually the Zero Credit Switch. But don't assume a given switch is the zero credit switch until you short it and see if you get free games. On Chicago Coin machines, at least the ones I've seen, there is only one switch, and that's the Zero Credit Switch.

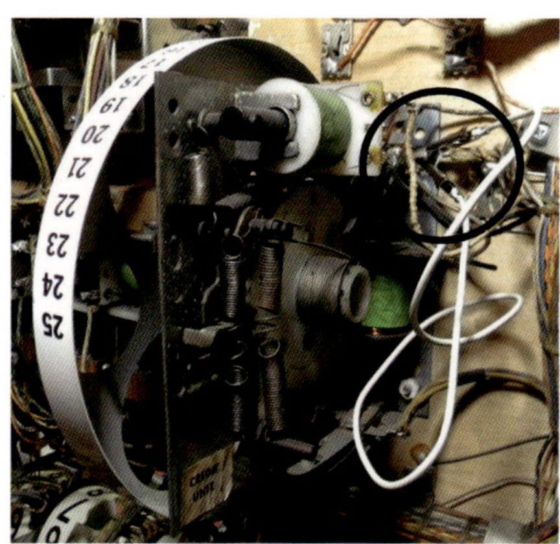

Credit wheel and zero credit switch with jumper (circle).

One of those two or three switches is the zero credit switch, the one you want to permanently close. Make sure the credit wheel is at zero, with no free games showing. Take a jumper wire with alligator clips at each end and short (close) one of the switches. Try to start the game. If the game does not start short a different switch. When you have the right switch the game will start. That's the zero credit switch. You will not damage the machine if you short the wrong switch, so feel free to experiment. If you are good at reading schematics, instead of guessing which switch to short, locate the zero credit switch on the schematic (usually near the transformer), note the color of the wires, and short the switch that has the same color wires.

You can take a pair of needle nose pliers and bend the switch blades together or solder them together, or what I usually do, use a jumper wire to connect the switch solder lugs. Put one alligator clip on each of the solder lugs to the switch. That way you can easily remove the alligator clips any time you want to get people to drop some quarters in your machine.

People who want to change the game back and forth, free play to coins, and don't want to constantly open the back door and play with the switch, can solder a pair of wires (a thin lamp cord works well) to the two lugs of the zero credit switch and run the wires through the inside of the cabinet to a location near the coin door, then put a toggle switch on the end of the wires. You can reach in the coin door and throw the toggle switch to select coin operation or free games. I've also seen toggle switches mounted inside the backbox near the credit wheel, thus not requiring a lengthy run of wire through the cabinet, but that is practical only if you have easy access to the back of the machine.

On a very few machines, this free-play adjustment will not work if the machine was turned off in the middle of a game. When you turn the machine back on, you will have to activate the coin switch, not the start button, to start a new game. Most machines do not have this problem.

If you do have easy access to the back of the machine, you can get free plays by opening the back door of the machine and hand operating the credit step up solenoid until the credit wheel gets to the maximum number of credits, maybe fifteen or twenty games. And there you go . . . until the wheel runs down to zero and you have to do it all over again. A bit of a nuisance, but it requires no new wiring.

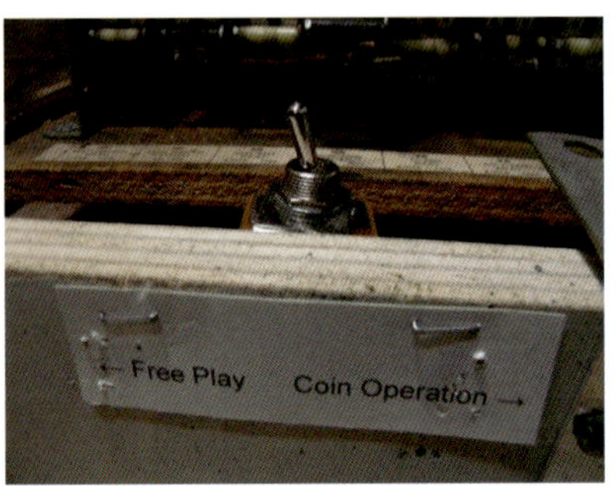

This switch is mounted just inside the coin door, with a long pair of wires running to the back of the machine, up into the backbox, soldered to the zero credit switch.

CREDIT WHEEL OUT OF ALIGNMENT

It is uncommon, but sometimes a credit wheel can get loosened and then re-tightened in the wrong position, so the zero on the credit wheel is not aligned with the actual zero position when the zero position switch opens. If you operate the credit unit's reset solenoid by hand, stepping down the numbers, you can watch the metal pin on the rotating wheel (the wheel with the gears) come around and open the zero position switch. At that point the credit wheel should show zero (or on some credit wheels, just a blank space). If it doesn't, the credit wheel needs to be loosened, rotated to zero, and re-tightened.

Now I'm sure this is obvious, but the rotating geared wheel is *not* the credit wheel. The credit wheel is the large white wheel with numbers on it. You don't adjust the geared wheel, you adjust the credit wheel.

The illustration of a credit unit shows the pin mounted on the geared wheel opening the zero position switch and the credit wheel showing zero. This is how the credit unit should be aligned. The unit in this illustration has a bracket framing the number on the credit wheel, making it easy to see what the credit number is; in this illustration, zero. If the credit unit on your machine does not have a bracket framing the number on the credit wheel the number is usually aligned at 3 o'clock, the same location as the zero in the illustration.

Sometimes the credit wheel will not show any numbers, just white paint. The numbers are on the back side of the wheel, hidden from view. This can happen if the credit wheel got loosened as explained previously. More likely this happens if the number of free games won has exceeded the numbers printed on the wheel: If, say, the wheel goes up to 15 games and you have won 16 games. Most machines have a second pin on the geared wheel that stops the wheel from rotating beyond a certain number of games. Sometimes the pin hits a bracket that will not let it rotate any farther. Sometimes the pin opens what's called the "limit switch" that shuts down the rotation. That pin can be unscrewed and remounted in a different hole in the geared wheel to reduce the maximum number of free games. Another possibility: That pin, or the pin that opens the zero position switch, could be missing and needs to be replaced.

HALF-MOON CREDIT UNIT

In the mid-1970s, some Gottlieb pinball machines were manufactured with a different design for the credit unit. Instead of a credit wheel with the number of credits appearing behind a window in the backglass, Gottlieb created an elaborate contraption where a plastic cover resembling an arch (or a half moon) rotated, exposing the number of credits. The "half-moon credit unit," as it is commonly called, had a different switch arrangement than the credit wheel. The switch closest to the frame of the half-moon mechanism was the zero credit switch. Permanently closing this switch would give you free play.

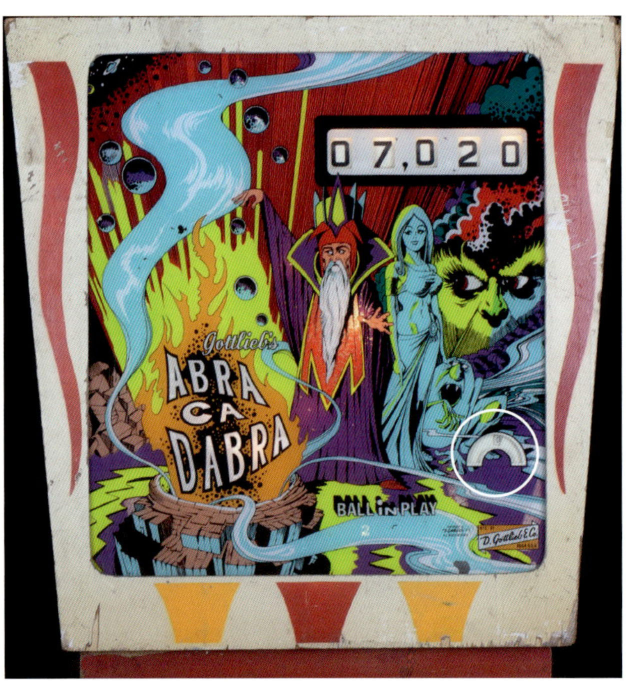

Gottlieb **Abra Ca Dabra**, 1975, with a "half moon" credit window.

ANOTHER WAY TO GET FREE PLAY (ALL MACHINES)

If your electronic machine does not have a free play setting, or if you took one look in the backbox of your electro-mechanical machine and said *not me*, you have another way to set the game for free play. You can wire a new switch on the coin door that will give you free plays.

Inside the coin door, on the bottom of the coin slot hardware bolted to the door, you will see a small wire switch, one switch for each coin slot. These are the switches the coins hit, and any of these switches should start a game or give you a credit towards a game. Turn on the machine—don't start a new game—and operate one of the switches by hand. You should get a game started, or the credit wheel stepped up to 1. If not, the switch is broken or out of adjustment, or the wires are disconnected. Short the switch wires together momentarily and see if you get a game. If no success try the other switch. And if still no success, postpone this idea until you find out what's wrong with the machine. See "Game Won't Start" in the "Repairs: All Machines" chapter.

If you did have success when you operated a switch, here's what you do. Buy a spring-loaded push button "momentarily on" switch at a hardware or electronics store. You push and release the button and the switch makes a momentary contact. Wire this switch in parallel to one of the coin switches mounted inside the coin door, whichever switch is working. Parallel wiring: one wire to each side of the coin switch. Use the new switch to get your games. Each time you push the button you start a game or get one credit (one play).

Mount the new momentarily on switch inside the coin door, or if possible mount the switch on the coin door. You might try removing the coin return button from the coin door and see if the switch will fit into the mounting hole. Don't drill a hole in the coin door, you'll ruin its value.

Three-wire switches: Most coin switches have two wires, a simple open-close. Some electro-mechanical machines have a three-wire switch on one of the coin slots. The three-wire switch is actually two switches and both are operated by the coin: one closing and the other opening. You cannot start a game just by closing the open switch. If you have a three-wire coin switch, look at the switch on the other coin slot. Most likely it will just have two wires. That's the coin switch you want to wire. If both coin slots have three-wire switches, well, things are more complicated. You will either need to examine the wiring schematic to figure out how to rewire the switch or go back to the other options for free play.

Electronic machines: Some electronic machines have a diode mounted on the coin switch. This diode protects the electronic components from high voltage spikes. If your machine has the diode, be sure not to disconnect or bypass the diode when wiring the new switch.

AND ANOTHER WAY TO GET FREE PLAY (ELECTRO-MECHANICAL MACHINES)

The credit wheel is operated by two solenoids: the step up, adding credits (plays); and the step down, subtracting credits. Hand operate the step up solenoid to 1 play and then cut or unsolder one of the wires to both solenoids. The wheel will no longer rotate up or down, so the switches activated by the wheel will stay in the 1 play position. Any time you want to go back to coin operation just reattach the wires.

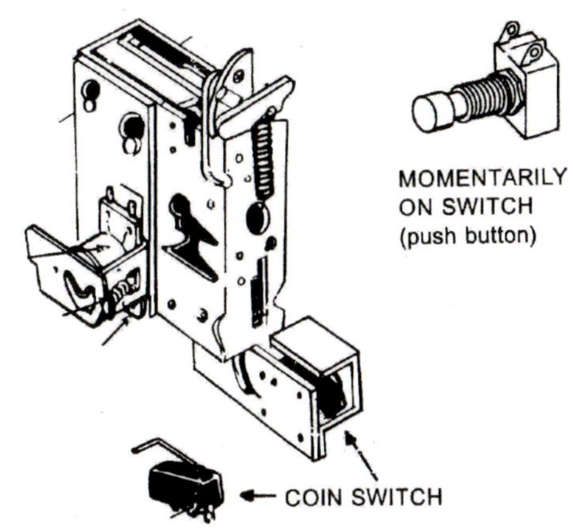

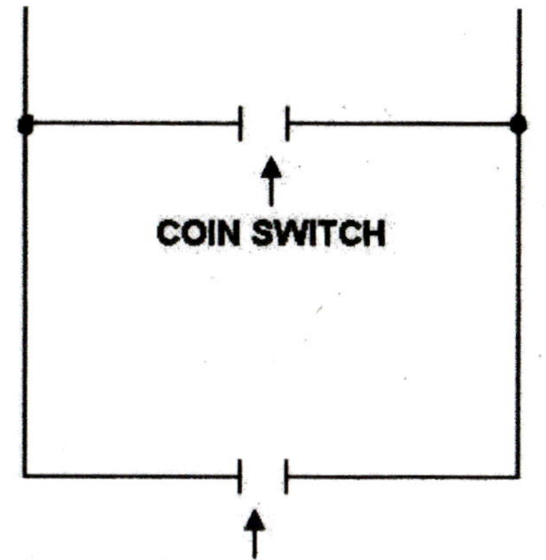

NEW MOMENTARILY ON SWITCH
(wired to original coin switch)

BATTERIES (ELECTRONIC MACHINES)

Most electronic pinball machines have a battery or set of batteries on the CPU board inside the backbox. Electro-mechanical machines, home models, and most cocktail table machines do not have batteries.

The batteries provide backup power to the memory chips so that the chips can store high scores, bookkeeping information, and special settings and adjustments such as free play, when the machine is turned off. On older electronic machines the batteries are not necessary to power or play a game; the machines will work with the batteries removed. Newer electronic machines often require the batteries to be installed or the game will not play.

Some pinball machines use AA alkaline batteries, usually two or three on one battery holder. Some pinball machines have built in battery charging circuits and use rechargeable nicad (nickel-cadmium) batteries. Some nicad batteries look identical to regular AA batteries. Some nicad batteries are small cylinders or squared-edge tubes about 1" long soldered to the CPU board. And some machines use lithium (lithium-ion) batteries, which are usually flat, round, and silver, about the size of a quarter, and soldered to the CPU board.

When alkaline batteries get too old they will eventually leak. And just the way leaking batteries destroyed your flashlight or alarm clock, battery electrolyte can destroy the CPU board on your pinball machine.

Nicad batteries don't usually leak with age, but if the charging circuit in your pinball machine overcharges a nicad battery—something that can happen in older machines—the battery can overheat and leak corrosion onto the CPU board. Many pinball shops recommend converting nicad machines to alkaline machines (see following).

Lithium batteries rarely leak but it's still possible, and still worth taking precautions.

There is no advance warning when a battery is about to leak. One day it looks fine and the next day it's a mess of corrosion. Be on the alert when your machine starts showing incorrect high scores, different replay levels, or no more free play. This is a warning that the batteries are dying. On newer machines, if the machine is suddenly behaving erratically or not working, the problem may simply be a failing battery.

You can't always tell if a battery is good by pulling it out and checking the voltage. Your multi-meter may show good voltage, but when the battery is actually working ("under load") the voltage on an older battery can drop enough to cause failures.

CHANGING BATTERIES

To be safe, change batteries once a year. If there is more than one battery, change all of the batteries at the same time, and use batteries from the same package. If you mix brands and manufacturing dates you can shorten the life of the batteries. When you replace the batteries, tape a note to the batteries (or to the back of the machine) with the date you replaced them.

If your pinball machine uses alkaline batteries, replace with alkaline batteries. Don't buy the cheap "heavy duty" batteries; they don't last very long. You can use rechargeable NiMH (nickel-metal-hydride) batteries instead of alkaline batteries, but my own experience is that the NiMH batteries have a short life and have to be removed and recharged too often.

If your machine used rechargeable nicad batteries do not use the newer NiMH batteries. They do not charge properly in the old nicad charging circuits. Either buy new nicads or convert the machine to alkaline batteries (see next paragraph). But **do NOT put alkaline batteries in a nicad charging circuit. The batteries will explode!**

It is a good idea to relocate the batteries off the circuit board, and mount the batteries somewhere in the backbox where they will not damage the CPU board if they leak. For AA alkaline and nicad batteries, pinball suppliers sell an inexpensive battery remount kit: a "remote" battery holder with mounting hardware and wiring. The kits usually include a diode that blocks the recharging circuit (batteries will no longer recharge) so you can use regular alkaline batteries instead of nicads.

If batteries corrode when mounted in the remote holder kit, make sure there is no corrosion on the wires. Even after removing the damaged batteries the corrosion can continue down the wires and work its way into the circuit board. Get a new kit.

If you put a machine in storage, or you know you won't be playing it for a long time, remove the batteries.

If you change batteries when the machine is turned off you will lose all the high game scores and special settings. The machine will revert to the original default factory settings. If you change batteries when the machine is turned on (not

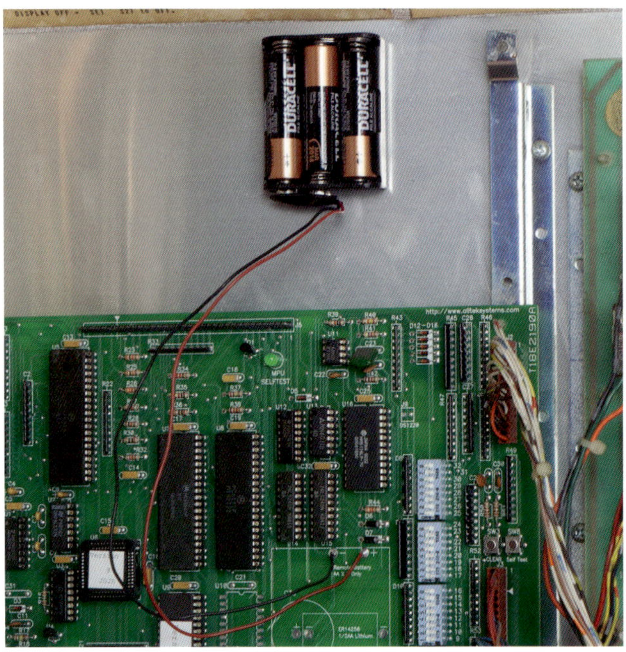

CPU board with the original battery holder removed from the board and replaced with a "remote" battery holder and wires long enough to mount the battery holder in a safe place in the backbox, so a leaking battery can't damage the circuit board..

just plugged in, but turned on and powered up) you will save the settings. Be careful not to reverse the polarity of the batteries and not to let any static electricity cause any damage. See “Static Electricity” in the “Repairs: Basics” chapter.

“FACTORY SETTINGS RESTORED”

When you first turn on a machine after replacing batteries, often you will get a message “Factory Settings Restored” or something similar. The message appears briefly and then the machine goes to game over (“attract”) mode.

If the machine keeps reverting to the “Factory Settings Restored” message, most likely one or more of the new batteries are not making good contact.

BATTERY ELIMINATORS

Some pinball suppliers sell battery eliminators: electronic modules that replace the battery circuits in pinball machines and eliminate the need for batteries, and eliminate the problems batteries cause. The modules are not simple plug-in units. They require careful installation, not terribly challenging but possibly beyond the “comfort zone” of many pinball machine owners.

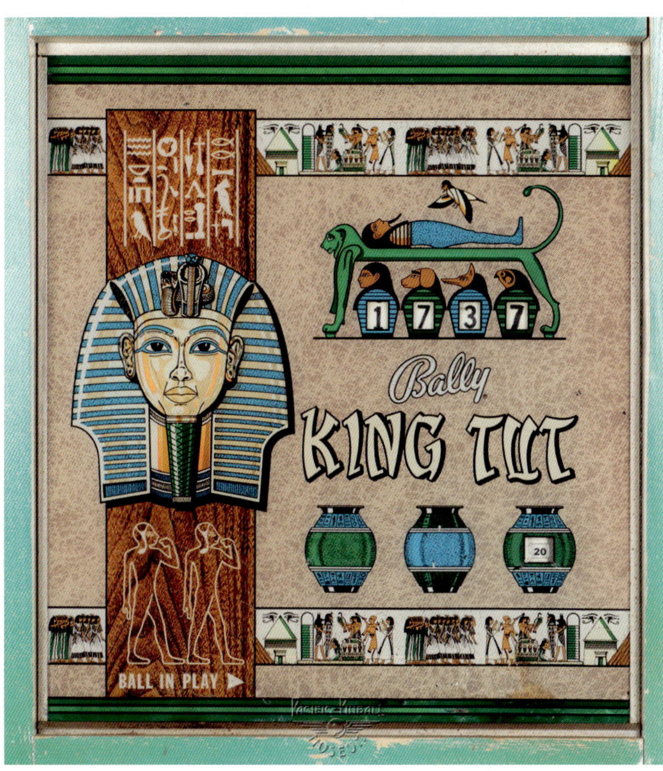

Bally **King Tut**, 1969. Donated to the PPM by Michael Schiess. Photography by RobPerica.com

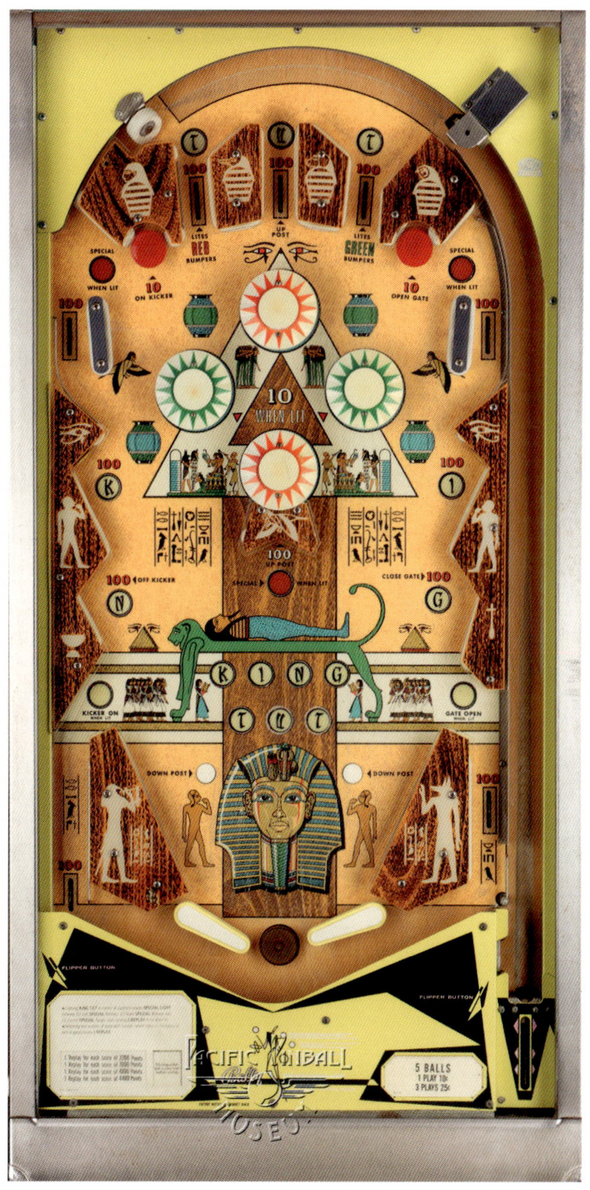

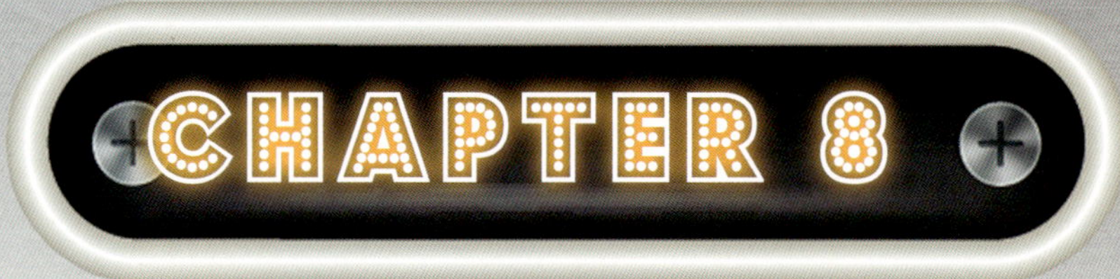

REPAIRS: BASICS

An old pinball machine = Fix me.

—**Joel Cook**, *The Pinball Lizard*

THE JOY OF PINBALL . . . REPAIR

There was a time not that long ago when *things*—appliances, electronics, even toys—were built to be taken apart, repaired, and put back together. The concept of just throwing away something that stopped working and buying a replacement was, quite frankly, ludicrous. And now look at any garbage dump: toys, TVs, computers, cell phones. Nobody wants to fix it, nobody knows how to fix it, and besides, the new model with the awesome purple case just came out.

So here's a toast to one "thing" that continues the honorable tradition of value and is worth fixing and keeping: the pinball machine. Most old electro-mechanical machines can be brought back to life simply by cleaning or replacing some inexpensive switches, coils, and rubber rings. Replacement circuit boards are available for many electronic machines and are often just a straight swap-out.

And you can do so much of it yourself. One step at a time, one circuit at a time, one flipper at a time, and that shiny steel ball is once again flying off a pop bumper, roaring up a ramp, and scoring another jackpot.

The repair help in this manual is separated into five chapters. "Repairs: Basics" is what you need to know to make repairs, and what tools and supplies you will want to have on hand. "Repairs: Flippers" is not only the most common repair, it is probably the easiest for any pinball owner to take on. The remaining three Repairs chapters are broken out by the two eras of pinball machines: electro-mechanical and electronic. Repairs that are common to both eras, other than flipper repairs, are covered in "Repairs: All Machines." Repairs that apply only to electro-mechanical or only to electronic machines are in separate chapters. If you don't know the difference, read "Electro-mechanical machines" and "Electronic machines" in the "Buying a Pinball Machine" chapter.

EVERY MACHINE IS DIFFERENT

I've gotten the same basic question at least 500 times: Some target—or switch, or button, or kicker, or eject hole, or whatever—does not work. What's wrong?

People are looking for an easy answer to what they think is a one-size-fits-all problem. But every machine is different, and every switch does something different. The answer to the problem may in fact be easy, but it's a different answer for different machines, and different manufacturers, and different technologies. A book that covers every problem on every machine would have several thousand pages!

Still, there are many problems—easily solvable problems—common to all, or almost all, machines, and I'm going to try to cover all of them in the Repairs chapters of this manual. The circuitry of a pinball machine is complex if taken as a whole, but is often simple if taken one circuit at a time, one switch at a time.

Being able to fix your own pinball machine can be as enjoyable and as satisfying as playing the game.

PROBLEMS AND SOLUTIONS

This manual covers both electro-mechanical and electronic troubleshooting and repair. Some adjustments and repairs are common to both electro-mechanical and electronic machines, while some apply to electro-mechanicals or electronic machines only. Some problems are easy to solve, and some can be as mysterious as the Sphinx in Egypt.

If there is a short somewhere, or if something isn't working, sometimes it is obvious—sometimes you can see that a coil burned, a component melted, or something is very hot or even smoking. More likely there will be no visible sign of the problem. Then it is a matter of how much you want to follow circuits, read schematics, and test circuit board voltages.

This manual will help you with the easiest repairs, ones not requiring a thorough knowledge of electronics and ones not requiring elaborate testing equipment.

DON'T FORCE ANYTHING

Don't force anything; you'll break it. If something isn't going in, or isn't coming out, if something is stuck, if something doesn't fit, or if something is jammed, stop. I mean, STOP! Have a close look. Try to find out what's wrong.

GET ORGANIZED

The number one key to success, and to keeping your sanity, when working on a pinball machine is organization:

1. Have your tools in one place and continually return them to that place. You will save hours searching for a screwdriver you just had a minute ago. The best birthday present I ever bought myself was a rolling tool cabinet. I park it next to the pinball machine and all my tools are right there.
2. Have a can or a box for parts.
3. Take photos or make diagrams before disassembling a unit. Take photos or make a diagram of all parts you remove, where they came from, and what order they came off so you can figure out how to put the parts back together. Some people set up a video of what they're doing to play back when the parts get mixed up and you're hopelessly lost.
4. When reinstalling a component, if there's more than one screw, start all of the screws before tightening any of them. Quite often the component needs a little wiggle room to set exactly where it is supposed to be.
5. Keep notes on the steps you take and the results. The same problem may come up again. With notes you can go back and see what you did last time.
6. There is an ancient Latin saying, *Age quad agis*. "Do one thing at a time." Do only one repair at a time, make only one change at a time. Check to see if you've fixed the problem before trying something else. That way, if you do solve the problem, or if you cause a new problem, you'll know exactly what you did. If you make several changes or repairs or adjustments at the same time, you will not know what caused or cured the problem. Multitasking = multi-problems.
7. One saving grace of pinball machines is that components often come in twos or more: flippers, pop bumpers, slingshots, roll-overs, scoring units. So if one isn't working, or one is hopelessly disassembled, you can examine the other to see how things should be. Don't take them both apart at the same time.

SAFETY FIRST

Pinball machines have 110–120 volts inside of them and exposed wiring. If you touch the wrong wire you can get a nasty shock. You could kill yourself. Whenever you work on a machine, unplug the machine first. **Don't just turn the machine off. Unplug it from the wall.** If you must work on a machine that is plugged in or turned on, know your circuits before touching any wires or connectors. Remove any loose jewelry to prevent accidental contact with wires. Do not work on a machine when you are wet or standing in water. Water conducts electricity right through your body.

CIRCUIT VOLTAGES

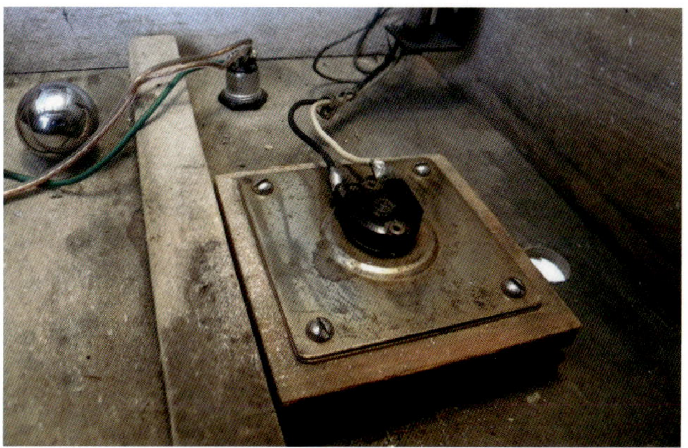

The wiring to the on-off switch is just inside the coin door. On some machines (such as this one) the switch wires are exposed, and they are carrying a full 120 volts. Keep the coin door locked. Unplug the machine before touching wires.

Most light circuits are low voltage, typically 6 volts. Most electro-mechanical circuits and coils are under 30 volts, which is not likely to shock you, but some circuits are 50 volts, and dangerous. And some electro-mechanical machines from the 1960s had 120 volt circuits—full line voltage!—on the coin door and for a few of the bigger solenoids, such as those resetting a long bank of relays.

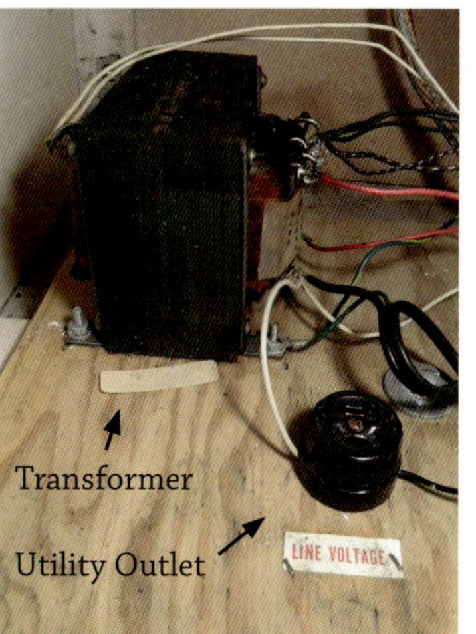

The power to the on-off switch—usually a toggle switch or push button switch mounted on the bottom of the cabinet—is a full 120 volts. So is the power to the main fuse, power to the transformer, and power to the utility outlet found in many machines, so you can plug in a work light or soldering tool. This is line voltage, and extremely dangerous.

Electronic machines: Electronic machines have operating circuits that generate as much as 100 volts. Be especially careful around capacitors, particularly the fat round ones on power supply boards. Capacitors store electricity and can give you quite a shock even after the machine is turned off and unplugged.

EXAMINE THE WIRING

Some machines may be 30, 40, or even 50 years old, and may have been altered or rewired by someone. Old insulation may be frayed or disintegrating. Check the AC power cord and plug carefully. Follow the power cord into the machine (unplugged from the wall of course), follow it to the main fuse, from there to the transformer, to the on-off switch, and to the utility outlet in the cabinet. That's all full 120 volts. If there is damage or a loose connection, replace or repair before plugging in the machine. See "Replacing AC Power Cords."

If there is a ground wire (green wire attached to a 3-prong plug) do not bypass the ground. If the ground prong is broken off, which it often is, replace the plug. Or instead of replacing the plug, it's often cheaper and much quicker to buy a heavy duty extension cord, cut off the female plug, and replace the entire power cord. See "Replacing AC Power Cords."

Some machines have the ground wire disconnected inside the machine. Reconnect the ground wire. The ground wire sometimes connects to the frame of the transformer, sometimes elsewhere. The ground wire is always green.

Safety for cleaning: Do not use flammable cleaners. Rags soaked with combustible cleaners and solvents, if left exposed to the air, can ignite from spontaneous combustion. Last year a friend of mine saw his workshop burn to the ground when, in the middle of the night, a rag burst into flames.

Safety for your kids: Keep the coin door and backbox locked. If the locks or keys are missing install new locks. Don't let curious kids and prying fingers get into trouble.

Safety for your health: Cleaning and waxing products, solvents, solder, and lubricants are often toxic if breathed or absorbed through the skin. Protect yourself.

SHORTS / SHORT CIRCUITS

Throughout this manual are discussions about problems with short circuits and about testing a circuit by "shorting" it.

Any time you complete an electrical circuit—connecting two ends or leads or connections together when they are not normally connected—you have created a short circuit, i.e., "shorted" it. In a short circuit the current travels a shorter distance than it should, which is where the term comes from.

When a switch is stuck closed, with the switch blades constantly touching, it is "shorted" or "short circuited." A coil that is stuck in the on position (activated) when it should be deactivated is shorted. Most important, in a short circuit the electrical current generates more heat than it should. This can cause all sorts of problems, usually minor ones like damaging a component, or not so minor problems like starting a fire.

If you think you have a short circuit—if a component melted or burned, or if a fuse blows repeatedly—do not turn on the machine until you can locate and repair the problem.

Most short circuits in pinball machines are somewhere in one of the machine's operating circuits, not involving high voltage and not dangerous. But if there is a short in the 120 volt line coming from the wall outlet, this is a situation that needs immediate attention. One quick test you can make if you suspect a wiring short in the 120 volt line:

1. Unplug your machine from the wall.
2. Connect one lead from a test light (continuity tester) or a multi-meter to one side of the plug on your machine's AC power cord (the plug that you unplugged from the wall).
3. Touch the other test lead to metal parts of the machine: the front door, side rails, lock down bar, etc. There should be no continuity. The circuit should be open and the test light should not light.
4. Test the other prong of the power cord plug. You should again get an open circuit. If you have a three-prong plug, the round middle prong is supposed to be ground. This is not the prong we are testing here.

If your machine passes this test it is not, repeat *not* a sure guarantee that you don't have a dangerous short in your wiring, but it is an indicator that your machine is most likely not shorted. But if your machine fails this test—if you do get continuity from either side of the line plug to metal parts on the machine and if the test light lights—be warned. You have a serious, life threatening electrical problem. Do NOT plug in the machine until the short is located and repaired. And **DO NOT work on any 120 volt circuit if you are not trained to do so.** Hire a professional electrician or pinball repair technician.

SHORTING A SWITCH TO TEST IT

While short circuits cause problems, sometimes you want to create a temporary short circuit to locate problems. Some testing and troubleshooting involves "shorting" a switch: temporarily closing the switch to observe the results. The easiest way to short a switch is to attach a short piece of wire to the two solder lugs of the switch using alligator clips at each end of the wire. This is often called a jumper wire, and the procedure is sometimes called jumpering or jumping it. Don't confuse this jumper wire with the jumper wires that connect a series of relays in some electro-mechanical machines.

STATIC ELECTRICITY PROBLEMS (ELECTRONIC MACHINES)

Anyone who has ever touched something metal and seen a spark or gotten a shock knows what static electricity is. Static electricity can damage components in electronic pinball machines. You can fry solid state components just by touching them if static electricity is present.

To prevent static electricity, you should be grounded when working on an electronic machine. If your machine is properly grounded (see "Safety First"), if the machine is plugged into the wall (see "Safety First"), and if your house wiring is grounded, you can ground yourself by touching any ground on the machine. If the machine has a ground strap—usually a flat, braided, and uninsulated wire mesh that runs between the main cabinet and the backbox—touch the strap. If there is no ground strap, the metal frame at the back of the backbox is usually grounded. Avoid wearing nylon or rayon clothing, as these fabrics generate static electricity easily.

Electro-mechanical machines cannot be damaged by static electricity.

WARNING! COIN DOOR OR START SWITCH SHOCK

On many old machines from the 1960s and earlier, there was a full 120 volts to the start button, the coin switches, and the anti-cheat switch on the coin door. Those machines usually had a label inside the front door warning about the voltage at the door, a label that may or may not still be on the door. A short or touching the wires could give you a dangerous shock. The start switch was insulated from the door to protect you from shock, but that 50-year-old insulation may be crumbling away.

If you do get a shock when you touch the start switch or the coin door, unplug the machine immediately. Check for frayed or loose wires on the start switch, or replace the start switch. Remember, this is not just another pinball circuit. You are dealing with lethal voltages. Be extremely careful.

When one of these old machines comes into my shop, even if it is working fine, I remove as much of the 120 volt circuit from the coin door as possible. I disconnect the wires and bypass the anti-cheat switch and the coin circuits. I replace the insulation between the start switch and the coin door. Sometimes I replace the old metal start switch with a new plastic switch that will not conduct electricity.

However, I am a trained, experienced repairman. As I repeat throughout this manual, **do NOT work on any 120 volt circuit if you are not trained to do so.** Hire a professional electrician or pinball repair technician.

FLIPPER BUTTON SHOCK

Like everything else on the outside of a pinball machine, the flipper buttons should not give you a shock. And they rarely do, but it has happened to me twice on old electro-mechanical machines that came through my shop.

In one case, the flipper buttons were metal and able to conduct electricity, the insulation separating the buttons from the flipper switches had worn away, and the 30 volts going through the switches went right into the flipper buttons and right into my fingers. Replacing the insulation solved the problem.

Many flipper buttons are plastic and unable to conduct electricity. If you can get a plastic replacement flipper button (available from some pinball parts dealers) it offers one more layer of protection from shock.

In the second case of flipper button shock, one of the flipper switches—the switch mounted inside the cabinet behind the flipper button—was mounted slightly crooked, so one of the switch blades touched the flipper button's metal mounting hardware, sending a shock. Realigning the switch solved the problem.

REPLACING AC POWER CORDS AND PLUGS

New pinball machines have removable universal power cords, but most pinball machines have permanent power cords that need to be wired correctly when replaced.

There are several kinds of AC power cords. All AC circuits have a "hot" side and a "neutral" side. Three-wire circuits also have a ground wire. Round cords have an outside insulation that conceals insulated wires inside the cord. The wires in the cord have color coded insulation: black and white. A three-wire round cord will also have a green wire.

If the wires are installed correctly, the black wire is the "hot" side and connects to the brass or gold-colored screw in the line plug. The white wire is the "neutral" side and goes to the silver-colored screw. The green wire is the safety ground and always goes to the green screw on the round middle terminal on the plug.

On a flat power cord, the ribbed side is the white or "neutral" wire and the smooth side is the black or "hot" wire. If the flat wire has no rib or similar marking you need to determine which side of the wire is which by tracing the wire from the plug to the components in the machine.

On both round and flat cords, the "hot" wire—the black wire on the round cord, the smooth (not ribbed) wire on the flat

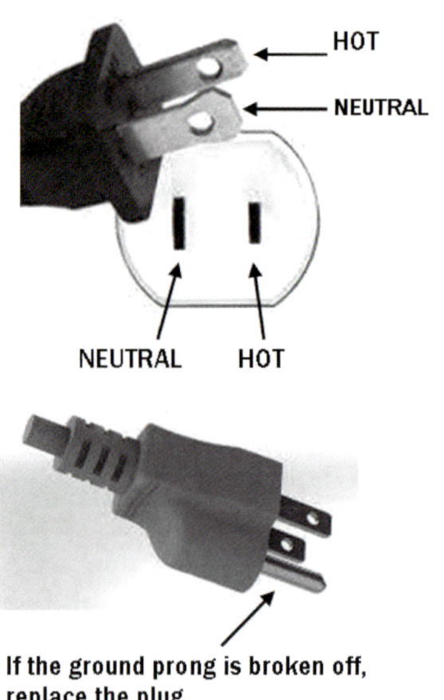

If the ground prong is broken off, replace the plug.

cord—is the switched wire, the wire that goes to the power switch.

Modern electric plugs that plug into wall outlets have wide and narrow prongs. The narrow prong is hot, the wide prong is neutral. On the wall outlet of the house the smaller slot is hot, the larger slot is neutral.

However, if someone has previously rewired the machine, or your house, there is no guarantee they got the polarity correct. Don't rely on the wiring until it is checked out. If you are unsure about your house wiring, get a professional electrician to test the wiring.

LONG CORDS AND EXTENSION CORDS

If it's a long run from your pinball machine to the nearest wall outlet, long cords and extension cords are usually okay. If the distance is more than about ten feet be sure the wire is heavy duty.

After you first plug in the machine and finish your first game, carefully feel the outside of the plugs and connections. You should not feel any heat. If the wire is warm, either use a shorter wire or a heavier gauge wire. **If the wire is hot, shut everything down!** Something is wrong.

Where people have serious problems and risk starting a fire, is when several appliances are plugged into the same outlet or extension cord, particularly anything generating heat, such as a portable electric heater or strings of Christmas lights.

AGAIN, BE WARNED: There are many wiring issues in a pinball machine that you can work on in complete safety, but working with 120 volts is not one of them. **Do not work with high voltages if you are not trained to do so.** Get a professional electrician or a pinball repair person to help you.

TOOLS

You'll enjoy working on your pinball machine a lot more if you have the right tools for the job. Nothing causes more aggravation and wastes more time than trying to use a tool that doesn't fit the purpose.

Splurge on tools. Buy whatever tools you need, and buy quality. Cheap tools are cheaply made. If you cannot afford quality new tools check out yard sales.

Most of the tools you need to adjust and repair pinball machines are common hand tools. Some tools are unique to pinball and can be ordered from a pinball parts dealer.

Regular (non-specialized) tools include:

- Set of magnetized screwdrivers, regular and Phillips. Magnetized screwdrivers will save you hours of searching and cursing, looking for lost screws. You can magnetize screwdrivers with a magnetizing tool that you can get at any hardware store.
- Set of socket wrenches and a set of combination (open-box) wrenches. Pinball machines use American (SAE) measurements, not metric.
- Set of L-shaped hex (Allen) wrenches
- Long nosed pliers
- Wire cutter/ stripper/ crimper tool
- Soldering gun or pencil and a desoldering tool (solder sucker; more on this to follow)
- Magnet, preferably the kind that is mounted on a hand-held extending rod.
- Clamp-on work light
- Flashlight
- Jumper wire, 1 or 2 feet long, with alligator clips at both ends. You can easily make one or buy one at any hardware or electronics store.
- Small wire brush
- Small smooth-cut metal file, such as an automotive point file or a burnishing tool, although a pinball flex file works better (see following).
- Bench grinder with a grinder wheel, a wire wheel, and a buffer wheel is a very useful tool for smoothing rough ends of coil plungers and shooter rods, and for removing rust and polishing metal parts.
- Power screwdrivers and drills. Be careful if you are screwing down plastic parts with a power screwdriver or drill. In fact, I suggest you never use a power driver on plastic parts, and I wish I hadn't the other day when I cracked a pop bumper cap when the screwdriver kept right on going . . . If you do use a power driver, set the torque at the lowest setting. The lower the number, the lower the torque.
- Adjustable wrenches. Just about every tool chest includes an adjustable wrench (crescent wrench, monkey wrench),

but try not to use one unless there is no other option. A friend of mine, who owns a bicycle repair shop, calls adjustable wrenches "nut rounders" and forbids them in his shop. Because that's what happens too often. The wrench is adjusted a hair too big for a nut and suddenly the nut is damaged, won't loosen, and won't tighten.

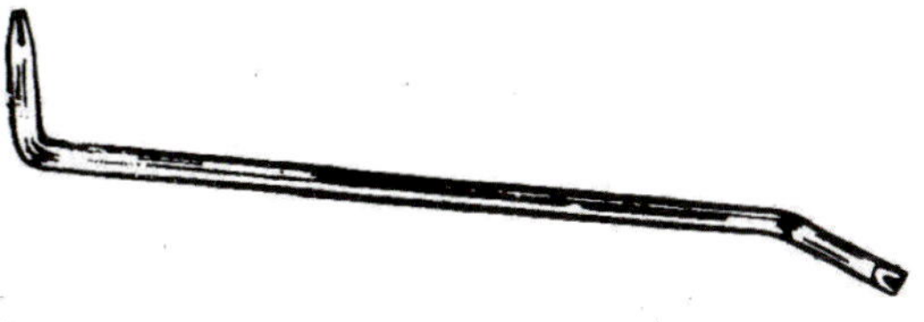

SPECIALIZED PINBALL TOOLS

Pinball parts dealers sell tools designed especially for pinball machines. None of these tools are absolutely necessary to maintain and repair a pinball machine. You can use conventional tools to do the job. But the specialized tools can sure make working on a pinball machine a lot easier. Most of these tools are inexpensive, and well worth the investment. A few of the most useful tools are:

- Switch adjusting tool (contact adjuster) for electro-mechanical machines and early electronic machines that used leaf switches.
- Pinball "flexstone" flexible file for electro-mechanical machines
- Light socket cleaning stick
- Lamp remover: Special pliers (expensive) or rubber tube (cheap) to grab light bulbs.
- Circuit board remover (electronic machines)
- Thin, flat wrench for removing playfield post caps.

TESTING EQUIPMENT

For simple troubleshooting and repairs you only need a minimum of testing equipment, and all of it is inexpensive or easy to make yourself.

MULTI-METER

An essential piece of equipment for all electrical work is a multi-meter, also called a multi-tester, a VOM (volt-ohm-milliammeter), or an ohm meter, even though it reads more than ohms. A multi-meter is the Swiss Army Knife of electronics. You can use it to check shorts, broken wires, fuses, lamps, switches, coils, and batteries. You can use it to check voltages and current draw. You can use it to test solid state components.

Multi-meters are either analog or digital. Analog meters have a moving needle that shows the readings. Digital multi-meters have digital readouts. Analog meters are fine for electro-mechanical machines and testing for breaks or shorts. But if you will be troubleshooting electronic machines—reading circuit board voltages and testing batteries—a digital meter will be more precise to read.

Some multi-meters have a very useful feature when testing for shorts, broken wires, burned lamps, fuses, and other continuity tests: an audible signal. Instead of having to look at the meter every time you test a circuit you can listen for the beep. I highly recommend getting a meter with this feature.

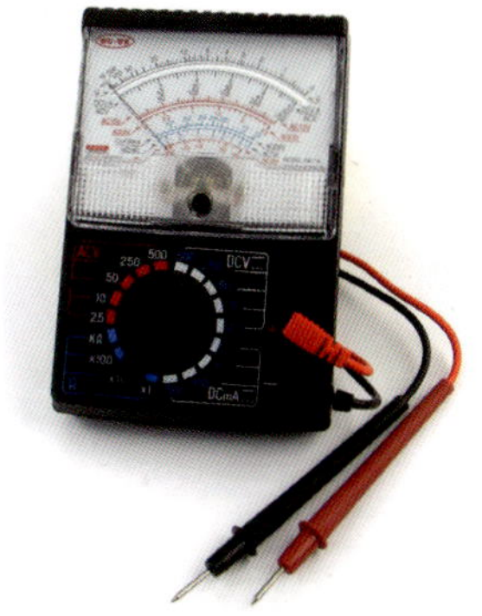

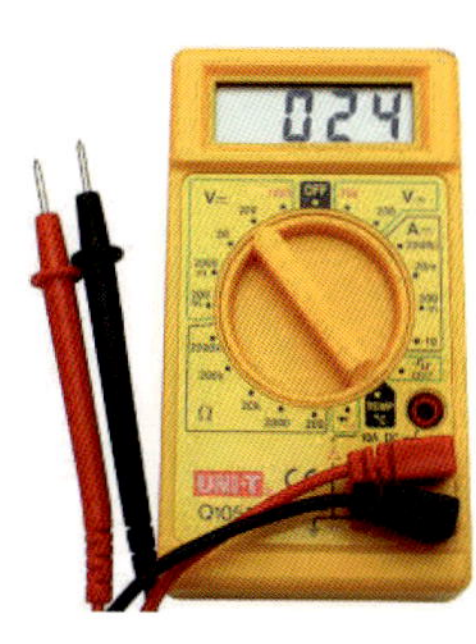

Analog (left) and digital (right) multimeters cost about $35 and are essential for any kind of electrical troubleshooting.

Using a Multi-meter

Testing for open or shorted components is the most common use of multi-meters for people doing pinball machine troubleshooting: Is a switch open or closed? Is there a break in the wiring? Is there a short circuit? Is a fuse or a lamp or a coil burned?

To make these tests, set the meter to read ohms (Ω). Touch the two test leads (test probes) to the two ends of the component or wiring you want to test. If there is an open circuit (open switch, disconnected or broken wire, bad fuse, no short circuits) an analog needle will stay at rest (left side of the meter window), while a digital meter will read 1.00. If there is a closed circuit (closed switch, wiring intact, a short circuit) an analog needle will swing all the way to the right, a digital meter will read 0.00. Clean, fast answers. However, you might get a reading somewhere in between: the analog needle in the middle, the digital reading somewhere between 0.00 and 1.00. Which we won't get into here. We're going to start out simple. Most of the tests you will be doing will be to check for open or closed circuits. Clean, fast answers.

Be aware that you may have to disconnect a component from the wiring in the machine to get an accurate test of the component. You want to be sure that you are testing the component, not the circuit.

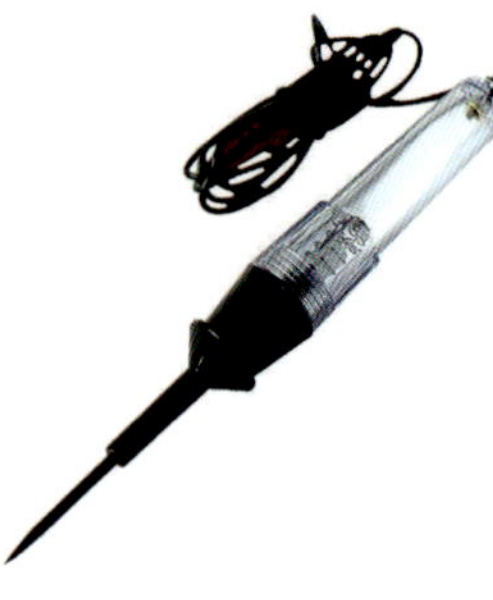

CONTINUITY CHECKER (TEST LIGHT)

If you are only going to test for continuity (shorts, lamps, fuses, broken wires, etc.), a continuity tester, also called a test light, may be all you need. This is a very simple tool resembling an ice pick with a pointed metal tip and a long wire with an alligator clip coming off the end of the handle. There is a battery and a light bulb inside and visible through the handle. If you touch the alligator clip to the pointed tip of the tester, you make a complete circuit and the bulb lights. So by touching the two ends of the checker (the pointed tip, and the alligator clip) to the two sides of coils, switches, fuses, light bulbs, and wires, you can see if the test light lights. If it does, you have continuity: the fuse is good, the coil does not have a broken wire, the switch is closed or shorted.

Unfortunately, I've found that the cheap continuity checkers sold at hardware and electronics stores do not work well and make poor contact. I don't know why they won't work, but they're a waste of money. Buy a good quality continuity checker or just get a multi-meter, which does the same thing and is much more useful.

By the way, don't buy an automotive test light. These testers look almost the same as a continuity checker but do not have a battery inside of them. They will not work.

LAMP TESTER

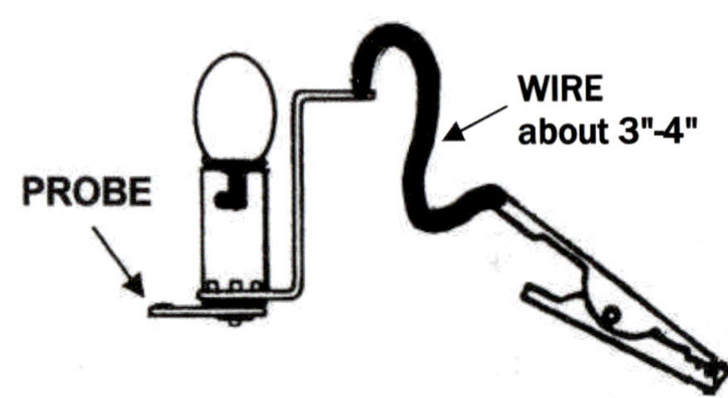

Homemade lamp tester easily and quickly tests lights, light sockets, wiring, and power to the light circuits.

I made a simple but very useful lamp tester by soldering a short piece of wire, about 4" long with an alligator clip at one end, to the base of a lamp socket (the mounting bracket). The solder lug on the lamp socket is my "probe." By attaching the alligator clip to one side of a lamp socket in the machine, or to the common wire running from socket to socket, you can quickly test the lamps, sockets, and the wiring. This is covered under "Problems With Lights" in the "Repairs: All Machines" chapter.

REMOTE START TESTER

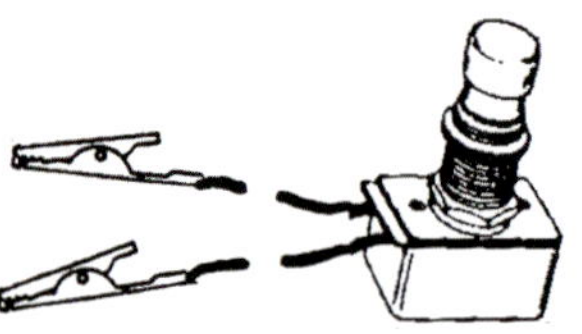

Another simple tool I made to help with repairs is a remote start button, so I can start a game from behind it or while watching components inside the machine. It's a double wire about 4 feet long (a thin lamp cord works well) with alligator clips on the two wires at one end and a momentarily-on push button switch at the other end. Push the spring-loaded button to activate the switch, release the button to deactivate the switch.

Connect the alligator clips to the two solder lugs of the coin switch inside the coil door (the switch the coin hits to start a game). That way, without having to reach in front of the machine to push the start button, I can just push the remote button and start a game. This is also one way to get free play, as discussed in "Free Play" in the "Components and Features" chapter. If the coin switch has three wires instead of two do not use this tool, it won't work.

SUPPLIES AND PARTS

Here is a list of supplies that I always have on hand:

- Isopropyl (rubbing) alcohol: 99% if you can find it, 91% is okay
- Pinball (coin) lubricant or a Teflon gel lubricant. Pinball parts suppliers sell different versions of this lubricant.
- Solder, rosin core, 60/40. This is standard solder for electrical applications. Do NOT use acid core solder; it will destroy the components. There is also coreless solder, solid solder that requires the addition of a liquid flux to get the solder to adhere. Coreless solder and flux will work for electrical applications, but there is no need to use this messy two-step process.
- Cleaners and wax for the playfield, cabinet, and plastics. Pinball parts suppliers sell a selection of cleaners, and for playfields, combined cleaner/wax.
- Glass cleaner, although a moistened paper towel is really all you need.
- Rags, paper towels, and a box of tissues.
- Chrome polish
- Brillo or SOS pads (steel wool soap pads) to remove rust.
- WD-40

Contact cleaner: What about contact cleaner? In my experience, spraying contact cleaner on pinball components does not fix problems or protect against future problems. I know, people swear by contact cleaner as a cure all for everything electronic. Which it isn't. Contact cleaner is useful on electronic components with difficult-to-access moving parts, such as volume controls or sliding adjustments on sound systems. But pinball machines rarely have problems that spraying contact cleaner will cure.

Which no one believes. I once had a customer bring a pinball machine to my shop for repair and told me that the last repair person—a computer repairman who had never seen the inside of a pinball machine—just sprayed contact cleaner all over everything and the machine started working again. (Perhaps. When I got the machine it had a melted coil, a broken switch, and a stuck ball count unit.)

But, yes, I do have a can of contact cleaner in the shop that I use on noisy volume controls and lamp dimmers on electronic machines. Different contact cleaners have different ingredients. Some leave a residue, some may not be safe for plastics, and I can't imagine that any contact cleaners are safe to breathe.

PARTS TO HAVE ON HAND

Unless you have several pinball machines you are regularly working on, it's not practical to keep replacement parts on hand. In every machine there are many different coils, switches, and electronic components, and no way to know which part might eventually need replacement. I suggest purchasing replacement parts as needed. If parts are inexpensive, when you need a part maybe buy two instead of one and start building up a collection of parts (that you'll probably never use).

Parts you will probably need:

- *Rubber rings:* Some people buy an extra set of rubber rings in case a ring breaks. But if the rings sit for a year or two they dry out with age, and lose some of their elasticity. I'd wait until a ring actually breaks. Or wait a year, replace all of the rings with new ones, and save the old rings for backup.
- *Shooter tips:* I'd have an extra shooter tip, because it will wear out. Shooter tips should be replaced immediately, so the exposed end of the metal shooter rod doesn't contact the ball and damage it.
- *Lamps:* Lamps burn out occasionally, so having an extra box of lamps is a good idea.
- *Balls:* If your pinballs don't get dinged or pitted or rusty, they should not need replacement. But it never hurts to have an extra ball, or set of balls, just in case.
- *Springs:* Pinball machines use all sorts of springs—different shapes, lengths, thicknesses, and tension strengths. Compression springs that, when pushed in, spring back out; and extension springs, that when stretched out, spring back in. The pinball manufacturers had specific springs for specific mechanisms, hundreds of different springs, many of which are no longer available. By gathering a collection of springs—new ones when you can find them, and used ones scavenged from anything and everything you are tossing away—you will usually be able to find a workable replacement for a broken or missing spring in your machine.
- *Miscellaneous hardware:* You are going to lose, or discover already missing, small screws, nuts, washers, and those little snap-on clips that keep parts from falling off whatever they're mounted to. It's always useful to build up a collection of hardware. All of it is available from any hardware store.

SOLDERING

Many pinball machine repairs require soldering. Good solder connections are essential to good working games. A poor solder connection means poor electrical contact, poor performance, or components that don't work. Loose or straggling wires and sloppy globs of solder can cause shorts and burned components.

SOLDERING GUNS AND SOLDERING IRONS

You can use either a soldering gun or a soldering iron (soldering pencil). Soldering irons take a few minutes to heat up, but they are on, and hot, all the time. Soldering guns turn on only when you click the trigger, and they shut off as soon as you let go. Soldering guns heat up in just a few seconds.

Soldering guns are perfect for electro-mechanical machines and for coil and switch connections. Most soldering guns have a light that helps tremendously in the dark corners of a machine. But soldering guns generate very high temperatures at the tip that can damage solid state circuit boards. If you work on circuit boards, use a low power soldering iron, about 25 watts. The best soldering irons have a temperature control. Use the lowest temperature setting that will accomplish the job.

As I mentioned under Supplies, use rosin-core solder, solder made especially for electrical connections, which is usually designated as 60/40 (60% tin, 40% lead). Do not use acid-core solder for electrical work. The acid in acid-core solder will destroy electrical components. Buy solder from a hardware store or electronic supplier. I've been told that some off-brand cheap solder is of poor quality, so don't save a buck by buying solder from someone at the flea market.

HOW TO SOLDER

Soldering is easy with a little practice. For soldering coils and switches, the wires should first be connected securely: twisted together, or bent and crimped around the lug. The physical connection should be as good as possible before applying solder. There should be no stray strands of wire sticking out, as they could touch the wrong thing and cause a short.

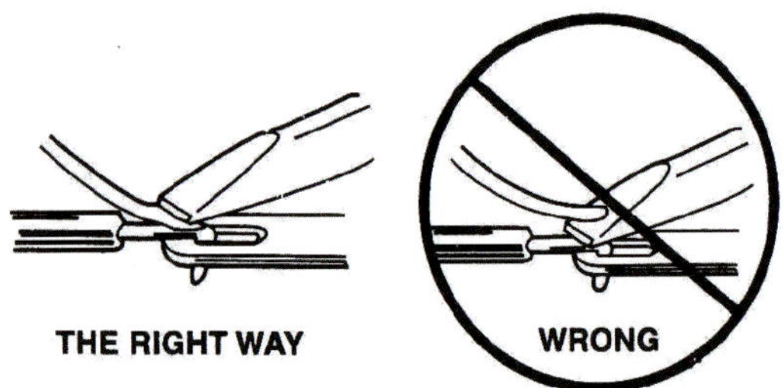

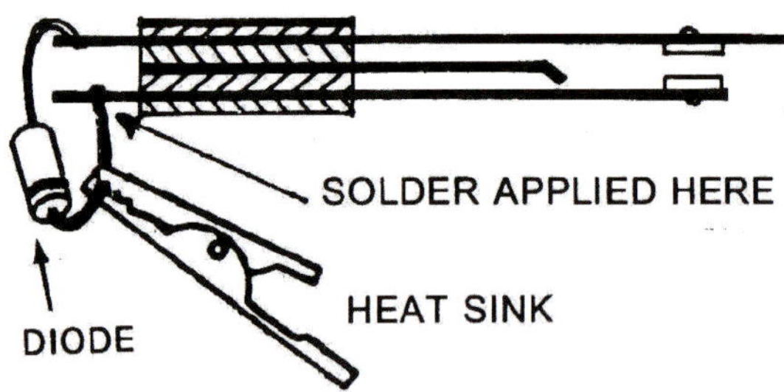

Alligator clip used as a heat sink to protect the diode. It clips to the wire between the diode and where you are soldering (lower switch blade).

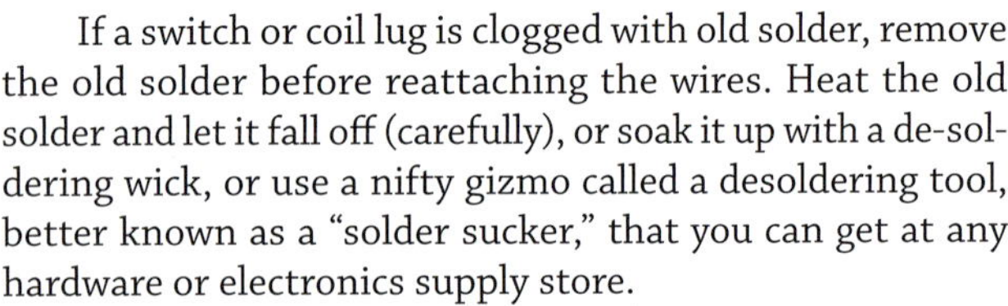

If a switch or coil lug is clogged with old solder, remove the old solder before reattaching the wires. Heat the old solder and let it fall off (carefully), or soak it up with a de-soldering wick, or use a nifty gizmo called a desoldering tool, better known as a "solder sucker," that you can get at any hardware or electronics supply store.

When soldering, apply the heat to the connection until it melts the solder into the wires evenly and smoothly. The melted solder should have a shiny, silvery look to it. If the solder balls up into a dull grey lump it wasn't hot enough, and the connection is likely to be poor. Reheat the solder until it melts. Try to apply as little solder as possible—enough to make a good, solid connection, and no more.

Don't put too much heat on a connection. You don't want to damage it or melt a part. Unfortunately, knowing how much heat is "too much" requires a little bit of experience. You are not likely to damage a coil or switch connection or a transformer lug with the heat from a soldering gun, as long as you keep the heat away from meltable parts.

Be especially careful around heat sensitive components such as diodes, capacitors, resistors, transistors, rectifiers, and ICs. Electronic machines have diodes soldered across coils and switches, and the diodes need heat protection. You may want to use a heat sink, which is a metal clip (an alligator clip will work if there is enough room) placed between the connection where you are soldering and the component needing heat protection. The clip absorbs much of the heat traveling down the wire from where you are soldering, keeping the heat from reaching and damaging the component you want to protect.

Look out for falling solder. I've burned the back of my hand more times than I care to count when some hot solder dropped right on me. Worse than my hand, solder can fall onto electrical components and short them.

If solder does drop, find it and make sure it did not land where it can cause trouble. This is the most fun when soldering components mounted under the playfield of an electro-mechanical machine. With the playfield propped up, gravity has a way of almost guaranteeing spilled solder, most likely falling right onto components mounted on the floor of the cabinet. When I'm soldering under the playfield, I keep a big piece of cardboard to cover the cabinet and catch dripped solder. (And then I never use it, because it's a bulky nuisance to keep putting on and taking off.)

And don't breath the fumes. They're no good for you.

SOLDERING CIRCUIT BOARDS

Be careful soldering circuit boards. Circuit boards and their components can be damaged by too much heat, and the connections are often so small you can easily short and damage the circuit board with one sloppy solder job.

If you don't have experience soldering circuit boards, get a junked CD or DVD player, or an old computer from a repair shop or the local recycling center, and practice your soldering on the boards.

The actual repair of a pinball machine may only take a few minutes, but finding the defective circuit and part can take up to several hours. It is quite possible for different problems to cause exactly the same symptom in a game. Worse, the problem might look like it is in one circuit when it is really on the other side of the game.

What everybody would like to have is a complete list of "If this happens, do this" procedures, an almost impossible undertaking. But with a logical, intelligent approach to isolating a problem down to a circuit area and finally to the defective component itself, you can repair nearly any problem.

—**Robert A. Hornick**, *Electronic Pinball Electronics*

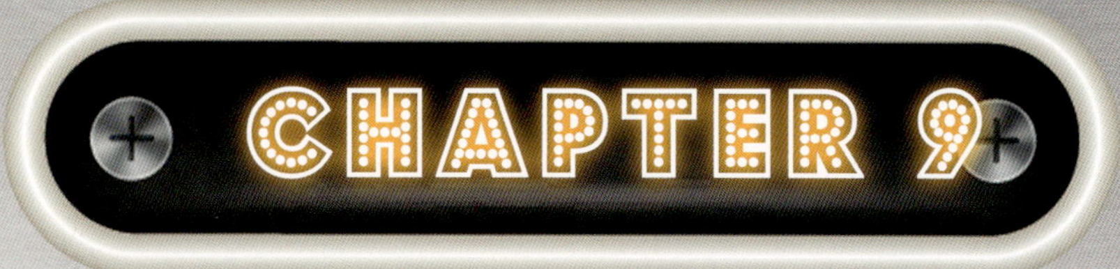

REPAIRS: FLIPPERS

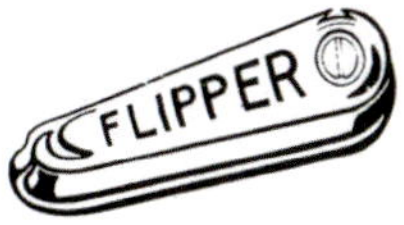

The flippers are the heart and soul of the pinball. The flippers are the contact between player and machine, the battleground where games are won or lost, where skill overcomes the laws of physics and gravity. The flippers are far and away the most important components of every pinball machine. Flippers are what make a pinball a pinball. In many countries pinballs are not even called pinballs They're called "flipper games" or simply "flippers."

Before flippers were invented, flipper-less pinball machines existed, and were loved. But when the first flipper pinball machine was invented in 1947, a game called Humpty Dumpty made by the Gottlieb Company, the world of pinball changed overnight. Flipper-less games were instantly obsolete. Pinball became a huge success. Flippers were the reason.

And weak flippers suck.

There are several reasons why flippers don't work, or don't work well. If you're going to try your hand at any repairs or adjustments, flippers should get first priority. They are easy to fix. You're not likely to cause any more damage. And, man, if you've got a lame flipper, you've got to fix it. You really have no choice.

DEAD FLIPPER (NON WORKING)

There are several causes for non-working ("dead") flippers, and most of them are easy to diagnose and easy to repair.

The flipper circuit is a simple one. A switch is wired to a coil. The switch closes when you push the flipper button on the side of the cabinet, allowing the electrical current to go through the coil. The powered coil becomes an electro-magnet that pulls in a metal rod called the plunger or piston, which is attached to a linkage that rotates the flipper.

There are five components to this circuit—power, the coil, the linkage (the mechanical operation), the switch, and the wiring—any of which could be the reason the flipper isn't working. We'll take them one at a time.

DEAD FLIPPER: NO POWER

If the game starts up and everything seems to be working except a flipper, there obviously is power to the machine. The flipper is probably getting power, and the problem is most likely in the wiring or one of the components in the flipper circuit. If neither flipper is working I'd suspect a lack of power

The Gottlieb Company called the flippers on their first flipper pinball machine Humpty Dumpty "flipper bumpers." The machine in fact had six "flipper bumpers." Although Humpty Dumpty, made in 1947, is the first pinball machine with powered flippers activated by solenoids, the first pinball flippers actually date to a 1935 pinball machine called Olympic Pins made by California Games in Los Angeles. Olympic Pins had six flipper "paddles" manually operated by handles on the sides of the cabinet.

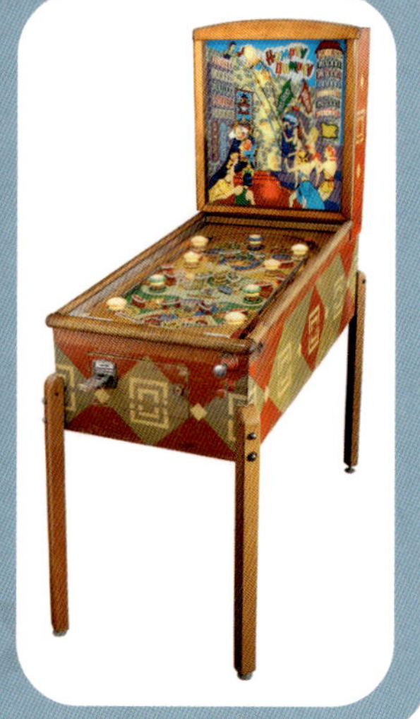

to the flippers. Are there other components on the playfield—targets, or pop bumpers, or slingshots—that are not working? If several playfield components are dead the problem is somewhere else in the machine, not in the flipper circuits.

DEAD FLIPPER: BURNED COIL

The most common cause of a non-working flipper is a burned flipper coil. Raise up the playfield so you can look at the coil that operates the flipper (remove the balls first so they don't come crashing down).

If the paper around the coil looks dark or burned, the coil is probably burned. Try moving the plunger by hand. If it will not budge, that's another indication that the coil is burned. The plunger, however, may be stuck due to broken linkage and not because of a burned coil, so check the linkage before replacing the coil (see following). If the coil is burned you'll need to replace it.

Before replacing the burned coil find out why it burned. If the plunger fell out of the coil, which can happen if the linkage breaks or comes apart, or if the coil fell out of its mounting bracket because of missing screws, the coil will overheat when it is activated. A missing plunger is almost guaranteed to burn out a solenoid.

If someone held in the flipper button for a long time, or if the flipper button switch (inside the cabinet behind the flipper button) is stuck closed, the flipper coil was constantly energized, and that would explain the burned coil. Repeatedly pushing the flipper buttons over and over, as many young kids like to do, can sometimes burn out a flipper coil. More on kids and flippers at the end of the chapter.

A broken bracket or connector on the flipper linkage could have jammed the end-of-stroke switch and burned the coil. These problems are covered below.

DEAD FLIPPER: BROKEN COIL

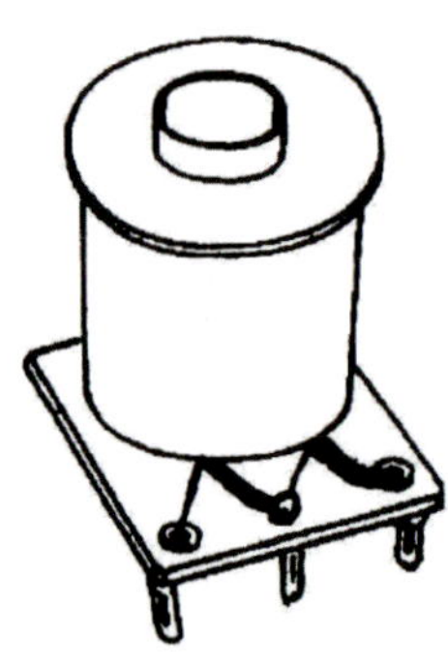

Another possibility is that the coil is broken, or has a broken wire. The coil will not be burned or stuck, it just will not work. This is not common, but it is a possibility.

The flipper coil has two wires, both rolled around the core of the solenoid. One of the wires is much thicker than the other. Flipper coils have three solder lugs. The thick wire is connected to one of the outside lugs and to the middle lug. The thinner wire is connected to the other outside lug and to the middle lug. The thin and thick wires connect together at the middle lug. Look at where the wires come out of the coil and attach to the solder lugs. Are any broken off?

You can test the coil to see if a wire is broken inside the coil. Before testing the coil, separate the two switch blades of the end-of-stroke switch mounted on the flipper mechanism and wired to the flipper coil. A closed end-of-stroke switch can complete a circuit and give you a false reading. A piece of paper, or electrical tape, or any insulator will work. End of stroke switches are covered below.

Next, put the test leads from your multi-meter or continuity checker on the two outside connectors of the coil—not on the middle connector. You should get continuity: The meter registers, the test light lights. There is no break in the wires. If you get an open circuit, or no continuity (test light does not light; meter shows no connection), the coil has a broken wire inside the coil and needs to be replaced.

Replacing a coil: There are probably a hundred different flipper coils. One size does not fit all. Any pinball supplier should know what coil to sell you just from knowing the name of the machine and the manufacturer. The coil in your machine will probably have a wire code or part number printed, though it might be a replacement that wasn't the one specified by the manufacturer. The machine's schematic, and possibly the instruction manual, will identify the correct coil. The "Coils" section in the "Repairs: All Machines" chapter explains how to replace a coil.

DEAD FLIPPER: BROKEN LINKS

The linkage is the mechanism that connects the coil's plunger to the moving flipper bat. You may find, as I have more than once, a broken bracket or connector on the flipper linkage, or parts that separated or fell off due to a missing screw. It might be as simple as remounting the parts and tightening screws. If a bracket or connector is broken, a pinball parts business may have a part to replace the broken one, or you may have to make one from whatever materials you can find. Look at the other flipper if you need help figuring out how the parts fit together.

Chances are that once the linkage is repaired the flipper will work fine. But as I mentioned above, if a broken linkage caused the coil plunger to come all the way out of the coil, the coil may have overheated and burned.

DEAD FLIPPER: FLIPPER BUTTON AND SWITCH PROBLEM

The flipper buttons are mounted on the sides of the cabinet. Pushing a flipper button closes a switch mounted inside the cabinet. The switch activates the flipper. A problem with the switch could be the reason a flipper isn't working.

Clean the switch. Make sure the switch closes and the switch blades make contact when you push the button. Check that the contacts (nipples) on the blades are intact. See "Switches" in the "Repairs: All Machines" chapter.

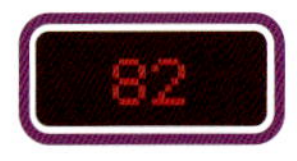

Some flipper switches have three blades instead of the usual two. These three-blade units are actually two switches, the flipper switch and a second switch that turns on the playfield lights or possibly has some other function. The middle blade is common to both switches. The blade closest to the cabinet, making contact with the middle blade, usually operates the flipper.

If cleaning and adjusting the flipper switch does not work, bypass the switch by shorting the two wires attached to the switch: Attach a jumper wire to the two solder lugs of the switch. If the flipper fires then something is still wrong with the switch. Keep playing around with it, or replace it.

If the flipper does not fire when you bypass the switch, the problem is probably in the wiring from the switch to the flipper. Read on.

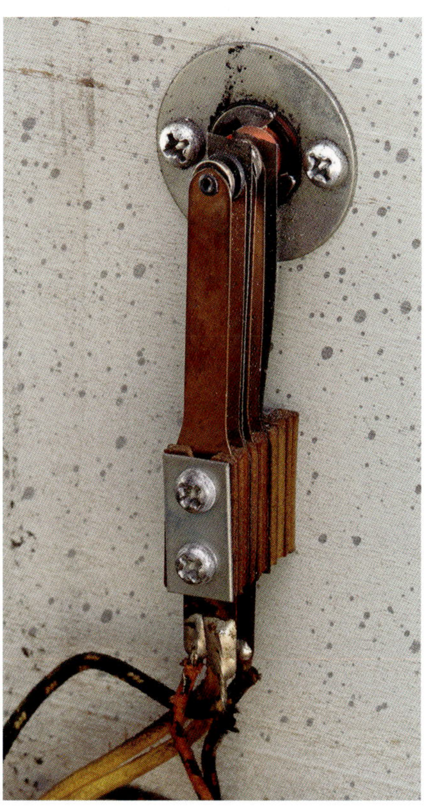

Flipper switch is actually two switches. Switch closest to the cabinet operates the flipper. The second switch turns on playfield and backglass lights after you turn on the machine. Most pinball machines did not have this two-switch arrangement; turning on the machine also turned on the lights.

DEAD FLIPPER: WIRING PROBLEM

If the flipper switch is working, if the flipper coil shows no obvious signs of damage, and if the linkage is ok, the next place to look is the wiring. Is there a disconnected or broken wire where the wires solder to the flipper coil? Is there a disconnected or broken wire on the flipper button switch mounted inside the cabinet? Both of these are easy resolder jobs.

DEAD FLIPPER: COMPARE FLIPPERS

If you haven't solved your problem, and if one flipper works fine, here is another test to try. Jump the wires from the two outside lugs of the working flipper coil to the two outside lugs of the non-working coil (it does not matter which lug is which) and push the flipper button of the working flipper. If the working flipper works but the dead flipper still doesn't work the coil is defective, even though our tests indicated otherwise.

But if the dead flipper suddenly does work when you jump the wires, the coil obviously is fine. The problem is in the wiring controlling that flipper. Recheck the connections. Sometimes I run new wires through the flipper circuit, right next to the old wiring, to see if there is a break in the old wiring. More than once the new wiring solved the problem, even though I still did not know where the old wiring broke or disconnected. There is nothing wrong with just abandoning the old wiring and leaving the new wiring in the circuit.

DEAD FLIPPER: WHAT ELSE?

Something as simple as a flipper circuit just can't have a whole lot of reasons why a flipper isn't working. But pinball circuits, even basic flipper circuits, are often puzzles challenging you to solve them. I have repaired more than 400 pinball machines and I still run into problems I had not previously encountered. For me, it's the most fun to keep trying different solutions. Fun, though sometimes frustrating. Sometimes no fun at all.

I do know that the problem is right where we've been looking. Somehow we missed it, again assuming that the problem is not widespread beyond the flippers, or for electronic machines, in the circuit boards (see following).

Come back tomorrow, or next week, and try again. Invite a pinball friend over and work on it together. At least you'll have someone to drink a beer with while you complain about the stupid machine. (Sorry, that's a bad joke. The problem is fixable and you will find it.)

Electronic machines: Most electronic machines have the same flipper wiring as electro-mechanical machines. The troubleshooting described applies to all pinball machines. But dead flippers on electronic machines could be caused by failed components on one of the circuit boards. The flipper coils and switches and connections are probably fine; they're just not getting power. This, unfortunately, is not an easy repair unless you understand electronic circuits. Read "Circuit Boards" in the "Repairs: Electronic Machines" chapter.

WEAK FLIPPERS

There are several possible causes of weak or poorly working flippers. It is usually not difficult to locate and fix the problem, or problems. It is not uncommon to find more than one reason your flippers are not working well.

WEAK FLIPPERS: OLD MACHINES

Some old flippers were not that powerful to start with. Machines made before 1970 had the old-style short flipper bats, 2" instead of the modern 3". These old flippers had less of an arc than the longer flippers and were not able to propel a ball quite as fast and powerfully as the longer flippers. The laws of physics were working against you. But just because the flippers may have been weak to start with, that doesn't mean you have to live with weak flippers. You can get old flippers working just the way you'd like. Here's what you do:

1. First, make sure that the flippers don't have any of the problems described in this Weak Flippers section. Get those problems fixed first and then see how the game is performing.
2. Next, try substituting more powerful flipper coils. Invest $30 in a pair of "hot" coils and see if you like the results (covered below).
3. Convert the coils to DC power. The old coils are AC (alternating current), which is not as strong per volt as DC (direct current). This is accomplished by wiring a bridge rectifier into the circuit, explained under "Converting to DC Power" in this section.
4. Something I've never tried, but might be interesting, is to swap the old 2" flippers for 3" flippers, if you can find ones that will fit your machine, that will not be so long they block the ball from draining, and that don't look out of place. Flippers can be removed and replaced in a few minutes, so there is nothing harmed if the new flippers don't work out. Just be sure to save the old flippers in case you want to switch back.

WEAK FLIPPERS: END-OF-STROKE (EOS) SWITCH

The end-of-stroke (EOS) switch is the *number one* cause of weak flippers, chattering flippers, vibrating flippers, and burned flipper coils. The end-of-stroke switch is something every pinball owner should know about, why it is there, and how to clean and adjust it.

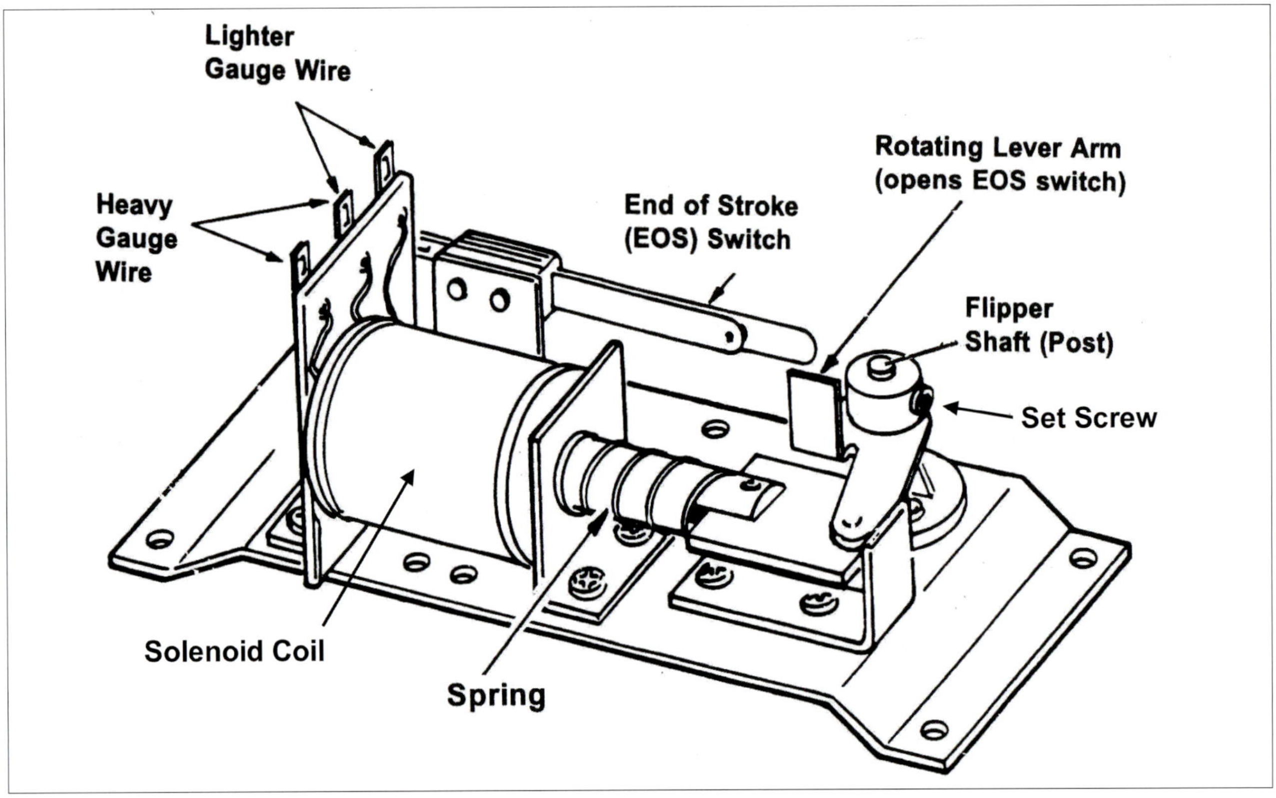

Underside view of a Bally flipper mechanism. When the plunger pulls into the coil, the flipper lever arm (which is attached to the flipper shaft) rotates. The lever arm makes contact with the end-of-stroke switch, opening the switch. Bally and other pinball manufacturers tried different flipper mechanism designs, but the basic idea is the same for all flippers.

All flipper mechanisms manufactured until the end of the 1990s included an end-of-stroke switch mounted on the flipper unit. Starting about the year 2000, Stern Pinball redesigned flipper operation and eliminated the need for the end-of-stroke switch.

The end-of-stroke switch protects flipper coils from overheating and burning out when a player holds in the flipper button for more than just a few seconds. The end-of-stroke switch opens at the end of the flipper's stroke when the flipper reaches the top of its arc. When the end-of-stroke switch opens, the current in the flipper coil is transferred through a longer winding, generating much less heat and protecting the coil from burnout.

But the longer winding also results in much weaker coil. That's why the end-of-stroke switch should not open until after the flipper has hit the ball and completed its movement. If the end-of-stroke switch opens too early, in the middle of its swing, the result is a weak flipper. If the end-of-stroke switch doesn't open at all (out of adjustment), it's like bypassing a fuse; the overheat protection is gone. If you hold in the flipper button you will fry the coil.

Locating and Testing the End-of-Stroke Switch

Here's how to locate the end-of-stroke switch. Lift the playfield and brace it up (remove the balls first). Have a look at a flipper coil and mechanism. Operate the flipper by hand. You will notice that as the flipper moves, a lever or bracket on the flipper lever arm rotates and opens a switch mounted next to, and wired to, the flipper coil. This is the end-of-stroke switch.

The easiest way to test the end-of-stroke switch is to short the switch so it is closed and stays closed. Put a jumper wire with alligator clips at both ends on the two solder lugs of the end-of-stroke switch so that the switch is in effect solidly closed. Turn on the game and operate the flipper. Don't hold the flipper button in or you'll burn out the coil. You can operate the flipper safely as long as you don't hold in the button. If the flipper is suddenly more powerful, if the problem went away, you found the culprit: poor contact on the end-of-stroke switch. From my experience, this is more often the cause of weak flippers than anything else.

Adjusting the End-of-stroke Switch

Adjust the end-of-stroke switch so that it is firmly closed and making good contact when the flipper is at rest. You want the switch to open as the flipper rotates, so that the end-of-stroke switch is about ⅛" to ¼" open when the flipper is fully up. But you don't want the end of stroke switch to open too early in the flipper's movement or it cuts power to the flipper coil too soon. You may have to play with the end-of-stroke switch adjustment, bending the switch blades back and forth several times before you get them to open at a good point. If your other flipper is operating with good power have a look at where its end-of-stroke switch opens and try to duplicate it. Sometimes it takes me several tries to get the end-of-stroke switch just where I think it should be. When you are finished adjusting the switch, clean the switch.

Cleaning the End-of-Stroke Switch

If the end-of-stroke switch is not making good contact when it is closed, it has the same effect as when it is open: It weakens the flippers. Poor electrical contact will also cause chattering, rattling flippers. I suggest cleaning and adjusting this switch on a regular basis, every two or three months. See "Switches" in the "Repairs: All Machines" chapter to learn how to clean and adjust switches. If the contact nipples on the switch blades are badly pitted or indented the switch should be replaced. This is not common. I've rarely had to replace end-of-stroke switches.

Once the end-of-stroke switch is adjusted and cleaned, fire up the game and try the flipper. If you still have a weak flipper, you haven't got the switch making good enough contact in the closed position, or the switch is opening too soon. Remember, when you shorted the end-of-stroke switch you found the problem. You just haven't yet solved it. On one machine I had in for repair, I readjusted one of the end-of-stroke switches six times—SIX TIMES—before I finally got it working correctly. Those thin metal blades can sometimes be very difficult to get bent just right.

Here's one more suggestion: If the wire connecting the end-of-stroke switch to the coil is very thin, replace the wire with a thicker wire, such as 18 gauge lamp cord. Thicker wire lets more current flow and might give more power to your flippers.

Electronic machine reverse EOS switches: Some pinball machines manufactured in the 1990s had end-of-stroke switches that close instead of open when the flipper is activated, just the opposite of the EOS switch explained. Since these switches are open during regular play they are not the cause of weak flippers. But check the switch anyway, to be sure it opens and closes when the flipper operates.

Some of the earliest flipper machines from the 1940s and 1950s used a completely different flipper system, called "impulse flippers." When you activated a flipper, right after it flips, power to the flipper was shut off. The flipper would return to rest, even if you were still holding the flipper button in. Before you could shoot the flipper again you had to release the flipper button. These old machines would not let you hold the flipper up to catch the ball and plan a shot. Many of these machines also had both flippers working off one solenoid. Either flipper button would activate both flippers at the same time.

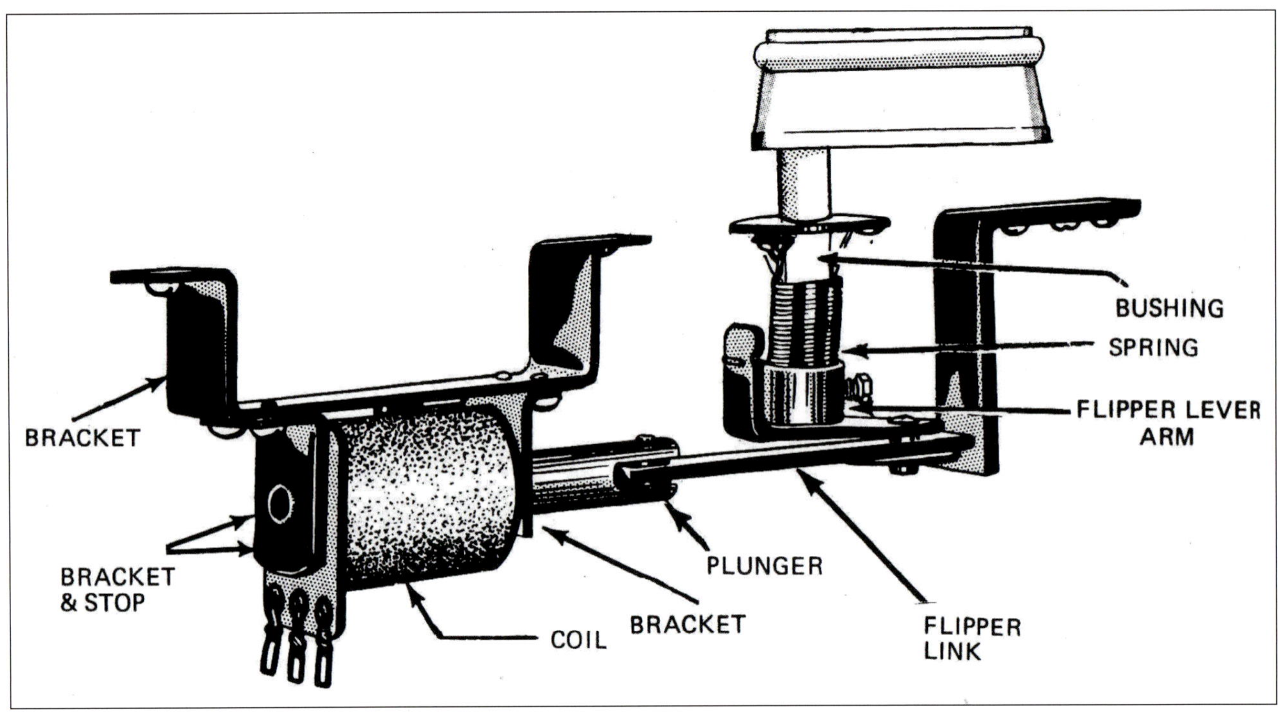

A Gottlieb flipper mechanism. This illustration shows the flipper lever arm, but does not show the end-of-stroke switch. The spring wound round the flipper shaft is held in place by a screw on the flipper bushing. Loosening the spring a turn, or part of a turn, may result in a more powerful flipper.

WEAK FLIPPERS: FLIPPER SPRINGS

The flipper spring returns the flipper to its resting position after you release the flipper button. If the spring is wound tightly, and if spring tension is strong, the spring can slow down and weaken the flipper, fighting against it. Flipper springs can be adjusted to make the spring weaker and the flippers stronger.

Some flipper springs are wound around the shaft (see the Gottlieb illustration). Sometimes you can unwind the spring one turn or part of a turn, depending on how the spring is mounted, reducing the tension on the flipper, in effect making the flipper faster and stronger. I've had some amazing flipper improvements by unwinding springs as little as ⅓ of a turn. There's no harm in experimenting with the spring. If loosening the tension makes the spring too weak, so it will no longer return the flipper fully to rest, put the spring back the way it was.

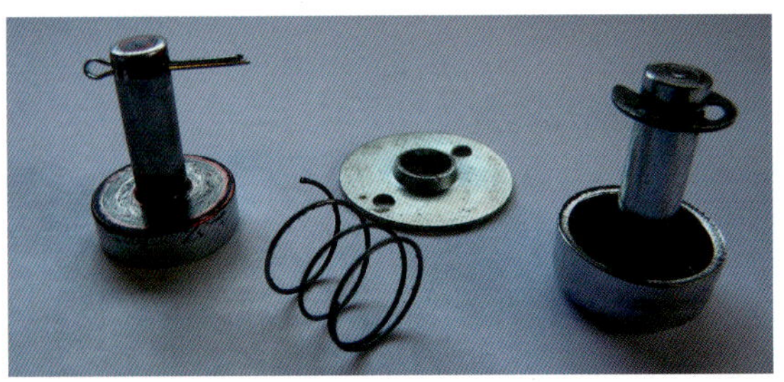

Two different types of flipper buttons and mounting hardware.

Some flipper springs are mounted on the coil plunger (see the previous Bally illustration), or hooked between the lever arm and the housing. If the spring is too stiff, too long, or too short, a different spring might make a difference. But do get a different spring to try out. Do not shorten or try to stretch the spring you already have. Keep your old spring in case the new one does not work.

WEAK FLIPPERS: MOUNTING HARDWARE

Check the brackets for loose or missing screws, on the mechanism and where the mechanism is mounted to the playfield. Loose brackets can wiggle or get out of alignment, rubbing moving parts and slowing down the flipper action. This suggestion also applies to all moving parts that are mounted to the playfield.

WEAK FLIPPERS: FLIPPER BUTTONS

The flipper buttons are mounted on the side of the cabinet. The flipper buttons activate switches that are mounted inside the cabinet, behind the flipper buttons, that in turn activate the flippers. The switch contacts build up carbon and get pitted, and make poor electrical contact, which will result in a weak flipper. See "Switches" in the "Repairs: All Machines" chapter to learn how to clean and adjust switches.

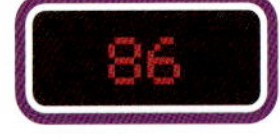

The flipper buttons are spring loaded and may stick if the buttons are dirty. Disassemble the button to clean it. It's an easy job. Remove the retaining clip or cotter pin and the spring from the button (don't lost them) and remove the button from the cabinet. Some flipper buttons don't have a spring. The flipper leaf switches act in place of the spring. Some flipper buttons have a metal bushing that slides onto the button shaft.

Put a little alcohol on a tissue or a paper towel, and clean the button. You may have to scrape off the dirt if it is caked on. Do the same cleaning on the hole in the cabinet where the button mounts. Get both the button and the cabinet good and clean. Don't use any lubricant.

Reassemble the button unit and check to see that it is operating smoothly, that it doesn't stick or jam, and that it springs back with no hesitation or sluggishness.

Pinball machines manufactured after 2000 use plastic optical switches that cannot be cleaned or adjusted. You can skip these suggestions.

Old Gottlieb machines: Until the late 1960s, Gottlieb machines used a different flipper system. The flipper button was not wired to the flipper. The button moved a long wire lever. The lever in turn pushed against a switch mounted to the underside of the playfield that activated the flipper.

It was a clunky system that Gottlieb eventually abandoned. If your machine has this set up, check that the lever mechanism is moving smoothly and that it closes the switches. The switches, mounted at the very bottom edge of the playfield, often got mangled when the playfield was raised and lowered, so check their condition.

You may want to replace the lever system with a new flipper button switch unit to get better flipper response. You will need to run wires from the new flipper switches to the wires that were attached to the original switches underneath the playfield. If you do rewire the flipper buttons, don't throw away the old lever mechanisms. You may someday sell the machine to someone who wants to convert it back to its original design.

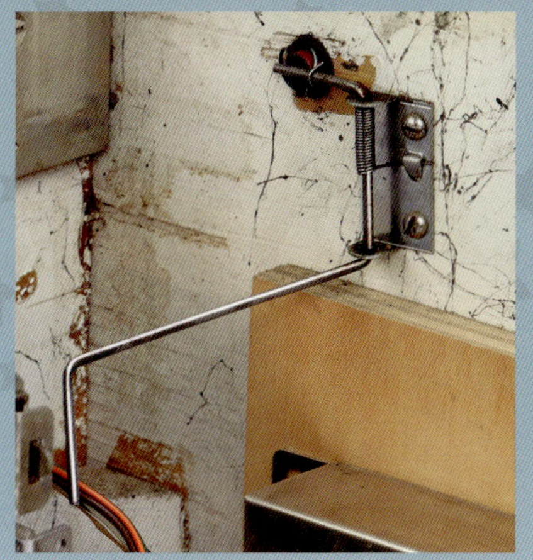

WEAK FLIPPERS: DAMAGED OR WORN PARTS

Flippers get so much action that it's inevitable that parts of the mechanism will wear down, loosen, crack, or break. An overhaul—a flipper "rebuild"—can work wonders, and it is actually quite easy to do.

Before you get going, consider the replacement parts you might need. It's likely that the coil sleeves should be replaced (see following), but other parts could be in good enough shape to clean and reuse. Or not. You can take apart one of the flipper mechanisms, see what's needed, and order parts, which means waiting about a week for the parts to arrive and then repeating the process for the second flipper. Or you could buy a "flipper rebuild kit" if there is one for your machine, which contains all of the parts you'll likely need. Even if it turns out you don't need the parts in the kit, you'll have the parts on hand if problems develop in the future. Flipper rebuild kits are covered below.

Again, disassemble and reassemble one flipper at a time. That way you will have the other flipper to refer to if you can't figure out how to put the disassembled flipper back together.

WEAK FLIPPERS: DIRTY OR DAMAGED PLUNGER

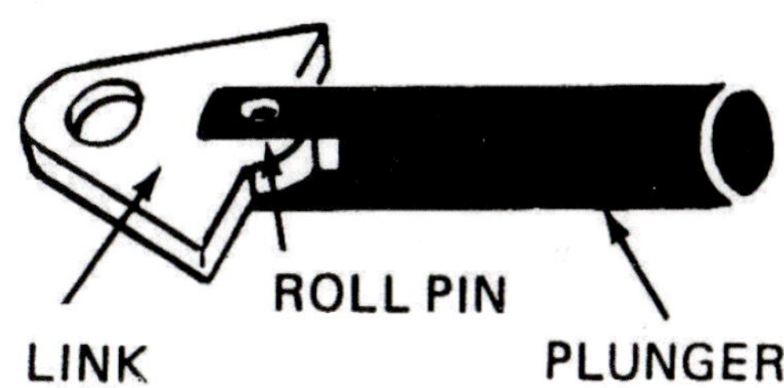

The coil plunger may be damaged. Disassemble the mounting bracket, remove the plunger from the coil, and clean the plunger. Rubbing alcohol works well. Remove rust or corrosion with steel wool or a wire grinding wheel. Do not use lubrication on the plunger.

Examine the bottom end of the plunger: the end that is always inside the coil. That bottom end of the plunger comes flying into the coil at least a hundred times every game, and it stops when it hits the coil stop (sometimes called a "core plug") mounted inside the end of coil.

Over the years, that metal plunger has probably hit the metal coil stop fifty thousand times. The plunger can get a mushroomed end or a ragged edge that will rub against the sleeve inside the coil, slowing the plunger and weakening the flipper. If the end of the plunger is damaged, file it smooth with a metal file or a grinder.

WEAK FLIPPERS: WORN LINK

On many flippers, there is a fiber link attached to the plunger. The link connects the plunger to the moving flipper bat. Different manufacturers used different systems. Over the years each manufacturer tried different ideas. If you look at the flipper mechanism and work it by hand, it will be obvious how the plunger is linked to the flipper.

The link is probably made of bakelite, a stiff, brown fiber material about 1⁄16" thick, sometimes rectangular, sometimes shaped like the illustration here. A few manufacturers used Plexiglas or polycarbonate plastic instead of fiber.

The link has a hole in each end with a pin through each hole. One pin attached to the plunger, the other pin to the flipper mechanism. After thousands of uses, the holes in the link often wore and enlarged. The too-large holes allow unwanted side-to-side movement of the plungers, causing them to vibrate as they pull into the coil. The vibration slows down the plunger, resulting in weak flippers.

If the hole is not worn, or is only slightly worn, there is no need to replace the link. If the hole is badly worn, you will have to buy or make a replacement. There are many shapes and sizes of flipper links. There is a good chance a pinball shop will have a replacement, but be prepared to tell them the description and measurements for the link.

If you cannot find a replacement you can make one using fiber material from a hardware or auto parts store, or you can use polycarbonate or Plexiglas plastic available at most hardware stores. Use the old link as a template.

To remove the old link, use a hammer and a punch to remove the pin that connects the link to the plunger.

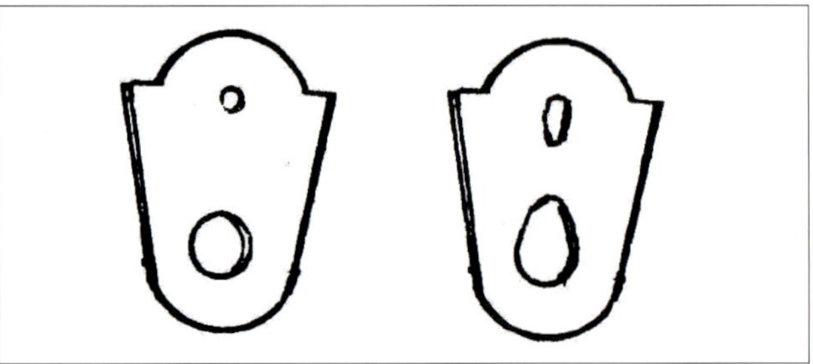

Flipper links, and links for other solenoid powered components, came in different shapes and sizes, but all links had a small hole at one end that mounted to the plunger with a pin, and a larger mounting hole at the other end--that often wore and expanded from years of use.

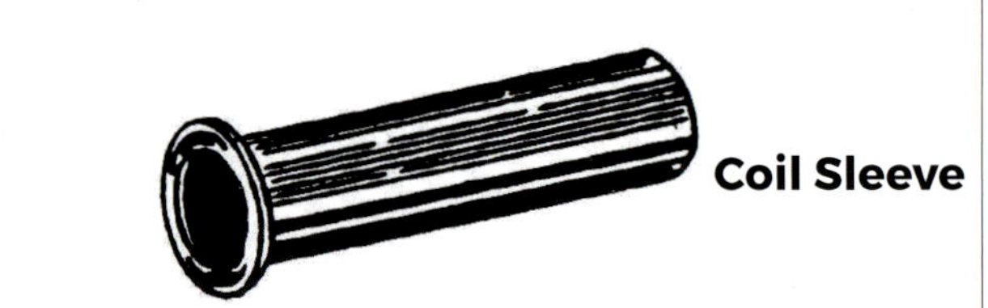

WEAK FLIPPERS: DAMAGED OR DIRTY COIL SLEEVE

When you have the flipper disassembled, remove the sleeve from the coil and clean it. Put some alcohol on a tissue and work it through the sleeve. If the sleeve is worn or damaged, replace it. A new plastic sleeve costs less than a dollar from any pinball supplier. Older pinball machines originally came with metal sleeves, but plastic sleeves are good replacements.

Coil sleeves should easily slide in and out of coils (except for some very old machines). If a sleeve is stuck and doesn't want to come out, the coil probably overheated at some time and partly melted the sleeve, permanently gluing it to the inside of the coil. The coil is most likely still ok, but getting the sleeve out may be a real struggle. The sleeve will often break in pieces, with part of the sleeve wedged inside the coil and not wanting to budge. And when you do finally get the sleeve out, the inside of the coil might be damaged enough that a new sleeve will not slide in.

If the sleeve won't come out, leave the sleeve in place inside the coil, as long as the plunger still moves through the sleeve smoothly and easily. But if the plunger action is hindered by the sleeve, take the easy way out and just buy a new coil and a new sleeve. Many pinball suppliers include a new sleeve when you buy a new coil.

On machines from the 1950s and earlier, and a few machines made in the early 1960s, the coil sleeves were built into the coil and were not removable. If a sleeve is damaged you will need to buy a new coil and a new sleeve.

Read more about coil sleeves in "Coils" in the "Repairs: All Machines" chapter.

WEAK FLIPPERS: DAMAGED COIL STOP

Coil stops get pounded by the plunger a hundred or more times every game, and they can get pretty beat up. The coil stop often gets a ragged or sharp edge that can grab the plunger as the plunger hits the coil stop, weakening the flipper. Remove the coil stop, clean it with alcohol, and if it has a ragged or sharp edge, carefully file or grind down the edge so it is smooth. If the coil stop is damaged or badly worn, not uncommon in a flipper, a new replacement costs about $8 or $10.

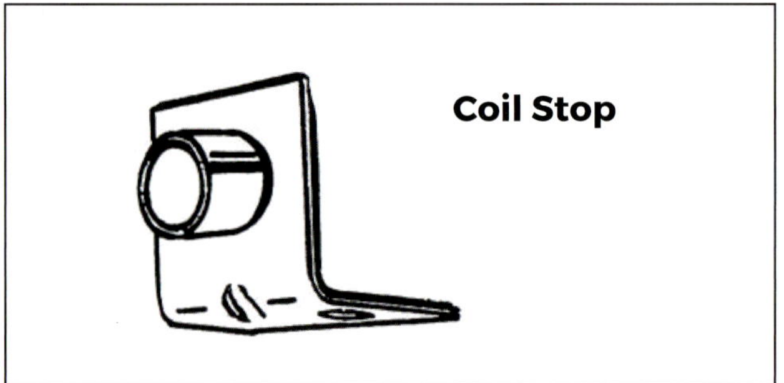

If you replace a coil stop, be careful to get the correct type of coil stop for your machine. In addition to different sizes and different screw placements, there are two basic types of coil stops: Those with solid steel cores used on electronic machines, and those with composition cores, including a visible band of copper (called "shaded" stops) used on electro-mechanical machines. The "shaded" coil stops help prevent coil buzzing.

Make sure the coil stop still has both mounting screws, tightened firmly. These screws sometimes work loose, which can affect the alignment of the coil stop and the flipper action.

WEAK FLIPPERS: CRACKED OR CHIPPED FLIPPER BAT

If the flipper plastic is cracked, it can give or flex when it hits the ball and result in a weaker flipper. If there is a chip broken off the flipper, the chipped edge could weaken the power of the ball when it contacts the ball. Replace any damaged flippers.

WEAK FLIPPERS: FLIPPER SHAFTS

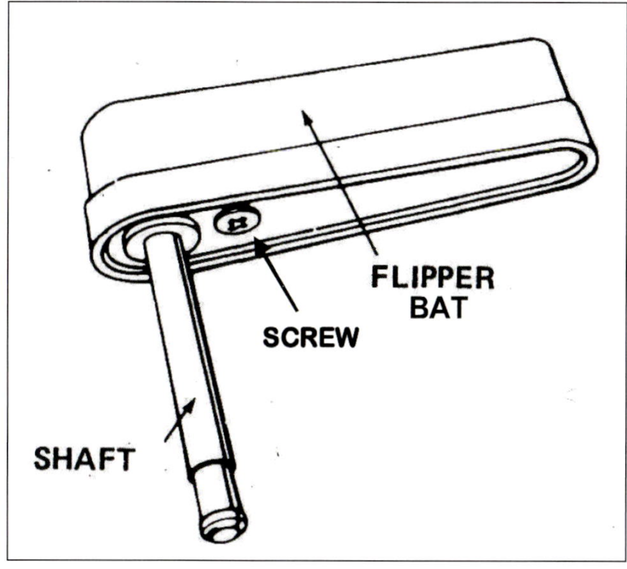

Some flipper bats and shafts are molded one-piece units. Some bats and shafts are separate parts held together with one or two screws that can work loose.

Sometimes a flipper is not working properly because the shaft of the flipper is not tightly fastened to the flipper mechanism under the playfield. Tightening the set screws and replacing missing set screws may solve your problem. This is covered under "Flipper Alignment."

Sometimes the set screws have been tightened and retightened so many times over the years that they've worn grooves in the flipper shaft and will not grab where you want them to. If this is the case, the flipper shaft should be replaced.

Most flipper shafts are smooth, round metal. Some flipper shafts had an indented grooved ring or channel around the shaft that the set screw fit into, making for a tighter fit, at least until the groove started widening and wearing out from years of use.

Most pinball parts suppliers carry a large selection of flipper shafts. Some flipper shafts and bats come as a unit, molded together. Some flipper bats screw onto the flipper shaft as two separate parts. When you order replacements be sure to specify what game you have.

Replacing a worn flipper shaft is an easy job. Loosen the set screws, pull the flipper and shaft out, put the new one in, set it where you want it (lined up with the other flipper), and tighten the set screws. Again, read "Flipper Alignment."

One more possible flipper shaft problem: If your flipper bat is attached to the shaft by screws, the screws (sometimes just one screw) can work loose and the flipper can wobble. The screws are underneath the flipper and not visible. To tighten the screws—in fact, to determine if there are screws to tighten—you have to remove the flipper from the mechanism.

WEAK FLIPPERS: FLIPPER BUSHINGS

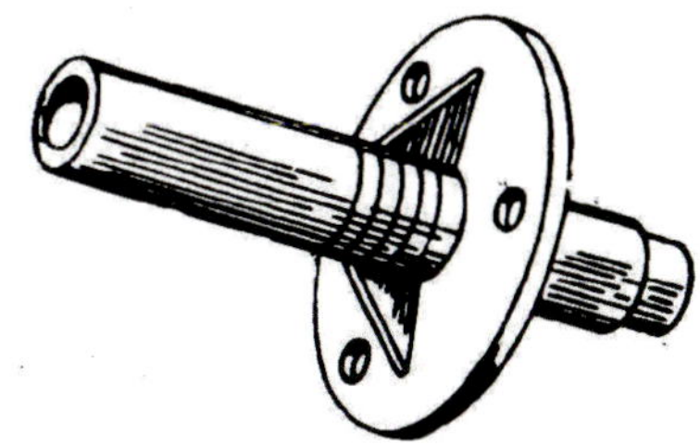

The flipper shaft passes through a plastic bushing—a hollow cylinder and flange with screw holes—that is mounted under the playfield. These bushings are often cracked or broken, resulting in a misaligned or wobbly flipper that will affect its ability to operate well. It is easy to replace the bushing: three wood screws. Replacement flipper bushings are available from most pinball parts dealers.

WEAK FLIPPERS: RUBBING FLIPPERS

Examine the flipper bats, shafts, and moving parts to see if anything is sticking, rubbing, or causing unwanted friction.

Make sure the flippers are not rubbing on the playfield. A lot of machines are designed so the flipper cannot rub on the playfield. There's a little lip on the flipper bushing that keeps the flipper from dropping too low. But a lot of machines don't have such a lip and the flipper could be set too low and is rubbing against the playfield. The flippers should sit a fraction of an inch above the playfield. Hand operate the flipper and observe if it rubs anywhere in its arc. If it does, loosen the flipper shaft (see Flipper Shafts) and raise the flipper just enough to clear the playfield. This will also prevent the flipper from scraping away the paint on the playfield if it isn't already too late.

WEAK FLIPPERS: PLAYFIELD TOO STEEP

Many people like to get the back legs of the pinball machine up higher than the front legs to get more roll, more speed on the ball coming down the playfield. I've even set pieces of 2"× 4" under the back legs for added angle.

That steep playfield angle may be the cause of weak flippers. The flippers may simply not be powerful enough to overcome a steeply angled playfield. Try lowering the back legs a little and see if you don't get improved flipper action.

If you like the steepness of the playfield, you don't have to sacrifice it to get better flipper action. This chapter includes changes you can make to the flippers so you can have your fast playfield and your hot flippers.

WEAK FLIPPERS: FLIPPER ALIGNMENT

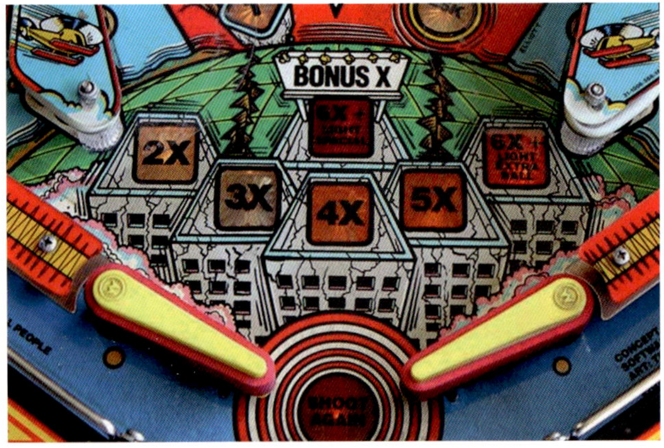

If the flippers are not aligned correctly—pointing too far down or too far up when at rest—they can send the ball in odd directions, often resulting in weaker flipper action.

Correct alignment varies somewhat from machine to machine. Many playfields have a little dot, about a quarter inch from the tip of the flipper, which is a guide to where the tip should be pointing. Some machines have a rail or rubber ring or other obstacle behind the flipper. On these machines, the flipper should be parallel to the rail, and as close as possible without touching the rail or rubber ring mounted behind it. And some machines have no guide to help you. You just eyeball the flippers and set them where they look about right, try them out, and see how they work.

On most machines, you adjust the angle of the flipper from underneath the playfield, where the flipper shaft is fastened to the flipper mechanism, usually by two set screws.

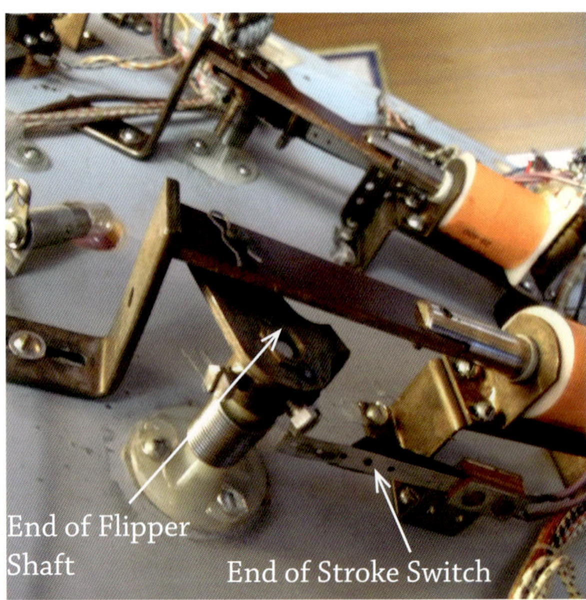

Flipper shaft and the mounting bracket on the underside of the playfield. On this particular game, the flipper shaft ends about ⅛" inside the bracket that holds the shaft in place. Different machines have different length shafts and different size brackets.

Loosen the set screws (don't remove them), adjust the alignment of the flipper, and retighten the screws. Only one set screw there? That is very common, and usually a reason the flipper slips out of alignment. There's almost always a second screw hole. Get another set screw.

Before you loosen the set screws, look at where the bottom end of the flipper shaft sits in the mounting bracket. Sometimes it's flush with the end of the bracket, sometimes it's sitting up or down a fraction of an inch. That's where you want the shaft to sit when you retighten the screws, so the flipper is not too high up (doesn't look very good) or too low (rubbing on the playfield). It can be a bit of a struggle, trying to hold the flipper bat on top of the playfield at the angle and height you want and tightening the screws underneath the playfield at the same time. A second set of hands is mighty helpful.

On some machines from the 1960s, the flipper has a screw on the top of the flipper that you loosen, adjust the angle of the flipper, and retighten. It was a much easier adjustment, but those screw threads stripped too easily and the manufacturers eventually abandoned that design.

After you retighten the flipper shaft, operate the flipper by hand and make sure nothing is grabbing, or scraping, or rubbing, and make sure the screws are tight and the flipper won't slip. If the flipper won't stay in alignment, if it slips when you push on it, you may not have tightened the set screws enough. Sometimes you really have to crank them down. Often the shaft is worn and will need to be replaced (covered previously).

WEAK FLIPPERS: ELECTRONIC MACHINES

Sometimes weak flippers on electronic machines are not caused by the flipper mechanisms, parts, alignment, or anything else in the flipper circuit. Defective electronic components on a circuit board can cause weak flippers. If you have one strong flipper and one weak flipper, you can switch the wiring between the two flippers and see if the weak flipper is suddenly strong and the strong flipper suddenly weak. That is a sure indication that the problem is in a circuit board. Also, if the flipper coils are warm or hot to the touch (and you haven't been holding in the flipper button) that is another sign that there is a problem in the circuit.

If the electronics is the problem, nothing in this chapter is likely to help. See "Repairs: Electronic Machines."

WEAK FLIPPERS: MACHINES WITH MORE THAN TWO FLIPPERS

If your machine has more than two flippers, have a look at how the flippers are wired together. Usually the right flipper button operates all the right flippers and the left flipper button operates all the left flippers. This is not always the case. Some machines have unusual flipper arrangements. Some machines have two sets of flipper buttons. When you operate the flipper buttons you will see which flippers are activated by which buttons.

Very often flipper coils are wired together so that they operate simultaneously. When you push in, say, the right flipper button, both right flippers are activated at the same time. However, you will sometimes find that one of the flippers, usually the upper flipper on the playfield, is activated not by the flipper button itself, but by a switch on the lower flipper.

Look at the end-of-stroke switch on the lower flipper. If there are two switches (three switch blades or four switch blades) instead of the usual one switch (two switch blades), most likely the second switch is normally open and it closes just as the end-of-stroke switch opens. This second switch, when it closes, activates the upper flipper. The upper flipper fires a fraction of a second after the lower flipper fires, since it is activated by the movement of the lower flipper. You can confirm this by tracing the wires (compare wire colors) from this second switch on the lower flipper to the upper flipper coil.

If the upper flipper is weak, the switch on the lower flipper—the switch that is, in fact, activating the upper flipper—may be the cause of the weak upper flipper. Short the lower flipper switch: Put a jumper wire across the two switch blades and see if the upper flipper improves. If so, clean and adjust the switch until you are happy with the upper flipper's action. I have sometimes permanently shorted (bypassed) this switch so both flippers are activated simultaneously by the flipper button with good results.

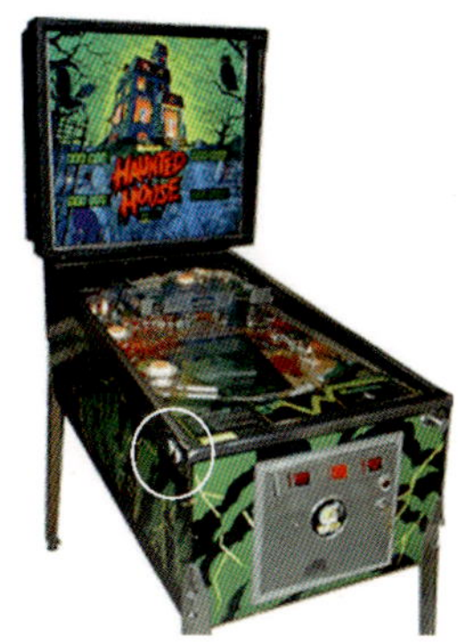

How about EIGHT flippers? Gottlieb's 1982 **Haunted House** had eight flippers—on three playfields: A main playfield with four flippers, an upper playfield with two flippers, and a "basement" playfield below the main playfield with two flippers. Haunted House had two sets of flipper buttons: one pair controlling the four flippers on the main playfield, the other pair controlling the flippers on the upper and lower playfields. It was a challenge to get your fingers on the correct flipper buttons.

WEAK FLIPPERS: A MORE POWERFUL FLIPPER COIL (ELECTRO-MECHANICAL MACHINES)

Your flippers may benefit from a more powerful coil. For starters, coils do not weaken over time. They do not degrade. Replacing an old coil with an exact replacement new coil will not result in a more powerful flipper. However, replacing an old coil with a different strength of coil can make a big difference in how your flippers perform.

Flipper coils on electronic machines are usually quite powerful to start with, so switching to a more powerful coil is not likely to make a difference. Flipper coils on electro-mechanical machines varied in power quite a bit.

First, check to see if the current flipper coils are the ones originally specified by the manufacturer. Someone may have replaced one or both original coils with weaker ones (or maybe stronger ones). Compare the coil number printed on the coil wrapper, or embossed on the flange holding the solder lugs, to the coil number specified in the schematic or the instruction manual. If the original coils were more powerful, switching back to coils with the manufacturer's specified ratings may give you the results you want. If you don't understand coil numbering—wire gauge and number of windings and how they affect coil strength—read "Coils" in the "Repairs: All Machines" chapter.

If the coils are unmarked or the markings are illegible, the coils are probably not original coils. Consider replacing one or both of them with coils rated the same as the original coils and see if there is an improvement. At about $15 per coil it's not a big investment in trial and error.

If the coils are originals or exact replacements, both Williams and Bally made different flipper coils of different strengths that were the same size as the original coils and are easy to swap out.

Gottlieb used one strength of flipper coil (A-5141) in most of their electro-mechanical machines. Gottlieb's unique numbering system made it impossible to know the actual rating of the coil, but it definitely was not very powerful. In response to demand for a more powerful replacement for the A-5141, you can now purchase a "warm" A-5141 (more powerful) or a "hot" A-5141 (even more powerful).

I have substituted coils from different manufacturers (such as putting Williams coils in Bally or Gottlieb machines), sometimes with success, sometimes not. The coils from one manufacturer did not always fit the mounting bracket of a different manufacturer. The operating voltages were sometimes different, possibly resulting in too-weak or too-strong action. Unless you already have a spare coil to try out, I would avoid cross manufacturer substitution. You are not likely to cause any damage, but you are likely to find you wasted your money on a coil that does not fit or work well.

Before replacing any coil with a more powerful one read "Coil substitution" in the "Repairs: All Machines" chapter. It explains the different coil numbering systems, how to determine the strength of different coils, and warnings about possible problems with replacement coils that might be too powerful.

An easier solution is to contact a pinball supplier, tell them what machine you have, and ask if they sell a more powerful coil for your machine.

REPLACING A FLIPPER COIL

Flipper coils have three connectors: three solder lugs. Each connector will have one or two wires soldered to it. It is important that the wires that were attached to the middle connector on the old coil go to the middle connector on the new coil.

Remove the old coil from the mounting bracket and let it hang by the wires. Attach the new coil to the mounting bracket. Unsolder or cut the wire, or wires if more than one, from one connector of the old coil—just one connector at a time—and solder the wire or wires onto the corresponding connector on the new coil. Then do the same for the other two connectors, one connector at a time. By cutting and soldering one connector at a time you will never have all these dangling cut wires and oops, which side do they go on? And if you paid no attention to this warning and now you don't know which wires go where, take a look at the other flipper to see how it is wired, particularly which wires go to the end-of-stroke switch mounted on the flipper mechanism.

DC POWERED FLIPPERS (ELECTRO-MECHANICAL MACHINES)

DC (direct current) powered solenoids are more powerful than AC (alternating current). Until the 1970s, all electro-mechanical pinball solenoids were AC powered. During the 1970s, manufacturers started converting flippers, pop bumpers, and slingshots to DC power. All electronic machines use DC power.

Williams started using DC power in 1972. Bally started using DC in 1976. Gottlieb did not use DC in any of their electro-mechanical machines.

Does your machine have DC power? Look around under the playfield for a bridge rectifier: usually a black block about 1" square and ¼" thick, with four wiring connectors sticking out of the four corners, a capacitor mounted nearby, and probably a fuse holder next to the bridge rectifier. Some older bridge rectifiers were circular, mounted on an oblong metal base.

If you find a bridge rectifier, the machine has DC power to one or more of the solenoids, though you will have to trace the wires or read the schematic to find out if the flippers are part of the DC circuit. Some machines had DC for the pop bumpers but not for the flippers.

If you are familiar with schematics, look on the schematic for bridge rectifier symbols. See "Wiring Symbols" in the "Tracing Circuits" chapter. Any coils wired to the rectifier are DC powered. If any of those coils are flippers, your flippers are DC powered.

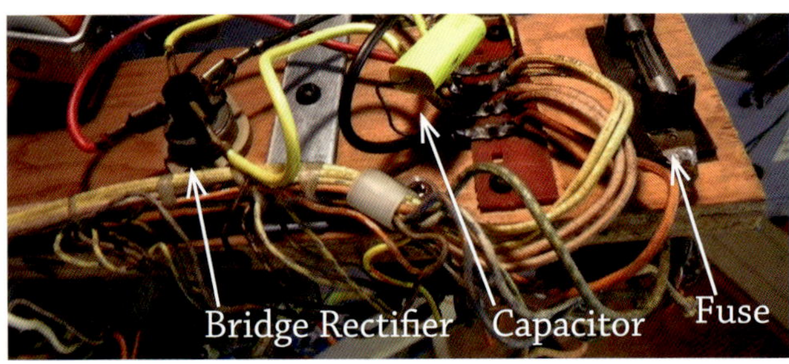

A Williams factory installed DC conversion, used in this machine (**Triple Strike**, 1975) to power pop bumpers and slingshots but, curiously, not the flippers.

Converting to DC

You can easily convert your flippers—and your pop bumpers, slingshots, and any other solenoid operation in your machine—to DC power.

For each flipper (or pop bumper, or whatever else you want to covert to DC), you'll need a small electronic component called a full-wave bridge rectifier, two bridge rectifiers for the two flippers. Get rectifiers rated at 4 amps or more, and at least 50 volts. You can buy bridge rectifiers at an electronics store or from some pinball suppliers and some hardware stores, for about $3 each. Be sure the bridge rectifier is "full wave" and not "half wave."

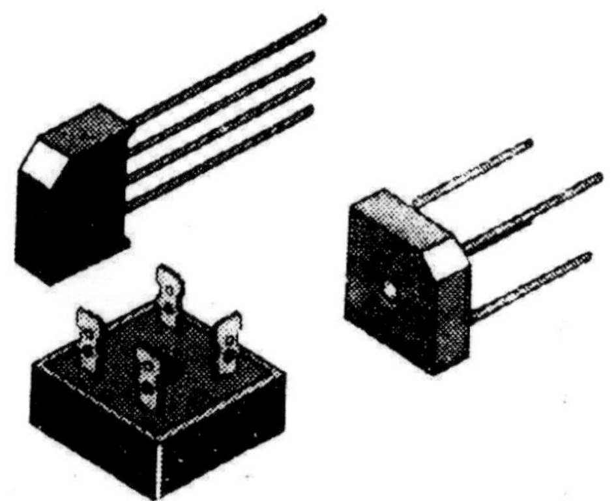

I suggest converting only one flipper and testing it out. See how well it works before converting the second flipper.

Bridge rectifiers come in different sizes and shapes, but all bridge rectifiers have four wires or four solder lugs. Two of the wires or lugs are labeled AC, one is labeled +DC, and one is labeled –DC, or possibly just + and –. Sometimes the AC connections will be marked with the symbol for alternating current ~ instead of being labeled AC. Examine your bridge rectifier and identify which connections are which.

Look at the coil you plan to rewire. Flipper coils have three connectors, but you ignore the center connector. Leave the center connector wired as is. You will be doing your rewiring on the two outside connectors. Coils other than flipper coils have two connectors.

Before you start, figure out how and where to mount the bridge rectifier. Some rectifiers are large and have screw holes so you can screw them to the underside of the playfield next to the coil. Some rectifiers are small and can be left hanging in the air, supported by their wires. You may have to do some creative squeezing and wire bending to find room for the rectifier.

Now, doctor, on to the operation. It's about a fifteen minute job, and the results are often amazing. Six steps:

1. On the coil, one or more wires will be soldered to each connector (each outside connector on flipper coils). Unsolder or cut off the wire or wires from one side of the coil—just one side. Leave the wire or wires on the other side of the coil attached for now.
2. Resolder the wire or wires to one of the AC connections on the bridge rectifier. It does not matter which AC connection. It is a good idea to clamp a heat sink to the wire to protect the rectifier from getting too hot. A heat sink is a small metal clip, such as an alligator clip, that absorbs some of the heat from the soldering tool. There is an illustration of a heat sink in "How to Solder" in the "Repairs: Basics" chapter.
3. Unsolder or cut off the wire or wires from the other side of the coil (again, not the center connector).
4. Resolder the wires to the other AC connector on the bridge rectifier.
5. Solder the + DC wire from the bridge rectifier to one side of the coil (either side of the coil, it does not matter).
6. Solder the – DC wire to the other side of the coil.

That's it. Operation completed. But . . .

Here's the warning: Be careful when you test drive your new hot-rod flippers and bumpers. Be sure we haven't created a Frankenstein monster by mistake. Too much flipper strength can be as bad, or worse, than not enough. Some old pinball machines were not built to handle powerful flippers. If the flippers or bumpers are too powerful they may shoot the ball much harder than necessary, and the ball may hit targets too hard. Some of those ancient targets are brittle and can break if hit by a fast-flying pinball. The ball may go flying up and hit the playfield glass, not likely to damage the glass but very irritating. If you find that the new DC power is too powerful, you can remove the bridge rectifiers as easily as you installed them. (If you do break a target, reproduction targets for most machines are available from pinball suppliers.)

It is also important to understand that this conversion is one bridge rectifier converting one coil from AC to DC. Each solenoid that you want to convert requires a separate rectifier. If you wire two or more solenoids to one bridge rectifier using the wiring I described, all of the solenoids will activate at the same time, which you certainly don't want to happen. It is possible to wire one rectifier to two or more solenoids, but the wiring is more complicated. The switches for each solenoid have to be included in the new wiring circuit, requiring tracing circuits and rewiring switches. It is much simpler to buy several bridge rectifiers—they are only a few bucks each—and use one for each coil.

Capacitor and Fuse

Pinball machines that have a factory-installed bridge rectifier will have a capacitor wired across the DC + and DC – of the bridge rectifier, and a fuse in the circuit. The capacitor and fuse are in the circuit to protect the bridge rectifier and to stabilize current flow. However, if your bridge rectifier is rated at 50 or more volts, it's not likely to have problems if installed without a capacitor or fuse. I do not install capacitors or fuses in these circuits and I have not had any problems.

Still, an ounce of prevention is worth a pound of cure. If you want to install a capacitor as part of the circuit, buy an electrolytic (polarized) capacitor, one that is rated between 1.0 mfd and 3.0 mfd (microfarad, also uses the symbol uf or mf) and 100–200 volts. The capacitor has plus and minus sides. If there is a printed or embossed band around one end of the capacitor, that is the + end. The plus side of the capacitor is wired to the DC + side of the bridge rectifier, the minus side to the DC – side of the rectifier. The fuse should be ten amps, wired between the DC + on the bridge rectifier and the coil. (If this ounce of prevention sounds like a pound of confusion, go back and read the last sentence of the previous paragraph.)

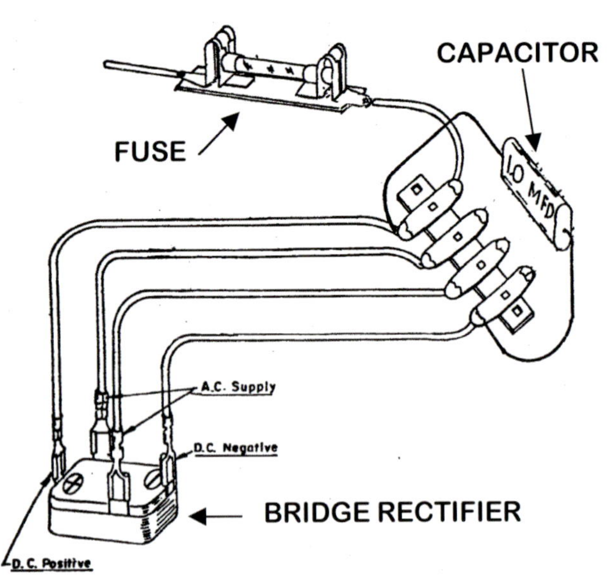

BRIDGE RECTIFIER WIRING DIAGRAM

The diagram below is drawn to make the wiring diagram easy to understand, but the illustration of the bridge rectifier is not anatomically correct. The two DC connections (+ and –) may or may not be next to one another. They may be at opposite corners, or on outside edges, or some other location. But they will be labeled. You will not have to guess which wire is which.

This illustration shows one wire going to each of the AC connections on the bridge rectifier, but there may be more than one wire going to one or both AC connections. If two wires were originally soldered to one of the coil connectors, those two wires should be resoldered to one of the rectifier's AC connectors.

It does not matter which original wire (or set of wires) go to which AC connector on the rectifier, and it does not matter which DC connector on the rectifier goes to which side of the coil. Just be sure the + and – DC leads from the rectifier go to the outside coil connectors of the flipper coil, not to the middle connector.

For coils other than flipper coils (coils with two connectors), simply ignore the wiring for the middle connector and the end-of-stroke switch.

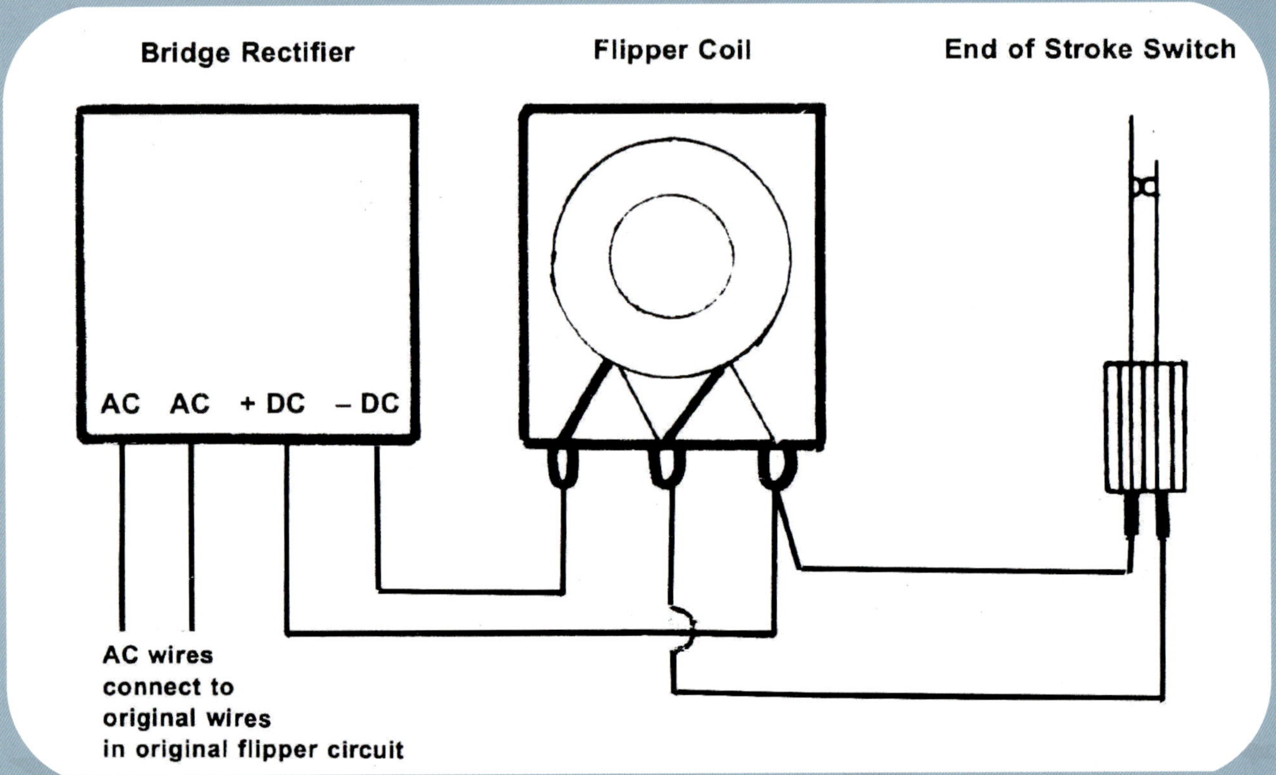

Bridge Rectifier

Flipper Coil

End of Stroke Switch

AC AC + DC – DC

AC wires
connect to
original wires
in original flipper circuit

DC TOO POWERFUL?

So, what do you do if the DC conversion results in a flipper that is too powerful? First, remove the bridge rectifier so you don't cause any (more) damage and go back to the original wiring. (Don't remember how it was wired? Look at the other flipper.)

Next, go back and double check everything in the flipper mechanism: All of the moving parts. The brackets, anchors, and links that hold and guide the moving parts. The coil sleeve. The mounting screws. The switches. The springs. It only takes one little wobbly or cracked part, easy to miss the first time through, to drag down the entire operation. See if things improve. If not, maybe try an even stronger coil if one is available. Talk to a pinball supplier and see if they have a suggestion for a different coil.

I know these suggestions are repeating stuff already covered in the manual, but sometimes a second reading, and a second go through, helps. I was recently struggling with a sluggish flipper that would not respond to anything I tried. I decided to start all over. I tore the mechanism down a second time, and only then discovered a cracked bushing (the plastic sleeve the flipper shaft goes through). The crack allowed the bushing to wobble just a tiny bit, but it was the source of the problem. I replaced it with a new bushing (a five-minute repair!) and immediately had a "happening" flipper. Either I missed the damaged bushing the first time through or forgot to check it.

And if you've gotten this far and still can't solve your problem, short of going down to New Orleans and getting some voodoo gris-gris to sprinkle on it, I guess you just live with it, or as I've done more times than I care to admit, you KEEP TRYING. Every pinball problem is solvable. Eventually.

NOISY FLIPPERS

When you hold the flipper button in, sometimes the flippers buzz loudly, which is very annoying. There are several causes of flipper buzz, many of them the same causes of weak flippers.

Vibrating Plunger: Chattering flippers, or flipper buzz, is usually caused by the plunger vibrating against the coil stop. A common cause of this problem is poor contact on the end-of-stroke switch. See "End of stroke switch". If the plunger is not smooth on the bottom, or if the coil stop has a sharp, rough, or jagged edge, this will often lead to buzzing or humming, as well as weak flipper action. See "Dirty or damaged plunger" and "Damaged coil stop."

Loose Mounting: A loose coil or a loose coil stop will cause a buzzing or humming sound. Check for loose or missing mounting screws. If the screws are tight but the coil is loose in its mounting bracket, you should adjust the coil mounting hardware so the coil is not loose. Some coil stops and coil mounting brackets have slotted screw holes. Loosen the screws, slide the brackets to get a tighter fit, and retighten the screws.

If the holes are not slotted try adding a washer. Use a large washer, or a special pinball flex washer, available from the pinball suppliers, that will fit over the protruding plastic sleeve where the plunger moves in and out of the coil. The washer will fill in the loose space and hopefully wedge the coil tighter in its mount. Do not put a washer on the coil-stop end of the coil.

Magnetized Coil Stops: Some coil stops are slightly magnetic. If they lose their magnetism, they can cause the plunger to buzz. Sometimes replacing the coil stop may

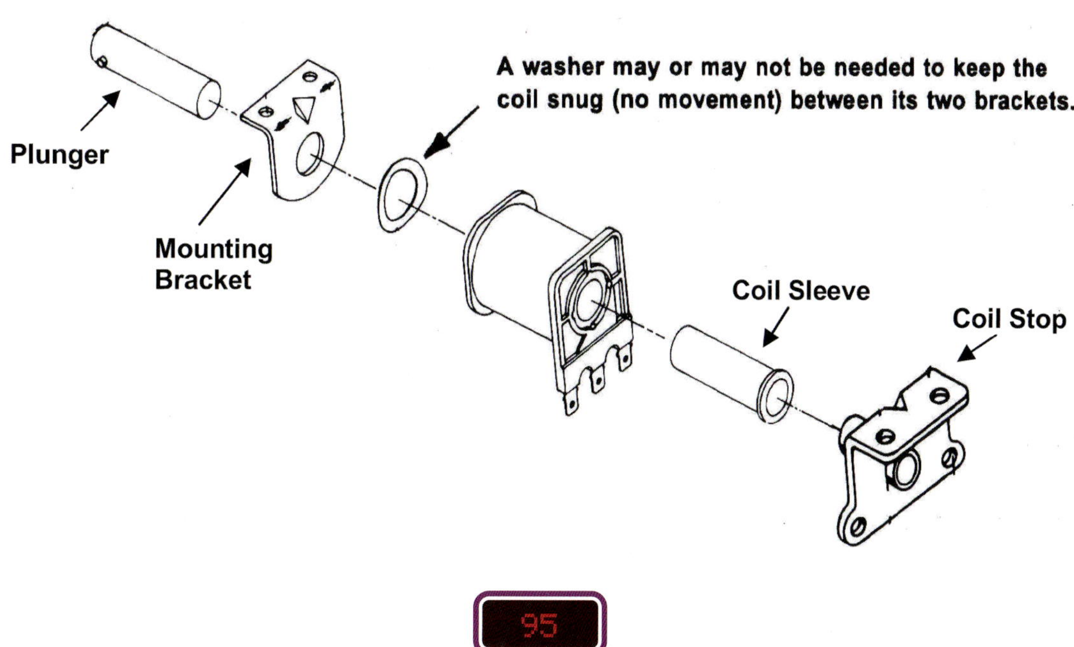

solve the problem. If you are unable to purchase a replacement coil stop, try swapping the coil stop with another one in your machine, one that is on a solenoid that doesn't get as much use as the flipper.

Electro-mechanical machines used a composition metal coil stop called a shaded stop, easily recognized by a band of copper that is visible at the end of the stop. This type of composition, due to its electrical properties, reduced buzzing flippers. If someone had replaced a shaded stop with a solid steel stop (used on electronic machines), that may well be the source of the noisy flipper. Buy the correct stop from a pinball supplier, or swap the stop with another one in your machine that gets less action and is not so irritating, and see if it solves your problem.

Worn Linkage between the plunger and the flipper mechanism can cause buzzing. See "Worn link."

FLIPPER REBUILD KITS

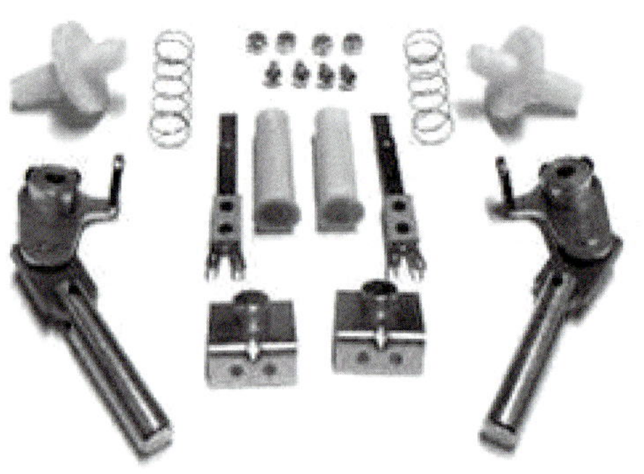

Several pinball suppliers sell flipper rebuild kits, a package of replacement parts to "rebuild" the flippers, for many different machines. "Rebuild" sounds impressive but really means replacing worn parts with new parts.

What is included in a rebuild kit varies from game to game, and from supplier to supplier. The kits typically include new plungers and links, end-of-stroke switches, coil stops, coil sleeves, bushings, springs, and mounting screws for two flippers. Rebuild kits for machines with three or four flippers sometimes include parts for the additional flippers and sometimes require a second kit.

The kits do not include the flipper coils, because coils, unless they've burned out, do not wear out. The kits don't usually include flipper rings. The kits also do not include the flippers themselves. If you have worn or damaged flippers you'll need to buy them separately. If your flippers are in good condition there is no reason to replace them.

When rebuilding flippers, the coil stops often need to be replaced. Many flipper rebuild kits include coil stops, but some don't. Compare kits from different pinball suppliers. If none of the kits include coil stops, order two when you order the kit.

Rebuild kits sell for anywhere from $30 to $60, depending on the parts included in the kit. Each kit is different. Flipper kits for one machine may not fit on a different machine.

I have not compared the cost of buying a kit versus buying the individual parts, but for anyone who has little or no experience and no spare parts at home, a rebuild kit is a worthwhile investment. If the rebuild kit does not come with instructions this flipper chapter describes each part and how to replace it.

KIDS AND FLIPPERS

Young kids who play pinball love to work the flippers, constantly. They'll push both flipper buttons repeatedly, over and over, almost non-stop. That can really heat up the flipper coils and damage them. What can you do?

Solution #1: You could tell the kids they can't use the machine. *Boo. Hiss. No fair.* (That isn't going to work.)

Solution #2: You could teach the kids how to use the flippers correctly. Maybe they'll pay attention. Maybe not.

Solution #3: You can check the end of stroke switch to be sure it is opening (covered previously). If the EOS switch does open, try adjusting it so it opens sooner during the flipper operation. This should result in a cooler, lower current flipper circuit that is less likely to burn out. But it also may result in a weaker flipper.

Solution #4: Replace the plastic coil sleeves with metal sleeves. Coils might survive some overheating, but the plastic sleeves inside the coils will sometimes warp and even melt from the heat, which seizes up the plunger and makes the flippers inoperative. All early pinball machines had metal sleeves, but starting in the 1960s, the manufacturers started using less expensive plastic sleeves, which work fine most of the time. You can still buy metal sleeves from many pinball suppliers.

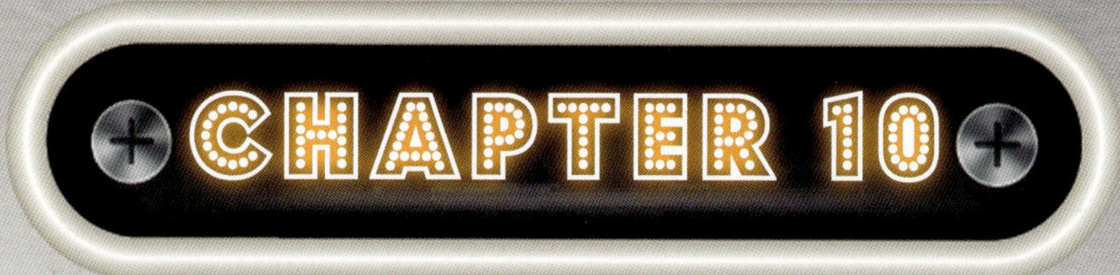

REPAIRS: ALL MACHINES

Most pinball problems have a simple solution. People reach too far into their troubleshooting. Remember the old saying, "When you hear hoof beats, think horses, not zebras."

—**Jim Tolbert**, *For Amusement Only*

IT'S BROKEN!

Now what?! For starters, relax. It's just a pinball machine. You're not stranded out in the desert with a car that won't start. The creek's not five feet and rising. You just have a broken pinball machine. Every pinball owner has been there, or will be there someday.

This chapter covers repairs common to all pinball machines, both electro-mechanical and electronic. Repairs that are only for electro-mechanical machines are covered in the next chapter. Repairs that are only for electronic machines are covered in the chapter after that.

GAME WON'T START CHECKLIST

When a game will not start, check the easiest and simplest problems first:

1. Is the game plugged in and turned on?
2. Is there a broken or disconnected wire from the on-off switch?
3. Are there credits on the machine? Or is it set for free play?
4. Is there a blown fuse or poor contact between a fuse and fuse holder?
5. Check the start button switch and wiring. Is there a broken wire? Does the switch close when the button is pushed?
6. Is a coin switch (the little wire switch on the coin mechanism) jammed? If the coin switch is stuck closed, it is usually because the wire is bent, and quite often the game will not start. See "Coin switches."
7. If you just worked on the machine, go back to the scene of the crime—whatever circuit or wires you were working on—and double check your work. Did you disconnect something? Did you wire something wrong?
8. Is the tilt light on? Is there a stuck or shorted tilt or anti-cheat (slam) switch? Every machine has two tilt switches (a few machines have three) and two or three anti-cheat switches. If any of the switches in the tilt circuit are activated the game will not start.
9. Check all switches on the playfield, including switches wired to targets, rollovers, the outhole, and kickout holes. If one of these switches is shorted or stuck closed, sometimes the game will not start.
10. Check all solenoids to see if any are activated. No solenoids should be activated before a game is in play. Some machines will not start if a solenoid is activated. Turn off the game immediately if there is an activated solenoid or the solenoid will burn out for sure. Don't confuse solenoids with relays. Solenoids have plungers that move in and out of the coil. Relays have solid cores and armatures. Unlike solenoids, some relays are activated all the time.

Electro-mechanical Machines:

1. Did all the score reels reset to zero? On some machines the game will not start until all score reels reset. See "Motor Runs Constantly" in the "Repairs: Electro-mechanical Machines" chapter.
2. Did all of the stepper units (rotating units) reset? Some games will not start until the units reset.
3. One desperate thing many people try—and it even works sometimes—is to, one by one, hand operate every relay on the floor of the cabinet. You are likely to find the relay that starts the game, and it may actually start the game. Sometimes that's all it takes to un-stick the switches and actually get the machine working. It's a long shot, but you are not likely to damage anything by trying. Just remember that there is exposed 120 volt wiring inside the pinball machine. Don't touch anything unless you know that it is safe to do so.

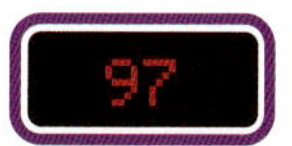

Electronic Machines:

1. The game may not start until the correct number of balls are installed. There is often a label underneath the lock-down bar stating how many balls are required, or the instruction manual will have the information or, on new machines, the digital display may tell you a ball is missing.
2. Some newer electronic machines will not start if the batteries are removed or dead.
3. Are you getting a message that says "Adjust Failure?" See Adjust Failure in the "Repairs: Electronic Machines" chapter.

If You Just Brought the Machine Home:

In addition check for:

1. Connectors that aren't plugged in and connectors that aren't snapped together.
2. Missing fuses.
3. Broken, cut, or dangling wires, and wires that have come off switches and coils. Old solder connections sometimes just break and the disconnected wire may be sitting right next to the solder lug, almost unnoticeable.
4. Empty brackets or empty spaces that might indicate a component is missing.
5. Missing circuit boards.
6. Any indication that something burned.
7. Moving parts that won't move, such as pop bumpers, flippers, solenoids, rotating (stepper) units, and other moving mechanisms.
8. Components and wires that are obviously replacements that may be wired backwards.
9. Amateur looking repairs. A lot of people who try to fix pinball machines don't know what they're doing. They cross wires, they try to bypass problem circuits, and they twist wires together without solder, making poor contact.
10. Electrical tape. Pinball manufacturers did not use electrical tape. Unwrap the tape and see what kind of patch job the tape is hiding.
11. Any other signs of previous repairs or alterations, particularly in electronic circuits, where someone may have replaced an IC, or a transistor, or a display unit, only to have it go out again.

FUSES: TROUBLESHOOTING BLOWN FUSES

A blown fuse may or may not mean a problem with the machine. You can change a lamp and touch the wrong wires, shorting the circuit and blowing the fuse. There's little likelihood that a brief short will damage a machine. That's what fuses are for. If you replace a burned fuse and everything works after that, you may have no problems at all.

If the new fuse blows immediately, or very soon, that is a sure indication that something is wrong in the machine. Do not continue to stick new fuses in until you locate the problem. This is especially important in electronic machines. If you keep putting in and blowing fuses, you are risking damaging the electronic components.

Each fuse in a pinball machine is for a different circuit or different components. The fuses are usually identified as to their purpose, what they are protecting. By knowing which circuit is blowing, you may be able to spot and possibly repair the problem.

A common cause of blown fuses is a short somewhere in the circuit. Maybe two wires are touching that shouldn't be touching, maybe a lamp socket got twisted and shorted, or maybe a solenoid is burned out and shorted. Repairing shorted wires and replacing shorted solenoids is easy to accomplish, if you can locate the problem. Sometimes it can take hours of hunting before you find the culprit. And sometimes it can take five or ten fuses—blown trying, and failing—to isolate the problem.

Many pinball suppliers have a large selection of low-cost fuses. Most hardware stores sell fuses, but the cost is often much higher than buying fuses from a pinball supplier, and the selection is limited. You can substitute a fuse with a slightly lower rating, such as a 7 amp fuse for a 10 amp circuit, and the fuse will probably handle the 10 amp circuit. But don't substitute a fuse with a higher ampere rating than the original fuse. This is not safe.

Unplug your machine before testing fuses. One or two of the fuses are in 120 volt circuits and are unsafe to work on if the machine is plugged in.

Never bypass a fuse. I thought it would be obvious to anyone that you should never bypass a fuse, but it's not obvious. I have seen machines where aluminum foil was wrapped around blown fuses. I have seen machines where nails were stuck in the fuse holders instead of fuses. Insanity. You could burn up your machine. You could burn down your home.

See "Fuses" in the "Components and Features" chapter if you are not familiar with fuses, how to identify them, and their location in the pinball machine.

Electro-mechanical Machines

Electro-mechanical machines usually have five or six fuses, each for a different circuit in the machine. Sometimes the fuses are labeled with the circuits they protect. If not, you will have to examine the schematic: Locate the fuses (almost always at the end of the schematic near the transformer and wall plug), note the wire colors for the different circuits, then compare the schematic colors to the wire colors in the machine. It really is not difficult and is a good way to introduce yourself to schematics. See "Schematics" in the "Tracing Circuits" chapter.

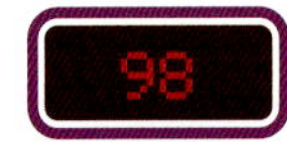

Main fuse: One fuse, usually a slow blow and usually mounted next to the transformer on the floor of the cabinet, is the main fuse. This fuse is wired directly to the line cord, the cord that plugs into an outlet in your home. If your machine has no power at all—no lights, no sounds when you turn on the machine, no game start, no anything—this fuse may have blown, though it is rare to have a blown main fuse.

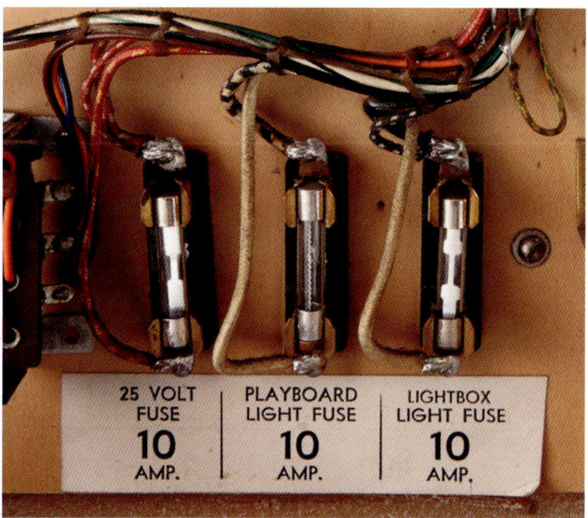

Set of three fuses: On most electro-mechanical machines, there is a set of three fuses in a fuse holder mounted on the left side of the cabinet or on the floor of the cabinet just inside the coin door. One of those fuses is probably to the on-off switch, and if so will have a full 120 volts going through it, so make sure your machine is unplugged from the wall. One of the fuses may be to the lamp circuits, or the lamps may be on a separate fuse elsewhere in the machine. One or possibly two fuses will be for the operating circuits: the flippers, pop bumpers, targets, and most everything else in the machine except the main power and the lamps.

If you are trying to locate a short, try unplugging the connectors between the main cabinet and the backbox. See "Backbox" in the "Disassembling and Transporting" chapter if you don't know how to unplug the connectors. If the fuse stops blowing, the short is probably in the backbox. If the fuse still blows the short is in the main cabinet. That will narrow down your search area considerably.

Lamp circuits: If the blown fuse is for the lamps, that fuse is wired to the lamps on the playfield and the lamps in the backbox. There might be as many as fifty lamps in the machines, a lot of tracing. Try disconnecting the backbox from the main cabinet to narrow your search.

Single function fuses: There may be a fuse mounted underneath the playfield, usually next to a long bank of relays with a large solenoid to reset the entire bank. The solenoid needs a lot of power and may even be a full 120 volts, thus getting its own fuse.

DC power: If you have a machine with DC power to some of the components (see the "Repairs: Flippers" chapter), there will be a fuse mounted on the underside of the playfield wired to a bridge rectifier and a capacitor, and sometimes on a separate wood panel. If this fuse is blown there is a problem in this rectifier circuit.

Electronic Machines

Electronic machines have many fuses, many more than electro-mechanical machines, and each fuse protects a different circuit or component. The blown fuse is your key to what failed. Each fuse has a part number: the letter F usually followed by three numbers, such as F102. The game instruction manual has diagrams of each circuit board and the fuses on each board, and what each fuse protects.

For example, if fuse F102 is blown, the fuse list in the instruction manual will tell you what circuit or component is protected by fuse F102. Using that information, you can examine the components wired to F102, such as a flipper coil or a string of lamps, and look for something obvious, such as a burned coil or shorted wiring. If you can trace circuit board schematics, the burned fuse will lead you to the defective component.

If you can't determine what's wrong, try this: In an internet search engine type in the name of your machine and the fuse number and see what turns up. Some pinball machines have a problem with one circuit or one component that shows up on many machines and that repeatedly blows the fuse. You may find the problem and the solution.

COIN SWITCHES

When you drop a coin in the coin slot, the coin goes through the coin mechanism, a delicate, almost clock-like apparatus that determines if the coin is legitimate and not counterfeit, or foreign, or the wrong denomination. Rejected coins are returned to the coin return slot. Accepted coins continue through the coin mechanism and land on the coin switch, a swinging wire contraption that starts a game. Or doesn't.

Coin switches are small and delicate, and they get mangled from people repeatedly operating the switches by hand to get free plays. If the switches are misadjusted sometimes the game will not start, and sometimes the game will constantly restart.

The most common and easiest solution is to bypass the coin switches completely and set the machine for free play. See "Free Play" in the "Components and Features" chapter.

If you want coin operation, remember that most pinball machines have two coin slots with two separate, identical coin operations. If one coin unit is broken or jammed, the other is a backup. So if one coin switch is damaged, see if the second switch is working. If there is only one coin mechanism (removable), move the coin mechanism to the working switch. And there you go.

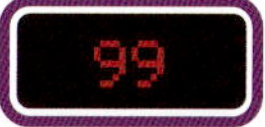

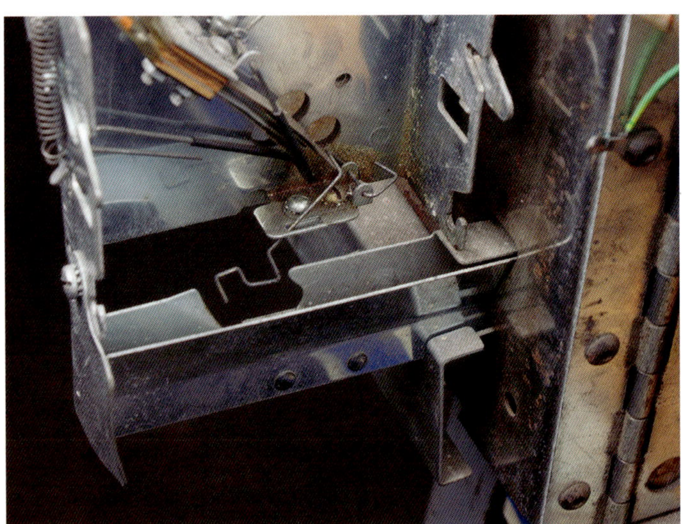

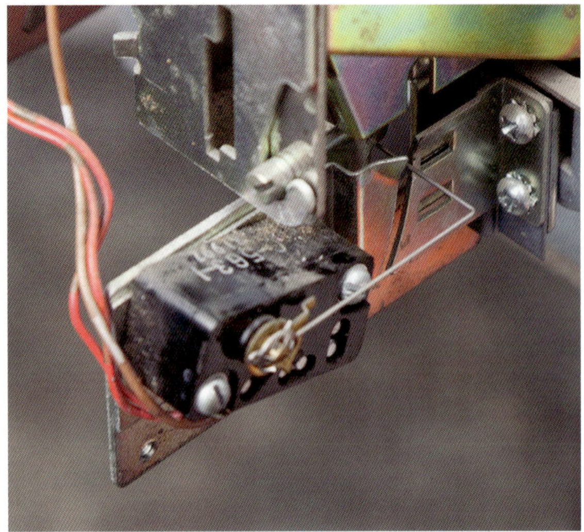

Coin switch on an electro-mechanical machine (left) and an electronic machine (right). The lever moves down when a coin lands on it. The lever, in turn, closes the switch that starts a game. The coin mechanism, which snaps in and out, has been removed to show the switch operation.

If both coin switches are damaged, you probably can fix them by bending the wire lever (that the coin lands on) so that the coin will operate it, and it in turn will close the switch that starts a game. You really have nothing to lose by trying. The thing is already damaged anyway.

Converting a 10¢ machine to a 25¢ machine: Many pinball machines made in the 1960s had both a dime slot and a quarter slot. One play for a dime, three plays for a quarter. You can convert the machine to give one play for a quarter. Cut the wires off the 25¢ switch and tape them off so they don't short out. Cut the wires off the 10¢ switch and solder them to the 25¢ switch. It does not matter which wire goes on which side of the switch. Now, a quarter in the 25¢ slot will give you one play, not three. (If either of the switches has three wires instead of the normal two, this rewiring will not work). Be sure to block off the dime slot. And be sure to unplug the machine from the wall before working on it: Many machines from the 1960s had 120 volt wiring on the coin door.

LAMPS

DIM, FLICKERING, OR NOT WORKING

Bright lights are part of what makes a pinball machine a pinball machine. It doesn't take much effort to get those lights working right. Usually.

I use the terms lights, lamps, light bulbs, and bulbs interchangeably.

NONE OF THE LAMPS ARE WORKING

If none of the lamps are working, or if a group of lamps are not working, check the fuse for the lamps. Fuse holders are labeled, showing which circuits they protect. If the labels are missing, the schematic identifies each fuse. The schematic shows which color wire goes to which fuse, so you can locate the actual fuse from the color of the wire. If the label and the schematic are missing, it's easy enough to check all of the fuses. Just be sure the machine is unplugged from the wall. Some fuses have 110 volts going through them.

See if the fuse is blown. Remove the fuse from the holder before testing the fuse; otherwise, you may get a false reading. If the fuse is not blown, check the fuse holder for a loose, broken, or corroded connection. Poor contact between the fuse and its holder is often the culprit. You can instead put a jumper wire across the two ends of the fuse holder and bypass the fuse (just for the time it takes to make the test), to see if the lamps light up. If they do, you have a bad fuse or a bad connection between the fuse and the holder.

If you have an electro-mechanical machine where the playfield lamps are working but the lamps in the backbox are not, try reseating the large wiring connectors in the backbox. See "Bolt on Backbox" in the "Disassembling and Transporting" chapter.

Many lamp circuits have a common wire that solders to every light socket in the circuit. These wires are usually bare, not insulated, and stapled to the underside of the playfield and to the backboard in the upright backbox. You can easily see these wires zig-zagging from one light socket to the next. Trace the wires and look for a broken or disconnected wire. Look especially along the edge of the playfield. I once found a broken wire, worn out from contact between the playfield and the supporting frame.

One way to test the circuit is to take a lamp that you know is working and test the wiring coming to the lamp sockets. See if you can set the good lamp across the wires coming to the socket. Does the lamp light up? If the lamp does not light, there is no power coming to the socket.

Instead of trying to manipulate a little lamp onto the wiring, if you built the lamp tester I described in "Tools and Testing Equipment" you can use it to test the circuit. Hook up the tester to the wires attached to the lamp sockets, the alligator clip on one wire (either one), the solder lug "probe" on the other wire. If the lamp tester does not light, there is no power coming to the socket.

Keep in mind that many lamps turn on and off as a game progresses, such as the ball in play light and the game over light. A socket that is not getting power may be working normally, but an entire string of non-working lamp sockets is not normal.

Also keep in mind that some pinball machines will not light the lamps until you push the left flipper button, and some old machines from 1960 and earlier will not light the lamps until you start a game.

If you've gotten this far and all of the lamps are still not working, is everything else in the pinball machine working? Or is the entire machine dead in the water? If so, your problem probably has nothing to do with the light circuits.

BURNED OUT BULB

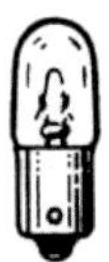

If a single lamp is not working, most likely the bulb is burned out. Just replace it. Pinball light bulbs usually have bayonet sockets or snap-in sockets. Some bulbs, particularly those in the backbox behind the backglass, are recessed in the wood panel and are difficult to remove. Try using a rubber shooter tip. It has just the right diameter to snugly grab the glass of the bulb so you can twist it out of its socket.

Or buy a short piece of 5⁄16" (inside diameter) flex tubing from a hardware store. All you need is two to three inches, so they might cut off a piece and give it to you. The tubing fits snugly over the glass bulb. The tubing, by the way, is clear plastic, and will "vanish" in your tool box. I wrapped a piece of colored electrical tape around it so I could find it. For real speed and efficiency, some pinball suppliers sell a rubber lamp remover similar to the tubing, and the Mill Wax Company makes a clever pair of pliers especially for pinball machine lamps.

To remove lamps with bayonet sockets, push the lamp in, turn counter-clockwise, and pull out. Lamps with wedge sockets pull straight out. Be careful not to bend the exposed wires at the bottom of the wedge. These wires contact the lamp socket, and if bent may not make contact.

LAMPS UNDER THE PLAYFIELD

To replace lamps under the playfield, lift up the playfield (remove the balls first) and remove the lamps from underneath. Do not remove the little plastic inserts mounted on the playfield above the lamps. Those inserts are glued in place, unless they're loose and falling out anyway.

Often lamps under the playfield are squeezed in very tight spaces. Usually you can bend the lamp sockets just enough so you can get the lamp out and a new one in. Sometimes lamps are under other mechanisms that may have to be dismounted to get to the lamps.

LIGHT SOCKETS

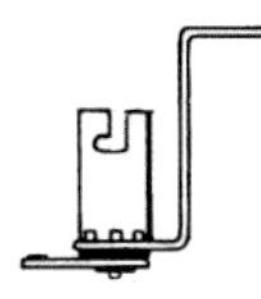

If you do have power to the light socket, and you do have a good lamp, next is to find out why the lamp will not light. Check the wires to the light socket to see if one broke off. If not, remove the lamp, stick a clean, new eraser tip of a pencil in the light socket—or buy a special socket cleaning stick from a pinball supplier—and rub the metal contact inside the socket. This will clean off dirt or corrosion that may be preventing good electrical contact. Also clean the metal nipple of the lamp. If the base of the lamp shows any corrosion or rust, replace it.

On some electro-mechanical machines, the light sockets have a little spring inside the socket. The bottom of the lamp contacts the spring to complete the circuit. With a pair of thin, long-nosed pliers, you can remove the spring to clean and adjust it, which may make the lights work much better.

If a lamp is loose in a bayonet socket, crimp or squeeze the side of the socket ever so slightly with a pair of pliers, just enough to grab the lamp firmly when you put the bulb back in.

Two warnings: (1) Turn off the game when using pliers on the socket. It's too easy to short out the socket when trying to fix it, and you can blow a fuse or, as happened to me once, burn out a dozen bulbs all at once. (2) Sockets are sometimes loosely mounted, so be careful not to rotate the entire socket when replacing a bulb or making an adjustment, which can short the wires.

WORN OUT LIGHT SOCKETS

If the previous cleaning and adjustments do not fix the problem, the light socket may simply be worn out. Light sockets are cheaply made. The little metal parts work themselves loose, or they corrode a little, or who knows. They just stop working, and that's that. And you usually cannot tell by looking if a socket is defective.

If you built the lamp tester as I described in "Tools and Testing Equipment" you can use it to test the socket itself. Hook up the tester to the wires attached to the socket, the alligator clip on one wire (either one), the solder lug "probe" on the other wire. If the tester lights, you know the problem is in the socket.

Sometimes you can bend or crimp the solder lug connection on the lamp socket and solve the problem. Sometimes you can solve the problem by putting a drop of solder where the round socket base attaches to its bracket; but be careful that the solder does not run and short out the socket. The best solution, though it is sometimes a tedious, time consuming job, is to replace the socket. Replacement light sockets are readily available. Lamp sockets come in many sizes and designs, so you'll need a good measurement and description when ordering a replacement.

DIM AND FLICKERING LAMPS

Dim and intermittent lamps are almost always caused by poor contact between the lamp and light socket, or by worn out sockets, with the same solutions as described.

LAMPS OPERATED BY ROTATING UNITS (ELECTRO-MECHANICAL MACHINES)

Some lamps are switched on and off by a rotating (step up, step down) unit. Lights that indicate the bonus, ball in play, which player is up (for multi-player games), and the match are controlled by the rotating contacts on a rotating unit. If a contact on the rotating unit is dirty or not pressing firmly enough the light will not work, or more likely, will flicker or work intermittently.

Rotating units are labeled, if the labels haven't fallen off. Locate the correct rotating unit, press firmly on the rotating contact points, and see if suddenly a dim or not working lamp comes to life. See "Rotating Units" in the "Repairs: Electro-mechanical Machines" chapter to learn how to clean and adjust the units.

If the labels are missing from the rotating units and you don't know which unit activates which lamps, try gently pushing down on the rotating metal wipers ("fingers") on the rotating units—so the nipple on the rotating wiper makes firm contact with the nipple on the stationary plate below the wiper—and see if any lamps turn on or go off. If a bonus light, or a player light, or a ball in play light suddenly lights, you've found the problem. And you can label the unit so you won't have to go hunting for it a second time.

LAMPS OPERATED BY SWITCHES (ELECTRO-MECHANICAL MACHINES)

Some lamps are switched on and off by switches mounted on relays. If these switches are not making good contact, or if the contact nipples on the switch blades are loose or missing, the lights will flicker or not work. Unfortunately, the only practical way to locate a problem switch is by studying the schematic, which is really not that difficult (see the "Tracing Circuits" chapter). All of the lamps are grouped together on one section of the schematic and the symbols easy to spot. Locate the lamp on the schematic and see what relay or other switch activates the lamp. Locate the switch in the machine. See "Switch Problems" in the "Repairs: All Machines" chapter to learn more about adjusting and repairing switch blades.

WIRING CONNECTORS

Electro-mechanical machines: Light problems in electro-mechanical machines are also caused by poor contact where the large wiring connectors plug into the backbox. Try wiggling the connectors, or pulling them out a fraction of an inch to make better contact. Or unplug the connectors, clean the pins with a wire brush, and reinsert the plugs.

Electronic machines: Light problems are sometimes caused by bad connections to circuit boards. If you can locate the connectors, check to see if they are loose, or burned, or corroded. Look for wires that may have pulled out of the plastic connectors. If you don't know which connection is which, see if you can determine if one of the circuit boards operates most of the lamps and check every connection to that board. Your machine's instruction manual will include a layout diagram of the circuit boards and what each circuit board does.

If the machine is using #44 lamps, the high current draw of these lamps is sometimes too much for the connectors and will burn them out. Switch to lower current #47 lamps. See "Lights" in the "Things to Know" chapter.

Electronic light problems are also caused by connector pins on the circuit boards—pins that the plastic connectors plug into. The pins can be corroded or dirty, making poor electrical contact. The pins may have broken solder joints. Look at the base of the pins for a crack in the solder. Resoldering the pins will solve the problem. See "Poor Connections" in the "Repairs: Electronic Machines" chapter.

COILS

Many problems in pinball machines are caused by non-working coils or weak coils, or by the mechanisms that the coils operate. Many of these problems are easy to diagnose and repair. Some of these problems are also covered in the "Repairs: Flippers" chapter.

TYPES OF COILS: SOLENOIDS AND RELAYS

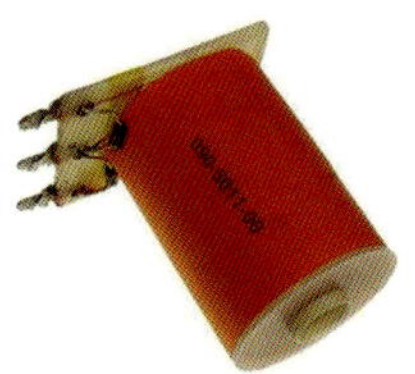

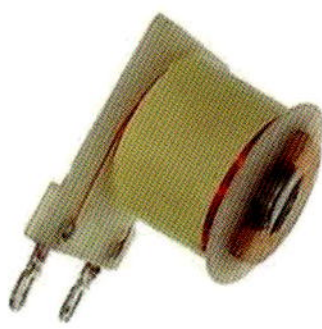

Left: Solenoid with coil sleeve. *Right:* Relay.

Pinball machines use two types of coils: solenoids and relays. Solenoids have hollow cores with plungers that move in and out. The plungers are activated by the electro-magnetism of the solenoid. Relays have solid cores. When activated, the electro-magnetism of the relay pulls a hinged metal plate, called an armature, against the coil. That plate opens and closes switches attached to the plate.

All pinball machines have solenoids. Flippers, pop bumpers, slingshots, kickout holes, and most other moving parts of the machine are operated by solenoids. All electro-mechanical machines have relays, twenty or thirty of them. Some electronic machines have a few relays, but many electronic machines have no relays at all.

HOT COILS

Hot coils (hot to the touch) may or may not be a problem. Some relays in electro-mechanical machines are activated all the time and can get very warm, darkening the paper wrapper. These relays often look burned but they aren't. Solenoids, however, are not normally on long enough to get hot. A hot solenoid is an indication of a short or a stuck switch. The solenoid will probably burn out or blow a fuse, or both.

> I knew a pinball repairman, a guy with years of experience fixing machines, who would troubleshoot coils by starting a game and then sit back and wait to see which coil started smoking! Definitely not something I recommend, but it sure worked for my friend.

RELAYS NOT WORKING

Relay coils rarely burn out, so if a relay is not working, first check the wires attached to the coil to see if a wire broke off. The wire inside a relay coil may be broken or shorted, unlikely but a possibility (see "Shorted or broken coils").

If the wiring is intact but the relay is not working, the relay is not getting power. This could be something simple, such as a switch not making good contact or having a broken wire; or something more complicated, such as a circuit with several components, any of which could be causing the problem.

Relays are labeled, describing what they do. Some descriptions are obvious, such as a 100 point relay or a tilt relay. Some descriptions are not so obvious, such as the ball index relay or delay relay. By knowing the name of the relay and what circuit it operates, you may be able to locate the components in the circuit, such as the 100 point targets that activate the 100 point relay. Or you may not, such as whatever it is that activated the ball index relay (usually a switch on another relay). Your ability to repair your pinball machine will grow quickly as you begin to understand circuits and their descriptions. And once you learn to read the schematic, it will be very easy to locate a non-working relay on the schematic and trace the wires to the switches that activate it.

SOLENOIDS NOT WORKING

A common reason why pop bumpers, slingshots, and other moving parts of a pinball machine quit working is due to burned out solenoid coils.

If you open up the machine and lift up the playfield, you can see all of the solenoid coils mounted under the playfield. Examine each solenoid. Each coil will usually have a paper wrapper around it. This wrapper is used to identify the coil—the part number and sometimes the manufacturer is printed on it—and to protect the wire winding. But the paper is also an indicator of the condition of the solenoid. A paper that is dark brown or black or looks burned is often an indication of a solenoid that has overheated or already burned out. Try moving the plunger inside the coil by hand. If the plunger will not move, the coil is most likely burned and will need to be replaced.

Solenoids rarely burn out from normal use. They usually burn out only if they've been activated constantly, stuck in the "on" position. This can happen to a coil if the switch that activates the coil, such as a slingshot switch or a target switch, is bent or stuck closed. This will cause the solenoid to be constantly activated, constantly on, and will burn the coil. The mechanism or linkage operated by the solenoid will sometimes break and jam the coil switch in the on position and burn the coil.

If you have a burned solenoid, first examine the switch that operates the coil and examine the mechanical parts attached to the coil. If you find a stuck switch or a broken part, you've most likely found the cause of your problem.

The wire inside a solenoid coil may be broken or shorted, unlikely but a possibility. See "Shorted or Broken coils."

SHORTED OR BROKEN COILS

A shorted coil is one where the coil's wiring has made electrical contact when it shouldn't, usually because of broken wires inside the coil. I should point out that in my experience, coils rarely short out or break a wire inside the coil. Still, it has happened, and it is fast and easy to confirm.

When a coil is shorted, the coil has lost most or all of its resistance. It is acting like a direct connection instead of a coil. The coil will most likely get very hot and burn and will probably blow a fuse. If you test the coil with a multi-meter you will get a closed circuit reading (no resistance), the same as if you touched the two meter leads together. An analog meter will read the maximum, the needle all the way to the right. A digital meter will read 0.00 or very close to it.

A broken coil means there is a break in the wire winding and the coil is dead. If you test the coil with a multi-meter you will get an open circuit reading (infinite resistance) the same as if the two leads from the meter were not touching anything. The needle of an analog meter will not move. A digital meter will read 1.00.

Sometimes the exposed end of a coil's wire, where the wire comes out from under the coil spool and attaches to the solder lugs, will break off. Repairing the coil can be a struggle. There usually is no excess wire to make a patch, and the wire coating, which is a form of insulation, will have to be scraped off to resolder the coil. Buy a new coil.

If you are testing a coil with a multi-meter, be sure that at least one side of the coil is removed from the circuit it is wired to, otherwise your meter could be reading the circuit and not the coil.

WEAK OR SLUGGISH ACTION

A solenoid or relay that is working but not strong enough to complete its task can be caused by several different problems, and can be repaired in several different ways.

Mechanisms: See if there is a problem with the mechanism. Relays have pivoting metal armatures, switches, switch actuators (the housing for switches), and springs. Solenoids have elaborate mechanisms: plunger, coil sleeve, coil stop, bracket, links, springs, screws, clips and washers, and the moving parts of whatever the coil is operating.

If any of these parts are dirty, worn, stuck, loose, missing or damaged, then clean, repair, or replace these parts. A coil should not have to fight its own teammates. Get everything in A#1 shape, then see how the coil is doing.

Springs: A spring with too much pull, or not enough pull, can slow down or weaken mechanical action. Loosening or tightening springs or using different springs can have a dramatic effect on the power of a solenoid or a relay.

The original springs might have stretched or might have been bent or altered. The springs might have worn out, what's called "metal fatigue." Or the original springs might be fine, the problem being in the mechanism the spring is attached to. A different spring or an altered spring will solve the problem. Sometimes old mechanisms are worn to the point where they need more spring assistance than they did when they were new.

A STRONGER COIL

If these remedies don't make your machine play any better, you can substitute a stronger coil. A coil with more power can really bring sluggish pop bumpers, kickers, and other solenoid-powered action to life. This applies to solenoids, not to relays. A stronger relay will do nothing to improve the machine and might cause problems.

And to repeat: Coils do not weaken over time and they do not degrade. Replacing an old coil with an exact replacement new coil will not result in a more powerful operation.

COIL POWER: WIRE SIZES AND WINDINGS

The power of a coil depends on the thickness of the wire and the number of windings around the coil.

The thicker the wire, the more powerful the coil. Wire thickness is designated by a standard wire numbering system used in all electrical applications. The lower the number, the thicker the wire. For example, a #27 wire will be thicker and create a more powerful coil than a #28 wire. Most pinball solenoid coils range from #22 wire to #30 wire.

The number of windings around the coil's core also affects the power of the coil. The fewer the windings, the more powerful the coil. A coil with 650 windings will be more powerful than one with 700 windings. Pinball solenoids typically have anywhere from 500 to 1,400 windings.

COIL IDENTIFICATION AND LABELING

All coils are identified by a number, or numbers and letters, printed on the paper wrapper or possibly on one end of the coil. Every number and letter on a coil has significance.

All pinball manufacturers except Gottlieb used a standard, easy-to-understand coil labeling system: a letter designation, followed by a number, followed by another number. The letter referred to the physical size of the coil or the way the coil mounted in its bracket. The first number referred to the thickness of the wire. The second number referred to the number of windings on the coil. For example, a coil labeled A-27-650 refers to an "A" type of size and mounting (coil's physical design), #27 size wire, and 650 windings.

The letter designation (the A in our example) is often interchangeable with other letter designations, of which

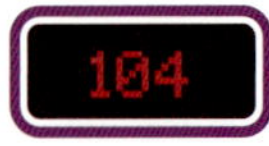

there were many, sometimes one letter, sometimes two or three letters. If the original coil called for an "A" coil, it's a good idea to get a replacement with the same letter, but if a coil with a different letter fits there's no problem using the other coil. An A-27-650 is the same coil as a B-27-650.

A third number in a coil label: Occasionally you will encounter a coil with an additional number right after the letter designation; for example, a coil labeled A-2-27-650. The first number (the 2 in our example) refers to the type of coil sleeve. You can ignore it for this discussion. An A-2-27-650 coil is the same as an A-27-650 coil.

"DC" at end of a coil label: Some coils have the letters DC added to the end of the coil's number. For example, an A-27-650 coil might be labeled A-27-650-DC. These coils were used on some mid-1970s electro-mechanical machines to indicate circuits that were converted to DC power, usually pop bumpers and flippers. Although the circuits were DC instead of AC, the coils are identical to coils without the DC designation. There is no difference between an A-27-650 coil and an A-27-650-DC coil. The coil, whether it has the DC appended to it or not, will work in both DC and AC circuits. The DC designation is meaningless.

MANUFACTURER

Many coils include the name of the manufacturer of the machine (Bally, Williams, Gottlieb, etc.), but many coils do not include a manufacturer's name. Generic (aftermarket) coils will work fine in any machine if the coil is the same number as the original coil.

No label: You may occasionally encounter a coil with the paper wrapper missing and no number. There is no practical way to identify an unmarked coil. If the coil is already installed in your machine chances are it is the correct coil specified in the wiring schematic, but there's no easy way to know for sure. If the coil is working fine, I see no reason to worry about it.

GOTTLIEB COILS

If you have a Gottlieb machine, you have a more difficult job deciphering coil numbers. Gottlieb did not use standard coil numbering. Gottlieb, bless their hearts, used a unique coil numbering system that did not disclose the thickness of the wire nor the number of windings. Gottlieb solenoid numbers were typically the letter A followed by a 4-digit or 5-digit number. You cannot tell from the coil number if a coil is more powerful, or less powerful, than a different coil. And the "A" did not refer to the coil size or type of mounting, it was just a letter Gottlieb used on their solenoids.

If a Gottlieb coil needs to be replaced, most pinball suppliers have coils with the Gottlieb designation, or know what substitute coils match Gottlieb specifications. If you are looking for a more powerful coil, some suppliers sell more powerful substitutes for Gottlieb coils. For example, for years Gottlieb used a flipper coil they labeled A-5141. It wasn't a very powerful coil, but I guess Gottlieb liked it a lot because you'll find it in most of their electro-mechanical machines. In response to demand for a more powerful replacement for the A-5141, you can now purchase a "warm" A-5141 (more powerful) or a "hot" A-5141 (even more powerful).

FLIPPER COIL LABELING

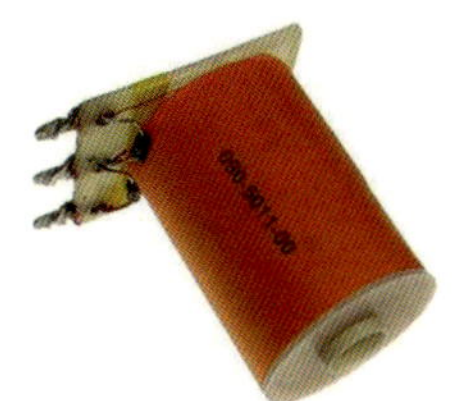

Flipper coils have two sets of windings, two separate wires wound around the core that are connected together at the middle solder lug of the solenoid. The thicker wire is the main (power) winding that is activated when you push the flipper button. The thinner, weaker winding is activated after the flipper completes its stroke, to prevent the coil from overheating if you hold in the flipper button.

Because of having two wires on the coil, flipper coils have two wiring designations. For example, your flipper coil might be labeled A-26-650/28-800. The first set of numbers, the 26-650 in our example, is the main coil winding (#26 size wire, 650 windings). The second set of numbers, the 28-800 in our example (#28 size wire, 800 windings), is the weaker second winding. The first set of numbers (the 26-650) is the wire size that indicates how powerful the flipper coil is.

COILS WITH DIODES

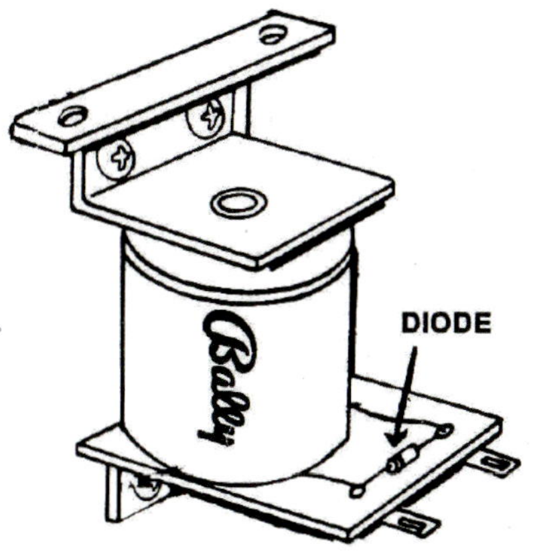

Electronic machines: Many (but not all) coils in electronic machines have a diode soldered to the coil's connectors. Flipper coils have two diodes. Diodes are electronic components that protect the machine's circuits from high voltage spikes. Most diodes on coils are cylindrical, about ⅛" in diameter and about ¼" long, with short wire leads.

Do not use a coil without a diode in an electronic machine unless the original coil also had no diode wired across it. If you use a coil without a diode when the coil should have one, you are likely to damage your machine.

Electro-mechanical machines: Do not use coils with diodes in electro-mechanical machines. The coils will not work and will probably blow a fuse, though they are not likely to cause any permanent damage. You can use the coils if you cut off the diodes first. Just take a pair of wire cutters and snip off the diodes. Other than the diodes, coils in electronic and electro-mechanical machines are interchangeable. One exception: Electro-mechanical pinball machines made by Chicago Coin (Chicago Dynamics) in 1976 had diodes wired across many of their coils, so replacements should also have diodes.

COIL RESISTANCE

All coils have resistance; that is, they "resist" or weaken the flow of electricity. The higher the resistance, the weaker the coil. The lower the resistance, the stronger the coil. Resistance is measured in ohms, and it varies from coil to coil. You can check a coil's resistance with a multi-meter, but the reading you get isn't helpful unless you know the different resistances of different coils and how to interpret the ratings. And fortunately it isn't necessary. It's much easier and simpler to rate coils by wire size and number of windings, information you don't have to determine on your own. It's printed right on the coil and in the schematic.

COIL SUBSTITUTION (MORE POWERFUL COIL)

You can buy replacement coils inexpensively, around $10 to $15 each, from any pinball supplier. If you are replacing a solenoid also buy a new coil sleeve, unless the coil already includes a new sleeve; ask the supplier.

Some coils are more powerful than others. If you want a stronger, more powerful flipper or pop bumper or slingshot or knocker, or any other pinball action controlled by a solenoid, you can substitute a more powerful coil.

First, make sure the solenoid that you already have is the correct one, not a weaker replacement someone put in there. Compare the number on the solenoid to the number specified on the schematic or in the instruction manual. If the number on the coil indicates that it is a weaker coil than the original, get a replacement coil that is the same rating as the original and see how that performs. If the solenoid is a stronger coil than the original and still not to your liking, you can try substituting an ever stronger coil, but maybe first double check the other possible reasons the coil is not performing well.

Coil substitution is not recommended for relay coils. Relays do not usually need to be powerful, as they are not operating any playfield action. Also some relays are activated for long periods of time and they can get hot, but they rarely burn out. If you substitute a more powerful relay you could burn it out.

The fastest, easiest way to find a substitute coil is to contact a pinball supplier, tell them the manufacturer and the coil number you have, which is usually printed on the paper coil wrapper, and ask if the supplier can recommend a stronger coil that will fit your machine and improve your machine's play. Pinball suppliers get this question every day of the week, so it's likely they will have a good replacement to recommend.

If you want to figure out for yourself what substitute coil to try, first read about coil numbering (above), and then read on . . .

CHOOSING A SUBSTITUTE COIL

Coil substitution is often a matter of trial and error. A substitute coil with a wire size one or two numbers lower, such as substituting a #26 wire for a #27 wire, might give you the boost you're looking for. A #26 wire makes a more powerful coil than a #27 wire. A #25 wire makes an even more powerful coil.

Substituting a coil with 10%–20% fewer windings, such as substituting a coil with 500 or 550 windings for a coil with 600 windings, will also increase your power. Substituting a coil that has both a thicker wire (lower number) and fewer windings will make it even more powerful. I suggest you don't go beyond an increase of 2 gauge sizes and a decrease of 20% of the windings, at least on your first try.

When substituting coils, also pay attention to the physical size of the coil. The thickness of the wire and the number of windings can affect the physical size of the coil. Slightly larger or smaller coils will often fit in the same bracket as the coil you want to replace, but sometimes not.

Experimenting with substitute coils requires you to have a selection of spare coils at home. Buying a bunch of different coils and trying them out is not really practical for most pinball owners. That's why I suggest just calling a pinball parts supplier and buy whatever they recommend.

GOTTLIEB COIL SUBSTITUTION

Gottlieb did not label their coils with the wire size and number of windings. This Gottlieb coil chart will help.

Gottlieb "Action" Solenoid Coils

Gottileb Number	Wire Size/Windings
A-1496	23-635
A-4893	22-535
A-5194	24-780
A-5195	26-1305

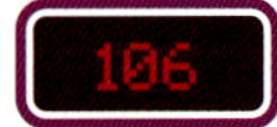

The chart lists the four most common "action" solenoids used in most Gottlieb electro-mechanical machines for pop bumpers, slingshots, kickout holes, knockers, outholes, and other playfield action where a good "kick" is wanted, and where a stronger substitute coil could make the difference between a listless playing machine and one with some punch. The chart shows the Gottlieb part number, always starting with an A, and the actual wire size (first number) and number of windings (second number).

You can use this chart to help decide what coils to substitute for the original coils. You can substitute a different Gottlieb coil or a coil from another pinball manufacturer or a generic coil (no manufacturer noted) as long as the replacement coil fits the mounting bracket.

This chart does not include relays, because a more powerful relay will not solve a weak relay problem and could cause more problems. The chart does not include solenoids for electronic machines because electronic solenoids are already quite powerful.

REPLACEMENT COIL WARNING

The manufacturers chose which coils to put in which circuits. The manufacturers may have had a good reason why they chose a less powerful coil. A stronger coil might power the ball too fast and hard, might break a target, might launch the ball into the air, or might affect the play of the game in a way you don't anticipate and don't like.

Be careful when you first test out a new coil. Remove the playfield glass so you can grab a ball that suddenly turns into Evel Knievel trying to jump ten school busses. Watch things carefully the first game that you play. If you don't like the new play, if the first target the ball hit snapped in half (oops...), then go back to the old coil or possibly a compromise between the two coils.

REPLACING A COIL

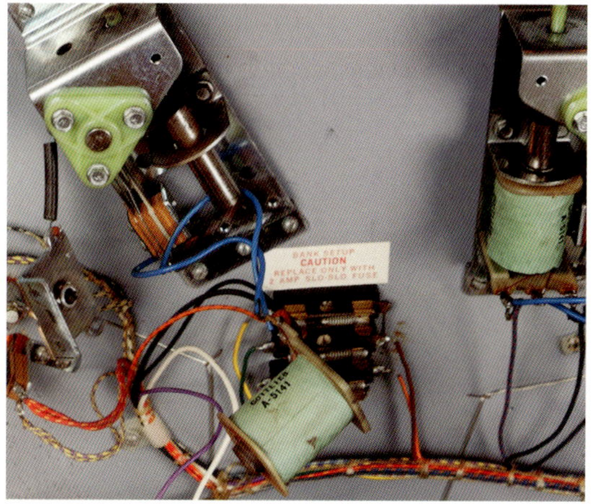

When replacing a coil, leave the old coil dangling by its wires while you mount the new coil, and then move one wire at a time from the old coil to the new one. That way, you won't get the wires mixed up, which goes where?

If you are good with a soldering tool, it is easy to replace a coil:

1. Turn off and unplug the machine.
2. Remove the old coil from its mounting bracket, leaving the old coil dangling by its wires. Don't unsolder the wires yet.
3. Mount the new coil.
4. The old coil has two solder lugs, three if a flipper. Work on one lug, one side of the coil, at a time. Each solder lug will have one or two wires soldered to it. Unsolder or cut off the wire(s) from one side of the old coil and solder the wire(s) onto the same side of the new coil.
5. Unsolder or cut off the wire (or wires if more than one) from the other side of the old coil and solder the wire(s) onto the new coil.
6. If it is a flipper coil, unsolder or cut off the wire from the center connector of the old coil and solder the wire onto the center connector of the new coil.

That's it. If the old coil is still in good condition, save it. It can be used again.

COIL MODIFICATION

Instead of getting a substitute coil to try out, some people like to modify the coils they have. They lop off about 10% of the windings. They unsolder the coil wire from one of the solder lugs, unwind 50 or 60 windings depending on how many windings the coil has, cut the wire, scrape the enamel insulation off the end of the wire, and solder it back onto the solder lug. A ten minute operation, a stronger coil.

I have modified coils this way when everything else I tried did not produce the power I was looking for, but I don't recommend it. You have to be careful unwinding the wire, count the number of windings you remove, make sure the remaining windings are still tight, and hope the result is satisfactory. If you don't like the result or if you botch the job, you can't put the coil back the way it was.

Do not modify flipper coils. Flipper coils have two separate windings, two separate wires that can overlap each other.

COIL SLEEVES

All solenoids have a plastic or metal sleeve mounted inside the solenoid. The plunger slides up and down the sleeve.

When a coil burns, the sleeve is usually destroyed along with the coil. The plastic sleeves will sometimes melt so badly that you have to struggle to get the plunger out. Replacement sleeves are inexpensive, but replacement plungers may be difficult to find. Keep working at the stuck plunger. It will come out eventually.

Most coil sleeves hold up well and don't often need replacing. I do recommend putting new coil sleeves on flippers

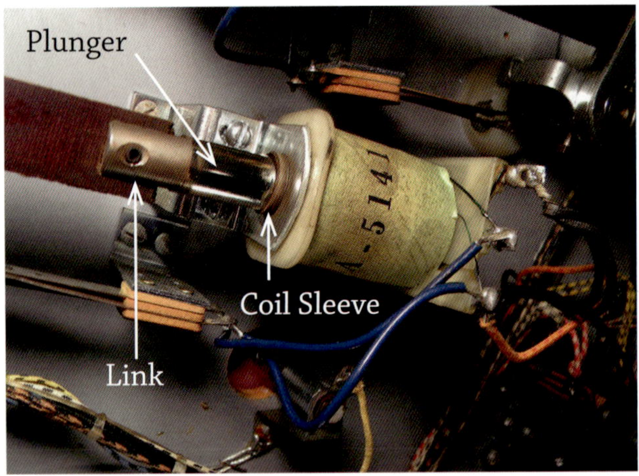

If a coil sleeve is too long the link could hit it, stopping the plunger from pulling all the way into the solenoid, stopping a flipper or other playfield action too soon.

(and pop bumpers if you are ambitious) once a year. Dirt and carbon build-up, and the scraping of the plunger after thousands of uses, damage the sleeves and slow down and weaken the plunger action. New coil sleeves may make your game more powerful and more fun to play. Sleeves are inexpensive. All pinball suppliers sell coil sleeves.

If the old sleeve is metal, there is nothing wrong with using a plastic replacement sleeve. Many suppliers no longer sell metal sleeves.

Almost all coil sleeves are the same diameter: ½" outside diameter to fit 7⁄16" diameter plungers, which is standard plunger diameter on most pinball machines.

Coil sleeves come in several different lengths. Different mechanisms, even within the same pinball machine, use different length sleeves. A replacement sleeve needs to be the same length, or very close to the same length, as the sleeve you are replacing. The replacement sleeve can be slightly shorter or slightly longer if it doesn't cause problems.

A coil sleeve that is too short may not fit the coil bracket. The sleeve needs to extend out from the coil far enough to go through the round opening in the bracket and hold the coil in place.

A coil sleeve that is too long may hinder operation of the plunger. The link attached to the plunger should not contact the coil sleeve when the plunger is pulled in.

Coil sleeves come in two different designs. Most sleeves have a lip at one end. The sleeve slides all the way through the coil until the lip stops it from going any farther.

Some sleeves have a lip a short length, maybe a quarter inch, from one of the ends. These sleeves are often referred to as having a shoulder or a flange. The sleeves stick out of the coil at both ends. You often find this kind of sleeve on knockers and chimes, where the plunger goes all the way through the coil, in and out both ends of the coil.

If the sleeve you are replacing is this second type, with the lip (flange) away from the end, be sure to mention it to the supplier. Most suppliers send sleeves with the lip at the end unless you specify otherwise.

REPLACING COIL SLEEVES

Here's how to replace a coil sleeve. Most coils are held in place with two brackets. Unscrew one or both brackets (depending on how the coil is mounted), slide the coil out, replace the sleeve, slide the coil back in, and reattach the brackets. Make sure the screws are tight. If either bracket is missing a screw, find a replacement. The tighter the bracket, the better the action. If you get confused, have a look at another flipper or bumper to see how it should be reattached.

Some solenoids from machines made before 1965 have built in metal sleeves that are not removable. The sleeve and the coil are one unit. If the metal sleeve is damaged, you will have to replace both the coil and the sleeve. Coils with built in sleeves are no longer made, but new coils and separate sleeves will work fine in old machines.

CHIMES AND BELLS

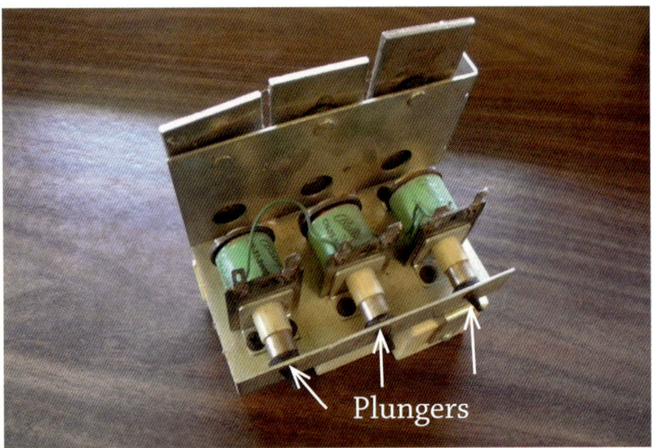

Underside of a Bally chime unit

Pinball machines made before 1970 had one or two bells that rang when you scored points. About 1970 bells were replaced by chimes, usually a set of three but sometimes only a single chime. The first electronic machines made in the 1970s still had chimes, but by 1980 all electronic machines had switched to electronic sounds. To learn more see "Chimes, Bells and Digital Sounds" in the "Components and Features" chapter.

HOW CHIMES WORK

When a chime solenoid is activated, a plunger shoots up out of the solenoid and hits the chime. The plunger falls back down to its resting position when the solenoid is deactivated, which should happen immediately.

Each chime is for a different score: 10 points, 100 points, 1,000 points. Switches and targets on the playfield activate the chimes. Those switches also score the points.

Each chime has a separate solenoid. A set of three chimes has three solenoids mounted on one frame. One side of each solenoid will have a common wire, usually black, attached to all three solenoids. The other side of each solenoid will have separate wires, one for each solenoid, each wire a different color.

NONE OF THE CHIMES ARE WORKING

If none of the chimes are working there is probably no power to the unit (the set of three chimes). Check that the common wire to all three chimes is connected. If the unit has a disconnect plug make sure it is connected. Try disconnecting the plug, clean the pins with a wire brush, and reconnect. Also check the fuse, though if the fuse is blown a lot more than the chimes will not be working.

ONE CHIME NOT WORKING

There are three possibilities why one chime is not working (and the other chimes are working):

1. One of the wires may be disconnected from the solenoid.
2. The plunger may be missing. If the unit had been previously disassembled the plunger could have fallen out. Look at the bottom of the chime unit. You can usually see the plungers sticking out from the bottom of the solenoids.
3. One of the plungers may be permanently up, jammed against a chime. This can happen if the solenoid is burned, the plunger held up by a melted or warped coil sleeve. This can also happen if the solenoid is constantly activated, holding the plunger up. Turn off the machine and see if the plunger drops.

Both of the stuck plunger situations are caused by a switch, probably on or under the playfield or mounted behind a target, that is shorted or stuck in the closed position. If you know which chime is stuck (the 10s, 100s, or 1,000s), check the switches that score those points. If you don't know which chime is which, check all of the switches that score points. You are likely to find a switch with the switch blades closed (touching), making electrical contact when they should be open. Adjusting that switch, and replacing the solenoid if it is burned, should solve your problem. See "Coils" to learn how to replace a coil.

MUFFLED SOUND

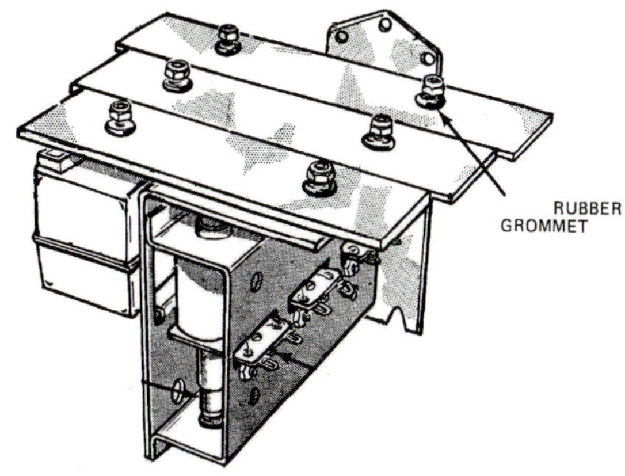

A dull chime sound is usually caused by disintegrated spacers, missing plunger parts, or too loose/too tight fastening to the cabinet. Or maybe all three.

Chimes are mounted with rubber grommets or rubber spacers that wear out and disintegrate after years of play, and muffle the ring of the chimes. Get replacement grommets from a hardware store, or use the small 5/16" or 3/8" rubber rings discarded when you put new rings on the machine.

If a chime has an indentation from being hit thousands of times by the plunger it could affect the sound. Turn the chime over and see if that makes a difference.

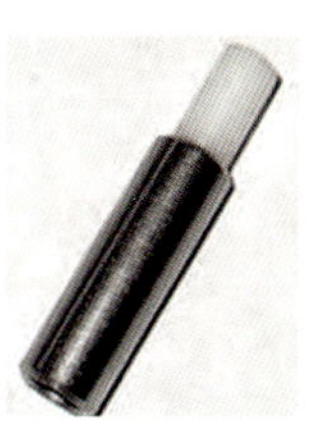

Many chime plungers have a plastic plug inserted into one end of the plunger. The plug stuck out about 1/2". If your plunger has a hole in one end where a plastic plug might once have been inserted, a new plunger might be all you need to get a good sound from the chimes.

Some chimes sound better when they're fastened tightly. Some chimes sound better when they're fastened loosely. Sometimes you can play around with the mounting adjustment to get a better ring, and well, sometimes you just can't.

BELLS

Some bells were operated by a separate solenoid and plunger, the same as chimes, with the same reasons they don't work and the same suggestions for fixing them. See "One chime not working."

Many bells were operated by a lever mechanism that was part of another mechanism, often a step-up scoring unit. As the step-up unit rotated, the lever swung up (or down), struck the bell, and swung back. If the bell is not ringing, the lever may not be moving because the rotating mechanism is not rotating. It could be because the lever arm

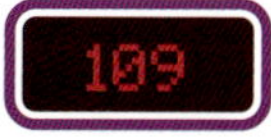

is jammed or stuck and in need of a good cleaning. It could be because the lever arm is missing, which I discovered on a machine that recently came into my shop. Someone had removed the lever arm to silence the bell. Which also explained the mystery why the wires to the other bell in the machine had been cut. I actually found a pinball parts dealer who had a replacement lever—for a mechanism that had not been manufactured in sixty years!

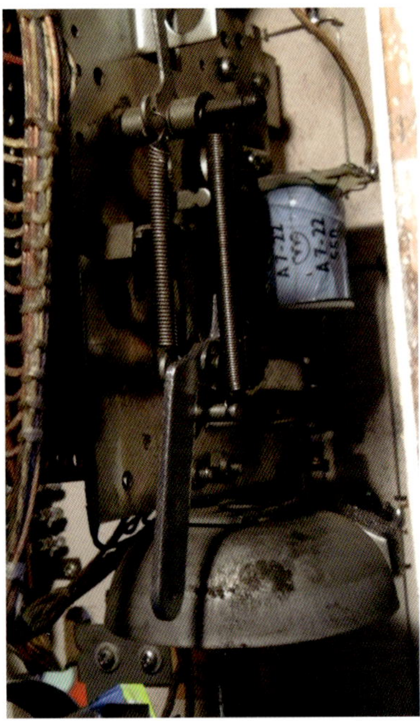

This bell, from Williams **Miss O**, is struck by a lever operated by the match unit, which moved one step (and rang the bell one time) when the ball hit the one-point bumper.

KNOCKER: NOT KNOCKING

When you win a free game, and in some machines when you get an extra ball, the loud bang or knock you hear is caused by a mechanism called the knocker. A solenoid launches a plunger up which hits a metal stop. It's a sound every pinball player loves to hear.

If the knocker isn't working at all there are a few quick things to check.

Are all the parts there? I've found machines where the coil was missing, where the plunger was missing, where the entire assembly—bracket, coil, and plunger—was missing. The knocker is usually mounted on the side wall of the cabinet or on the side wall of the backbox. If you can't find the knocker, look for cut wires or screw holes that might indicate where the knocker once was. Parts and entire knocker assemblies are available from pinball suppliers.

In many machines, the wires to the knocker are attached to the machine with a quick disconnect plug. I've often found knockers unplugged from the plug socket. I've also found knockers with the wires cut.

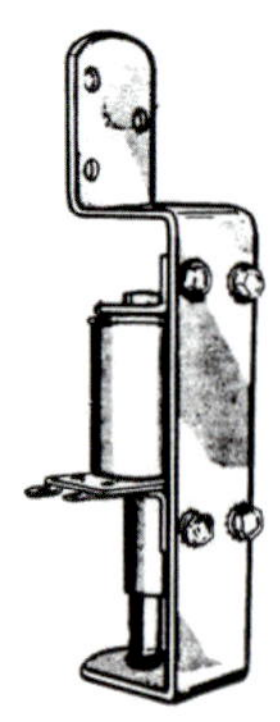

Replacing a missing knocker, reconnecting the plug, or patching the cut wires may be all that's needed. Or maybe not. There is a reason why the knocker is disconnected. Maybe someone just didn't want the knocker knocking. More likely the knocker coil may have burned or shorted, and the knocker was disconnected or removed because whoever owned the machine didn't know how to fix the problem. Does the solenoid look like it overheated or burned? Can you move the plunger by hand? If the plunger is stuck, the coil most likely burned and will have to be replaced. And it may be that easy a fix: a new coil.

Solenoids don't burn out in normal operation. They burn when they are activated continually, stuck in the "on" position caused by a short somewhere in the circuit, usually a switch that is stuck closed when it should be open. If you replace the knocker solenoid, and if the new solenoid is activated constantly, turn the machine off right away. Otherwise you'll just fry another coil.

The problem is most likely a stuck switch on the replay unit or credit wheel (the wheel that shows the number of free games). If the problem switch is not there, it is not difficult to locate the knocker on the schematic, trace the wires to the switches that activate the knocker, and then locate those switches in the machine. One of those switches will be shorted, causing the problem.

And if you can't find the replay unit, or don't want to trace circuits—*"not difficult?"*—disconnect the knocker again and wait until you find someone to help. The game will play fine without the knocker.

ADD-A-BALL MACHINES AND KNOCKERS

Some machines can be converted from replay to add-a-ball, where instead of winning a free game when you hit the Special, you get an extra ball. See "Add A Ball" in the "Buying a Machine" chapter. On some (but not all) of these "convertible" machines, the add-a-ball option disables the knocker. If the knocker works in replay mode and not in add-a-ball mode, most likely that's how it was designed.

WEAK KNOCKER

If the knocker is working but is not loud enough, there are a few quick things to check and a few remedies to try:

1. If the knocker's mounting bracket is not tightly attached to the cabinet, the knock is likely to be much quieter.

Tighten the bracket screws and replace any missing screws. If the screws will not tighten, or if the screw holes in the cabinet are worn, replace the screws with slightly fatter screws, or move the knocker and drill new holes.

2. Check that the coil sleeve isn't damaged. A damaged coil sleeve can rub against the plunger, weakening its movement.
3. Some knocker plungers, like chime plungers, have a plastic plug inserted into one end of the plunger. If the plastic plug is missing a new plunger might be all you need to get the loud knock back.
4. Replace the solenoid with a more powerful coil or convert the solenoid to DC. See "Coil substitution" in "Coils" and "Converting to DC" in the "Repairs: Flippers" chapter.

PLAYFIELD GLASS

If the playfield glass breaks, vacuum up the glass pieces carefully so they won't damage the playfield. Check in all the holes and slots in the playfield where tiny pieces of glass may have fallen through. The glass can fall into switches under the playfield, affecting their operation.

Replace the glass as soon as possible. A machine that sits without its glass is unprotected from dirt and objects, unprotected from curious hands and curious animals, and from scraping and scratching. Playfield components are easily damaged when not protected by the glass. Balls and bumper caps mysteriously disappear from machines without the playfield glass.

PLAYFIELD GLASS SHOULD BE TEMPERED

Tempered glass, also called safety glass, is a special heat-treated glass that, when it breaks, shatters into tiny pieces not likely to cut you as badly as window glass, which breaks into very sharp, pointed shards. Tempered glass is also much stronger than regular glass and is more likely to survive rough treatment. But be warned: Tempered glass is very weak at its corners. If dropped on a corner, or even just knocked on a corner, tempered glass can shatter.

All pinball machines came from the manufacturer with tempered playfield glass. However, there is a good chance that the glass on your machine was replaced sometime in the past, and the replacement glass may or may not be tempered. Only about half of the machines that come into my shop have a tempered playfield glass.

How can you tell if your playfield glass is tempered? Tempered glass often has slightly beveled or rounded edges and corners. If the glass has chipped edges, most likely the glass is not tempered. Some tempered glass has little dimples—indentations on the bottom or top end of the glass where tongs grabbed the glass to remove it from the kiln. Some tempered glass has an engraved or printed insignia in the corner called a "bug" that tells you it's tempered, but you rarely see this on playfield glass.

Playfield glass is usually 3⁄16" thick, thicker than regular window glass (although some window glass is also 3⁄16"). If your playfield glass is thinner than 3⁄16", most likely it is not tempered. Sometimes you can tell by tapping on the glass and listening. Tempered glass tends to have a more solid, less rattly sound. You can take the glass to a local glass store. They most likely can tell you if the glass is tempered.

Standard size tempered playfield glass for most pinball machines is available from nearly all pinball suppliers for about $35 or $40, but shipping can be expensive. If you live near a pinball supplier you can pick up a glass. If you plan to attend a pinball show in the near future, find out if any vendors will be there, contact the vendors, and see if they will bring a glass to the show for you. Call local route operators and see if any of them have an extra playfield glass to sell you.

A local glass company can custom order tempered glass for you, but it will be more expensive than pre-made playfield glass.

You can also buy playfield glass that is anti-reflecting and protects the playfield from UV light, reducing fading caused by sunlight. This type of glass costs about $200, an expensive, and for most people, unnecessary fix. If your machine is in direct sunlight you can simply cover it when not in use.

SIZE OF PLAYFIELD GLASS

Until 1990, almost all pinball machines from all manufacturers used the same size glass, 21" × 43". Some Gottlieb machines built in the 1980s and many machines built after 1990 used larger sizes.

Some Bally machines from the 1970s have the playfield glass mounted in a hinged metal frame that has to be disassembled to get the glass in and out. The frame is held together with screws, but after 40-some years it is also held together with rust and grime, and can be a struggle getting apart and back together. The glass in the frame was usually 21" × 41½", shorter than a standard playfield glass, which means that a replacement must be special ordered. You cannot cut tempered glass like you can regular glass. Again, you want glass that is 3⁄16" thick.

In the 1980s and 1990s, a few pinball machines were made with extra-wide playfields. These "widebody" machines also had a wider playfield glass, often 25".

I suggest that you measure the glass in your machine before it gets broken and confirm that it is or isn't the standard size. (And if you didn't get around to measuring the now-broken glass, don't worry. Every machine I ever had in my shop had the standard size glass.)

CLEANING THE PLAYFIELD GLASS

Clean the glass with a glass cleaner or just a damp rag. Avoid water. You don't want water running down into the cabinet. You also don't want spilled soft drinks or beer running down into the cabinet, so don't let your friends put their drinks on top of the glass.

PLAYFIELD: LOOSE COMPONENTS

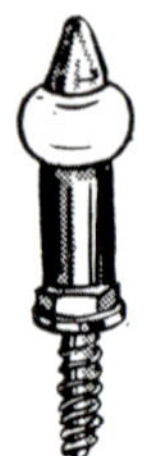

Pinball machines get banged around constantly and things get loose. And that always leads to problems. Components mounted to the playfield must be firmly secured. Tighten all mounting screws and replace all missing screws and nuts. And about once a year, recheck all of the mounting screws.

If a screw hole in the playfield is stripped and worn too large to hold the screw securely, you may be able to substitute a fatter screw, although for most playfield components this is probably not possible. You could also try a longer screw, if there is enough fresh wood a longer screw can grab.

A fix that usually works is to put some wood glue on a toothpick or a couple toothpicks and shove the toothpicks in the screw hole. Wipe off the excess glue so it doesn't stain or mar the playfield. Let the glue dry overnight, then cut the toothpicks flush with the playfield. You should now be able to mount the original screw in the hole, good as new. Instead of using a toothpick, mix some sawdust with white glue and fill the hole, let it sit overnight, and then mount the screw.

If patching the hole doesn't work, see if the layout of the playfield and the design of the component will allow you to drill a new hole a fraction of an inch from the old hole. Be careful that the two holes don't overlap, resulting in an even bigger hole.

If none of the above suggestions work, you could drill the hole all the way through the playfield and replace the old wood screw with a machine screw, tightening the screw with a washer and nut under the playfield. Before attempting this, look under the playfield to be sure you have access to get a washer and nut on the screw. There may be another component mounted directly under the screw preventing under-playfield mounting.

BALL SHOOTER (PLUNGER)

The ball shooter, sometimes called a plunger, will eventually acquire dirt or grime and will stick or not operate smoothly. It's a simple fix. Disassemble it, clean it, and put it back together. (If your shooter is a solenoid powered launcher or gun trigger you can skip this topic).

A simple cleaning with cloth and rubbing alcohol usually removes the dirt and grease. But first, make a diagram or take a photo showing what order all of the clips, spacers, washers, and springs come off, and go back on, the unit.

Shooter rod: If the shooter rod is rusted, it will usually come clean with a fine emery cloth or steel wool, or you can use a wire grinding wheel.

Don't lubricate the shooter rod. Lubrication attracts dirt and lint that will soon make the shooter action worse instead of better.

If the tip of the shooter rod is not smooth, if it's "mushroomed," or if it has a ridge or bulge or burr—usually the result of hitting a ball without the protective rubber shooter tip—file or grind the tip so it is smooth.

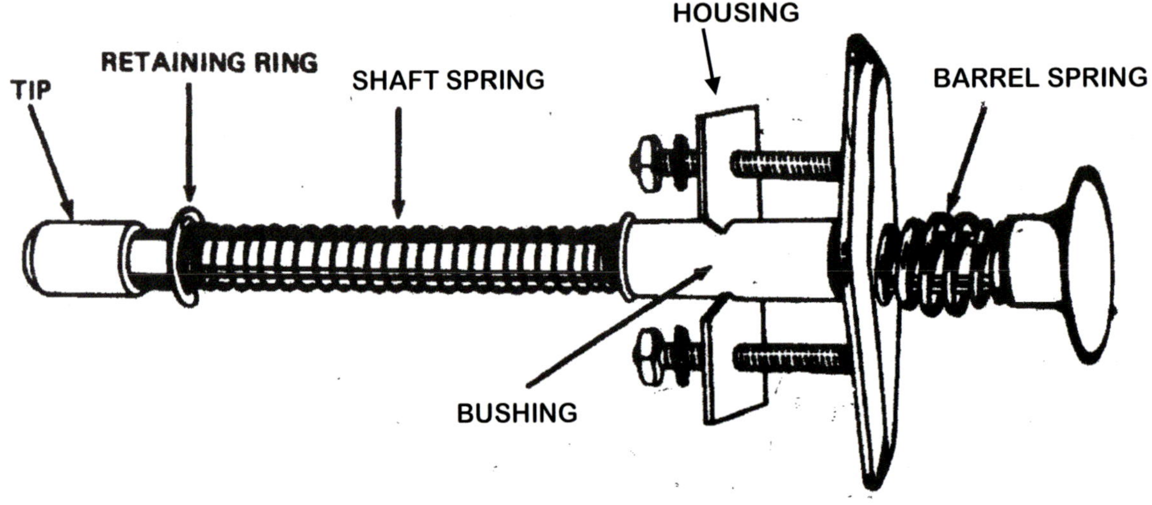

Shooter springs: If the springs are rusted or worn out you can get replacements from any pinball supplier. The long spring inside the cabinet is the shaft spring, which usually holds up quite well and rarely needs replacement. The short spring outside the cabinet is called a barrel spring, because it is shaped like a barrel, and it is often rusty and compressed.

Shooter bushing: The bushing is the mounting sleeve the shooter rod goes through. Clean the bushing with a tissue or a Q-Tip and rubbing alcohol. If the bushing is damaged and affecting the operation of the shooter, replace the bushing. If the shooter doesn't seem to go straight, if it wobbles or grabs a little when you pull it out to shoot a ball, a new bushing will probably solve the problem. On some machines there is a plastic, removable sleeve inside the bushing that can get damaged.

Rubber shooter tip: Most important, make sure the rubber shooter tip is in good shape. If it is damaged or missing, buy a new one. You do not want the metal shooter rod making direct contact with the ball. It will damage the ball, and the ball will damage your playfield. When you replace a worn shooter tip save the old one. It makes a great tool for removing hard-to-reach lamps.

ADJUSTING A SHOOTER

If the shooter is not centered in the runway, where the tip sticks out of the metal cover, you can fix this problem by adjusting the shooter housing (the bracket that holds the shooter in place). The housing is mounted to the inside of the cabinet with three or four screws that usually are in slotted holes, allowing for adjustment of the shooter. Loosen the mounting screws just a little, adjust the shooter, and retighten the screws.

If the tip of the shooter does not stick far enough out from the metal cover to make contact with the ball, or if the shooter rod seems too short, the playfield may be sitting in the wrong place. On many machines the playfield is held in place by two metal brackets mounted on the front edge of the playfield. If these brackets are bent or loose, they can set the playfield off from where it should be. Try straightening or replacing the brackets and see if your ball shooter is suddenly sticking out where it ought to be.

SOLENOID-POWERED SHOOTER

Some pinball machines have ball shooters that are solenoid powered, activated by a push button or a flipper button or some other mechanism such as a trigger on a gun. These machines do not have a hand-operated shooter, which takes away a little of the strategy and skill from the game. You don't get to decide whether you want to blast the ball up the runway, or maybe hit it just hard enough to land in a scoring hole or a slot at the top of the playfield. And yet the solenoid-powered shooter machines I get in my shop are always among the most popular.

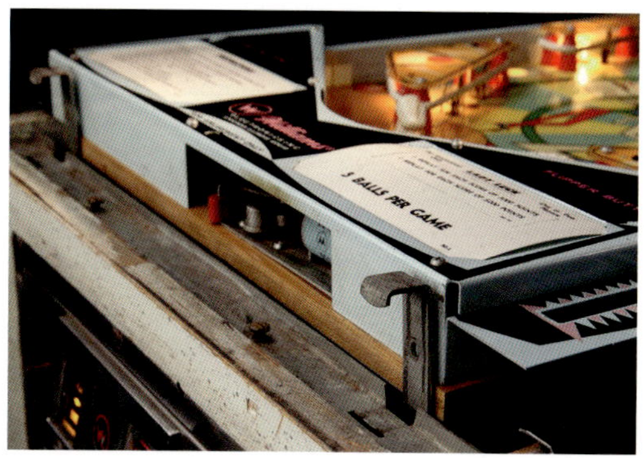

If the two metal brackets are bent or loose, the playfield might not align with the cabinet or the shooter.

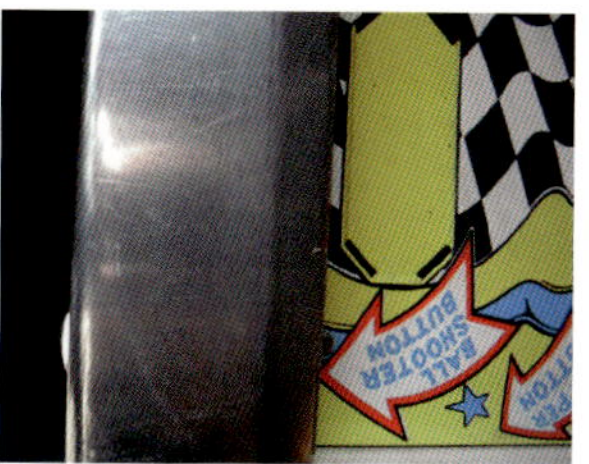

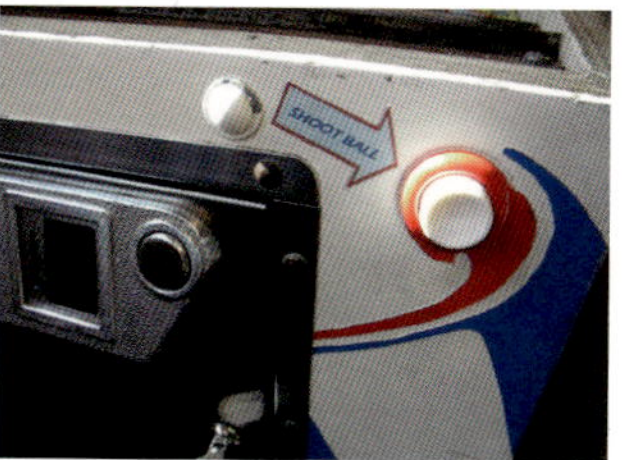

Players, not used to a shooter button, push the game start button (the silver button) by mistake and restart the game instead of shooting the ball, this despite the giant "Ball Shooter Button" arrow on the playfield. I added the blue arrow on the front of this Bally **"Twin Win."** It helped. Sometimes.

A solenoid-powered shooter will need more attention than most other powered action on the playfield. Just as no one wants weak flippers, no one will want a wimpy ball shooter. You should check the mechanism to be sure there is no dirt or gummy buildup that might slow the action, and check that the firing switch is clean and not pitted. And you might consider a stronger solenoid or a DC conversion to give the shooter more power.

TARGETS NOT WORKING

When the ball hits a target, the target closes a switch mounted behind the target. So when a target is not working, it's really the switch behind the target that's not working. The switch might not be closing, or it might not be making good contact, or it might be shorted (always closed), or it might have a broken off wire.

Sometimes a target will not score when hit by the ball, but will score if you operate the target by hand. This is sometimes caused because the metal switch blade, the one that the target is mounted to, is bending when the ball hits it, and the contact points are not making good contact. Try bending the top of the front blade (the blade with the target attached) just a little so that the contact nipples make better contact. If that doesn't solve the problem, buy a replacement target switch unit. It is inexpensive and easy to swap out. Many pinball parts suppliers sell replacement targets and switches.

If the switch is okay, the circuit or the components that the switch operates might not be working. There is probably a description on the playfield telling you what the target does—scores so many points, or increases the bonus, or opens a gate, or some other reward—or that information may be on the instruction card mounted on the apron (score card holder) at the bottom of the playfield. You'll have to locate the component that the target switch is supposed to operate, look it up in this manual, and see if you can determine what is wrong.

For more information about switches see "Switch Problems." Drop targets and roto-targets have their own unique problems (see following).

DROP TARGETS

Drop targets drop below the playfield when hit. Drop targets are one of the most popular features found on many pinball machines.

Drop targets often develop problems with age. They don't drop when hit, or don't drop all the way, or don't reset. Some drop target problems can easily be solved by cleaning, lubrication, spring adjustment, or some other minor repair.

Some problems will require disassembling the unit to replace worn or broken parts. Some drop target mechanisms are easy to disassemble and reassemble, but some, I tell you, can be a real job. Like every other repair in the machine, take your time, make notes and illustrations or take photos showing which parts go where, and keep a magnet nearby to find that little spring that just went flying across the room.

PROBLEMS WITH DROP TARGETS

If drop targets won't drop all the way or seem to get stuck when trying to reset, the mechanism may be gummed up and in need of a cleaning. Some manufacturers suggest putting lubrication on the back of the targets where they contact the switch mechanisms under the playfield, so the targets slide down more easily. A tiny dab of coin lubricant can sometimes make a big difference.

The plastic targets sometimes lose their rigidity and when hit may flex and not drop. Flexing drop targets also sometimes cause balls to launch into the air(discussed under "Airborne Balls"). You may be able to stiffen old drop targets by gluing a thin piece of stiff plastic or wood or metal to the

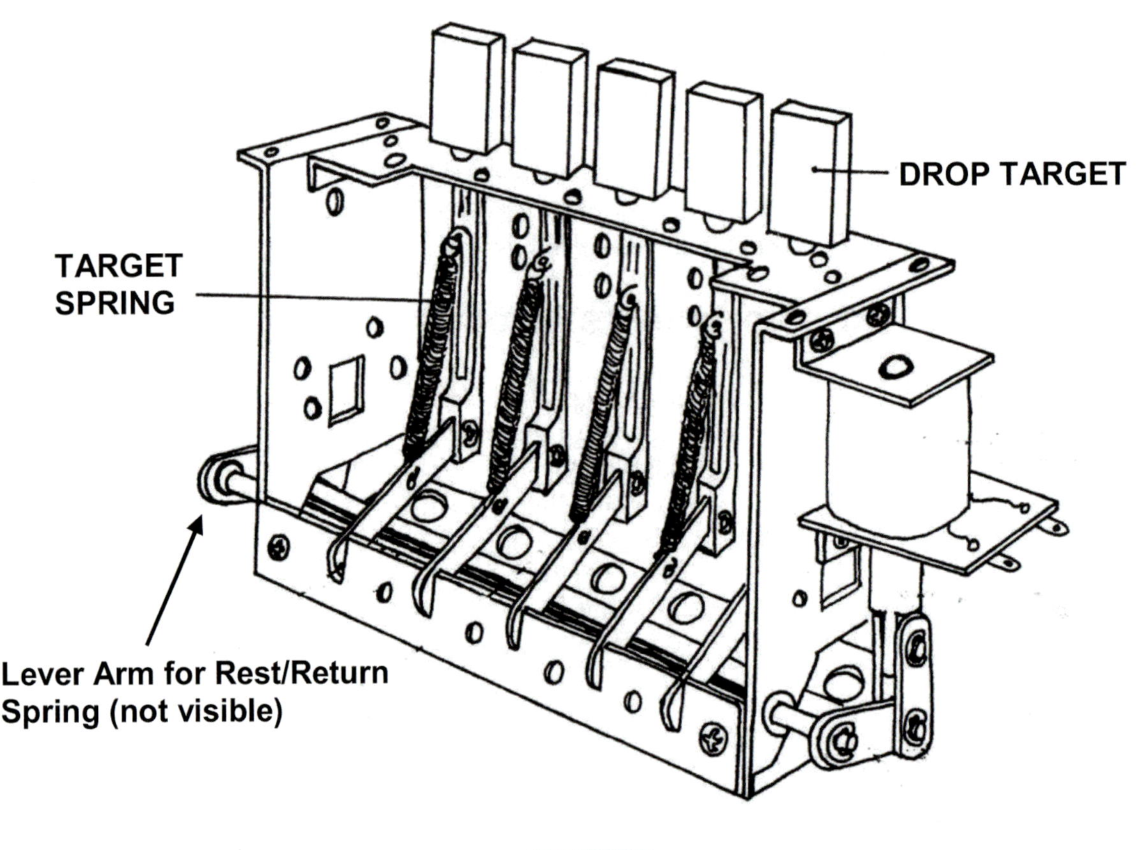

back of the target if there is enough clearance, though I have not had a lot of success trying this. A better solution is to replace the targets if replacements are available.

Drop targets often wear enough that they no longer go in a straight line. They are at a slight angle and friction keeps them from returning. They need to be replaced.

Replacement drop targets for many machines are still being manufactured. Try different pinball suppliers. If you can't locate a replacement, you can often find used targets for sale at pinball shows. Target designs and shapes varied from machine to machine, so take one of your old targets with you to compare.

SPRINGS

Spring tension, or lack of it, may keep drop targets from dropping or resetting. One of the springs may be broken or missing or disconnected, particularly the rest/return spring on the end of the unit. Some people stretch the springs to add tension and let the extra spring power compensate for what's wrong. That's a band aid approach, but it may fix—though not solve—your problem.

DAMAGED "TRIGGER" ARM/LEVER

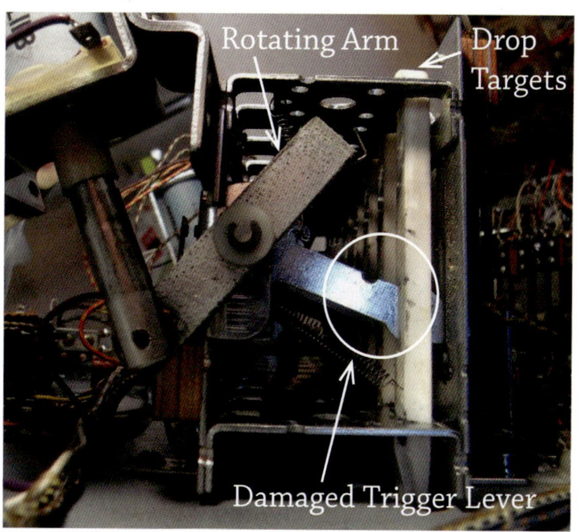

The rotating arm pushes the trigger levers down, and the levers pull down the drop targets. When indentation damage to the levers (circle) gets too large, the trigger levers no longer go all the way down. The drop targets do not latch into place and will not stay down.

Drop targets reset using metal "trigger" levers, pivoting arms that are lifted by a metal bar operated by a solenoid. After thousands of resets, those metal-against-metal collisions (the metal bar hitting the levers) often wear indentations in the arms, so much that the bar no longer lifts the arms all the way, and the targets will not reset. Sometimes you can jerry-rig some sort of patch on the arms so the targets will lift all the way. I've had some success doing this on some machines, but the patch usually does not hold very long.

The only sure way I've found to solve this problem is to replace the metal lever arms with new ones, if you can find them. Some pinball suppliers sell replacements. Some pinball suppliers actually salvage and sell parts from used machines when new parts are not available! You might also try to locate a used drop target unit or scavenge one off a parts machine, one that hopefully is not also worn out. Pinball shows are great places to find used parts and mechanisms.

OUTHOLE: BALL DOESN'T KICK OUT INTO PLAY

The outhole is where the ball lands after it leaves the playfield. Then a solenoid kicks the ball over to the shooter and back into play.

Sometimes when the outhole solenoid fires, trying to send the ball from the outhole back into play, the ball doesn't quite make it all the way and rolls back to the outhole, whereupon the outhole solenoid tries again to shove the ball into play. This might happen once or twice or more, before the ball finally gets itself in front of the ball shooter where it belongs. It's a real nuisance.

You can fix the problem. First try a simple solution. Put a level on top of the playfield glass. If the machine is out of level side to side, sometimes gravity will work against the ball, causing the kick-out problem. If this is not the problem—frankly, it's probably not—move on to the next step.

CHECK UNDER THE APRON (SCORE CARD HOLDER)

Unscrew and remove the apron (also called the score card holder), the large metal or plastic cover at the bottom of the playfield, to expose the metal rails that the ball travels over on its way from the outhole to the ball shooter. These rails are called the "trough." More than once I've found dirt or some object sitting in the ball's path that impeded its journey. Often I found the rails dirty or sticky or corroded. A simple cleaning made everything work fine.

If the ball kicks out fine when the apron is removed but not when the apron is on, most likely the ball is hitting the apron and being bounced back to the outhole. If the apron is buckled in a little, dented, or otherwise too low for clearance you found the problem. You can g-e-n-t-l-y bend it out again. But be careful; bending a metal apron can sometimes destroy the paint on it.

EXAMINE THE OUTHOLE MECHANISM

The outhole mechanism—the solenoid, brackets, and parts attached to the solenoid—may be causing your problem.

On some machines the outhole solenoid is mounted to the left of the ball and the plunger gives the ball a good hit, propelling it to the playfield. Check that the plunger isn't gummed up or dirty and that the spring isn't jammed.

On many Gottlieb machines, the outhole solenoid operates a lever in the shape of a hook mounted underneath the outhole that comes up through the bottom of the outhole to kick the ball back into play. This cumbersome mechanism often wears down and causes the "ball won't get to the shooter" problem. Sometimes the hook is not angled correctly, causing the ball to lift up as it kicks out, hit the metal apron, and bounce back to the outhole. With a pair of pliers try bending the hook slightly, one way or the other, and see if it solves the problem.

The lever (the hook) has a spring mounted on the mechanism. If the spring is weak or damaged, the lever will not have enough force to kick the ball all the way to the ball shooter. Tightening or replacing the spring may solve your problem.

On one old electro-mechanical machine when all else failed, I wired a bridge rectifier to the solenoid, converting it to DC power and giving it plenty of kick. See "Converting to DC Power" in the "Repairs: Flippers" chapter. You can, instead, substitute a more powerful coil. See "Coil substitution."

AIRBORNE BALLS

Some pinball machines have a problem with balls getting airborne and hitting the bottom of the playfield glass.

Flipper launch: If the flippers are causing the ball to launch, try moving the flipper rings up or down a little on the flipper if possible. If the flipper rings are regular round rubber rings, try different rubber rings, or a double set of rings. Make sure the flippers are not moving at an angle to the playfield, possibly caused by a damaged flipper bushing, which is covered in the "Repairs: Flippers" chapter. If the flipper height can be adjusted, try raising or lowering the flipper just a fraction of an inch and see if that solves the problem. Be careful, if lowering the flipper, that it doesn't rub against the playfield.

Another possible cause of airborne balls: The flippers may be too powerful. You might try weakening the flipper power a little. You can accomplish this by adjusting the end of stroke switch to open sooner. See "End of stroke switch" in the "Repairs: Flippers" chapter. If you installed a bridge rectifier (if you did a DC conversion) or a more powerful coil and suddenly have airborne balls, remove the bridge rectifier or go back to the original coil.

Drop target launch: Balls sometimes go airborne when they hit drop targets. Old drop targets can flex when hit, and can launch the ball into the air. The targets may need to be stiffened or replaced. See "Drop Targets."

Launch off of a playfield post: If the ball jumps up when it hits a post, try a thinner or thicker rubber ring on the post, try moving the rubber ring up or down, or maybe the rubber ring is missing and needs to be replaced. Make sure the post isn't bent or loose. You may need to experiment a bit to solve the problem (a comment I could add to almost every repair in this manual).

Launch out of a kickout (eject) hole: Keep Reading:

KICKOUT (EJECT) HOLES

Many pinball machines have kickout (eject) holes: holes in the playfield where the ball lands and is then ejected by a solenoid mounted under the kickout hole.

Don't confuse the kickout hole with the outhole. The outhole is where the ball lands after it leaves the playfield. Also don't confuse the kickout hole with a "gobble hole" found on some pinball machines, where the ball drops into and through the hole and winds up somewhere else on the playfield.

WEAK EJECT

If the ball doesn't seem to have much "oomph" when ejected, if it just kind of dribbles out of the kickout hole, check the eject mechanism for dirt or grime that may be slowing it down. Remove the solenoid from its bracket and check the plunger, the coil sleeve, and the coil stop for wear. See "Plungers and Coil Stops" in the "Repairs: Flippers" chapter; the same procedures apply here.

Try a different spring on the return mechanism —longer or shorter, weaker or stronger, depending on how the mechanism is designed—that might give less resistance and more power to the ejected ball.

BALL GOES IN THE WRONG DIRECTION

Sometimes a kickout hole will eject the ball in a direction that isn't where you want the ball to go, such as into a nearby obstacle, or into something that rebounds the ball right back into the kickout hole, or straight down between the flippers into the outhole. That's no fun.

You can change the ball eject angle. There is a metal lever under the kickout hole that comes up and ejects the ball. The top of the lever is shaped like a hooked nose. Use a pair of pliers to slightly bend the top of the lever to the left or right.

Go easy, just a fraction of an inch. This should change the eject trajectory enough to send the ball in a different direction. You may have to experiment a few times, bending the lever more or less one way or the other to get the results needed. Just go easy, and be careful not to damage the mechanism.

BALL WILL NOT STAY IN THE KICKOUT HOLE

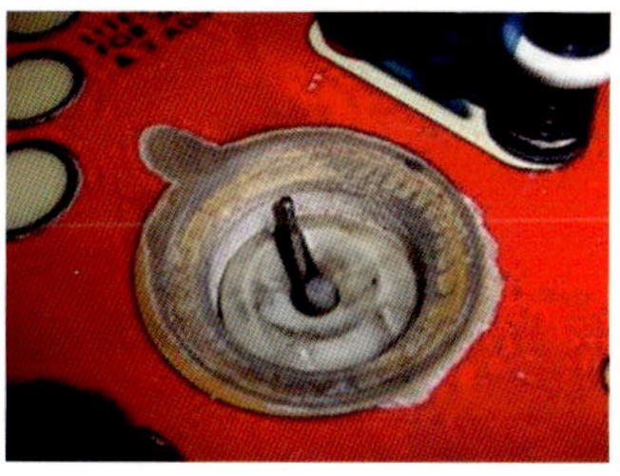

Kickout holes in playfields are often worn down around the edges, especially kickout holes that are dead center in the middle of the playfield where they see a lot of action. When the ball lands in the kickout hole, the ball does not stay in the hole like it should but comes flying out the upper end of the hole. If the ball will not land in the hole and stay until it is ejected by the mechanism underneath it, you've lost an important and special part of the game play.

I have never found a successful way to build up the missing wood around kickout holes. I've seen attempts people have tried, using wood filler or some other material, but the repairs usually didn't hold up to the pounding of a steel ball, or the unevenness of the repair caused the ball to move erratically, and it always looked like a bad patch job.

I did find a solution that usually works, depending on the design of the playfield, though some people might consider it radical. I drill a small hole and mount a metal playfield post and tiny rubber ring—similar to the illustration for "Playfield Components"—directly behind (above) the eject hole. When the ball comes flying out of the top of the hole, the ball hits the post and falls back into the eject hole.

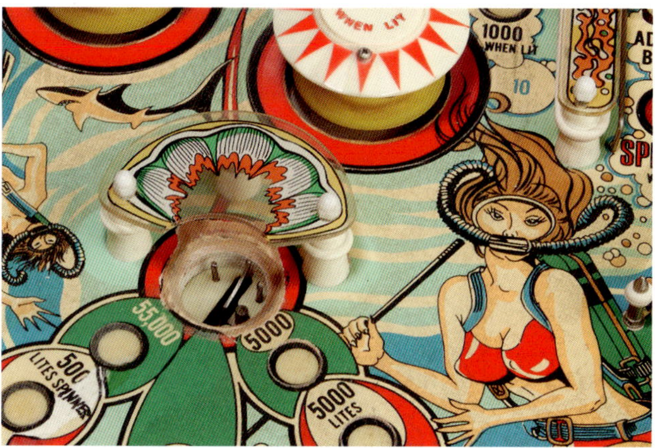

Plastic shield over the kickout hole on Stern **Stingray**, 1977.

For some machines, adding the post has proven to be an excellent solution to a major problem. For some machines, however, adding a post can alter the play of the game. If the playfield is designed so that the ball can enter the kickout hole from the top as well as the bottom, the post will block the ball from entering from the top. Study your machine and decide whether the "cure" is worse than the problem. If you try adding a post and find out that you made a mistake, you can easily remove the post and fill in the small screw hole.

BALL GOES AIRBORNE

I've had several machines in the shop where the ball, when ejected from an eject hole by the mechanism under the hole, goes airborne and hits the playfield glass. It is a real irritation. Quite often the cause is worn down wood around the eject hole which, as discussed under "Ball Will Not Stay in the Kickout Hole," is not easy to repair. You can usually find a way to solve the airborne problem without repairing the hole.

Quite often bending the eject lever will solve the problem. This is also explained under "Ball Goes in the Wrong Direction."

Sometimes installing a slightly weaker coil or putting a stronger return spring on the solenoid will give the eject mechanism a less powerful "kick" and keep the ball from launching into the air. This alteration, however, may weaken the ball action enough to adversely affect the play of the game. It never hurts to try. You can always reinstall the original coil and original spring.

Here is another suggestion that has worked for me when all else failed. On many pinball machines with kickout holes, there is a plastic shield, sort of like a roof, mounted over the eject hole. The shield is possibly just a decoration, possibly to protect lamps, or to cover posts near the hole. Whatever the intended purpose, that plastic shield might also stop the ball from launching up and hitting the playfield glass. If the ball hits the shield instead of hitting the playfield glass, then deflects down onto the playfield, it may alleviate the problem. If the plastic shield is not big enough or not centered over the hole, you may be able to slightly relocate the shield, or reverse the shield, and see if it helps.

By the way, if that plastic shield is cracked or broken in half—and a lot of them are—it may be an indication that too many airborne balls have already hit that shield and broken it. You may be able to find a replacement shield, someone selling a shield off another machine, but it's not likely. Now what? Glue the shield back together? Glue a second piece of clear plastic over it to stiffen it? Put a new and thicker piece of plastic in its place? You really have nothing to lose by experimenting. I admit, this is a real band aid approach to the problem, but as we used to say back in the Free Spirit days, Whatever works . . .

SWITCHES NOT WORKING

A ball hits a target and nothing happens. A ball lands in a kickout hole and nothing happens. A ball rolls over a trip wire and nothing happens. There is a good chance that the problem is a switch that is not working.

Every pinball machine has dozens of switches. Flippers, targets, bumpers, roll overs, spinners, motors, score reels, chimes, coin mechanisms, lamp circuits, and every other action in a pinball machine involve switches.

Most electro-mechanical pinball problems can be traced to switches that are out of adjustment, or are making poor contact, or are dirty or corroded, or have broken parts or broken wires. Or all of the above, especially if the machine has not been working properly for a long time. People will just ignore a problem, or not even notice the problem if a game is still playable. But then another switch stops working, and another, and what would have been a five-minute repair is now an entire machine in need of an overhaul. Whether you have one non-working switch or half a dozen of them, fixing the problem is not difficult.

All electro-mechanical machines used leaf switches: flexible flat switch blades separated by a spacer, with contacts (nipples) riveted to the ends of the blades. The blades, when pushed together, make electrical contact and complete a circuit that scores points, or advances a bonus, or rings a chime, or any of the dozens of other actions that are part of the game. Many of the early electronic machines also used leaf switches, but most electronic machines used micro-switches with one wire blade instead of the two-blade leaf switches. The information here on switch cleaning and repair applies only to leaf switches. Micro switches are covered at the end of this section.

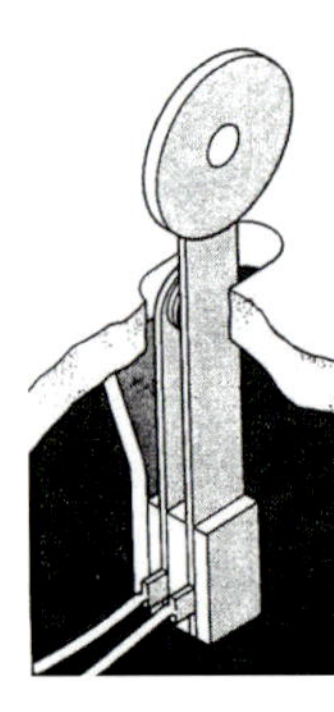

To confirm that you have a switch problem, first find the switch. For roll over switches and kickout holes, the switch is under the playfield. For targets, the switch is behind the target, usually above the playfield.

Manually close the switch blades so they make contact and see if they work. Or take a jumper wire or a small screwdriver, and short the switch: touch both solder lugs with the jumper wire or screwdriver. If the switch still doesn't work when you short it, the problem is not in the switch (see the suggestions at the end of this section). But if the switch starts working again, scoring points and doing what it is supposed to do, you've found your problem. The switch needs cleaning or adjustment, or repair.

RELAY SWITCHES (ELECTRO-MECHANICAL MACHINE)

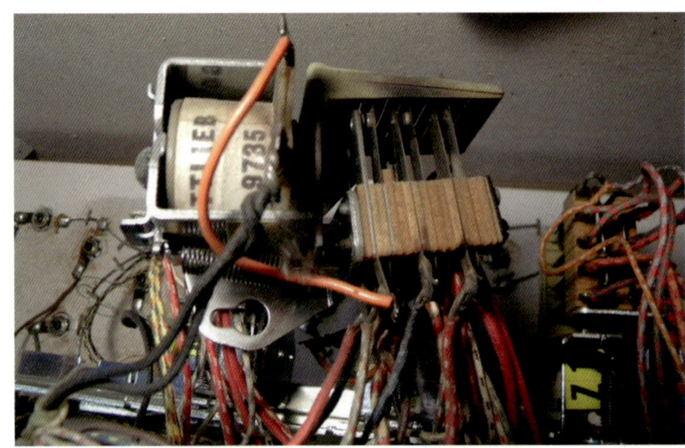

Many of the operations of electro-mechanical machines are controlled by the relays inside the machine. The relays and their switching circuits are often the source of problems.

A single switch on the playfield can accomplish several different operations. For example, a ball hits a target that closes a switch behind the target. Hitting that target, and closing that switch, might increase the score, ring a chime, turn on a lamp, or step up the bonus, all at the same time. What actually happened is the target switch did only one thing. It activated a relay. The relay closed or opened several switches mounted to the relay, and each of those switches activated a different circuit. One switch on the relay lit the lamps, one rang the chime, one increased the bonus, and one switch possibly activated another relay that in turn opened and closed more switches. A chain reaction that, taken together, seems overwhelming. But if you look at it one switch at a time and one circuit at a time (the circuit that the switch controls), you can isolate a problem, and troubleshooting gets a whole lot easier.

SWITCHES ON SCORE MOTORS (ELECTRO-MECHANICAL MACHINES)

The score motor, as it rotates, operates dozens of switches all at the same time. The switches are stacked on top of each other—hard to see, hard to get to. But fortunately, score motor switches are not likely to be the cause of switch problems. If you are not experienced in tracing pinball circuits, and if you are not good in working in tight spaces, score motor switches are best left alone.

You may, however, want to check the very top switches on score motors that have the switches mounted on top, which is how Bally and Williams machines are designed (Gottlieb machines have the switches mounted around the motor, not on top of the motor). Those top switches stick up high enough that they can get knocked by an arm or an elbow when making a nearby repair, and bent or shorted.

The top switches are easy to see in operation. With the machine turned off, you can rotate the score motor by hand and observe the top switches opening and closing, and you can easily adjust them. Still, if you are not sure what you're doing, best to leave the score motor switches alone.

CLEANING SWITCHES

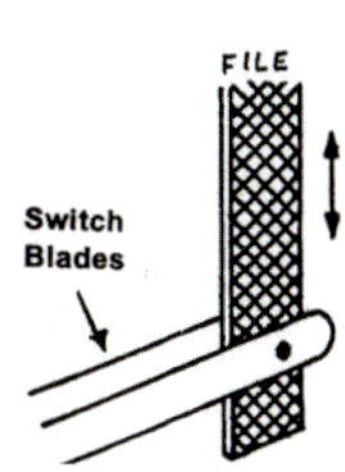

Switch blade contacts get dirty, corroded, pitted, and coated with carbon, causing switches to stop working or work intermittently. Poor contact can weaken the electrical flow, resulting in a poor playing game.

Leaf switches in electro-mechanical machines look identical to leaf switches in electronic machines, but there is an important difference in the composition of the switch contacts (the little nipples on the switch blades) requiring different cleaning methods.

Electro-mechanical machines: Use a special pinball flex file (flex stone, available from pinball suppliers), a burnishing tool, a metal fingernail file, or a flat metal "point" file, which you can get at an auto supply store. Put the file between the two contacts, apply light pressure to the contacts, and gently file the contact points (nipples) until clean. The contacts should look shiny. I've read suggestions to clean switch contacts with an old toothbrush soaked in alcohol, but I find it difficult to get a toothbrush in the tiny space between the switch blades. It's too easy to bend the switch blades with the toothbrush.

If the contact nipples on the switch blades are badly pitted or indented, or missing, the switch should be replaced. This is not common. I've rarely had to replace switches because of damaged contacts. They usually clean up.

Electronic machines: Do not use a file on electronic contacts because they have a very thin coating that is easily scraped off with a file. Clean switch contacts with a cotton swab dipped in alcohol. You can also clean switch contacts with a business card or similar piece of lightweight card stock. Put the card between the switch blades, squeeze together very lightly, and pull the card through.

Sparking: If a switch is "sparking" (an electrical spark between contacts when the switch is operated), try to file or adjust the switch so it no longer sparks. Sparking not only indicates less than perfect contact, but the spark itself can burn the contact nipples, making the electrical contact even worse.

ADJUSTING SWITCHES

Switches often go out of adjustment and will not open or close, or will make a poor or intermittent connection. If you have a switch that is not working, hand operate whatever component closes or opens the switch (a target or rollover, or push button, or some other mechanism) and watch the switch. Does it close, or open, all the way? If the blades are not contacting or making poor contact, bend them slightly until you get good contact. You can use a pair of long nosed pliers, or you can purchase a special switch adjustment tool called a contact adjuster, a most useful little metal rod with a slit in each end, from a pinball supplier.

A general guideline to switch adjustment: In the open position, the switch blade contacts should be 1/16" to 1/8" apart. When closing, the moving blade should touch the stationary blade and actually move the stationary blade a little bit (sometimes called "overwiping"). If switch blades are too far apart, they may not make good contact when the switch is activated. If the blades are too close together, the vibration of the machine may cause them to make contact or to "chatter" (repeated momentary contact). Different switches sometimes need different gaps. You may need to experiment with the gap size—the distance between the two switch blade nipples—until the switch behaves the way it should.

If you've never tried to adjust switch blades before, it may take several tries before you get it right. Sometimes a switch goes right back out of adjustment before the switch blade finally stays where you want it. Sometimes the blades will get all bent out of shape, and need to be straightened again, before you get them contacting smoothly.

After you adjust a switch and get it opening and closing correctly, sometimes the switch works for a short time, an hour or a day, and then goes right back to its out-of-adjustment condition. The metal had just enough spring from the old position that it reverted. I call it "metal memory" (although there probably is no such thing). Whatever the reason for the failed adjustment, a second adjustment usually holds.

"Fish paper" insulation: Some switches have a piece of insulating paper, called fish paper, between the switch blades to keep the blades from touching each other and shorting. This paper often gets bent and out of alignment, affecting the action and contact points. You can buy replacement fish paper from a pinball supplier or make your own. Take the switch apart, replace the fish paper, and put the switch back together. Or you can remove the fish paper—tear or slice it out, no need to disassemble the switch—and check that the switch blades are not touching and causing a short. I have found from my own experience that the switches, if carefully adjusted, do not need the extra protection that fish paper offers.

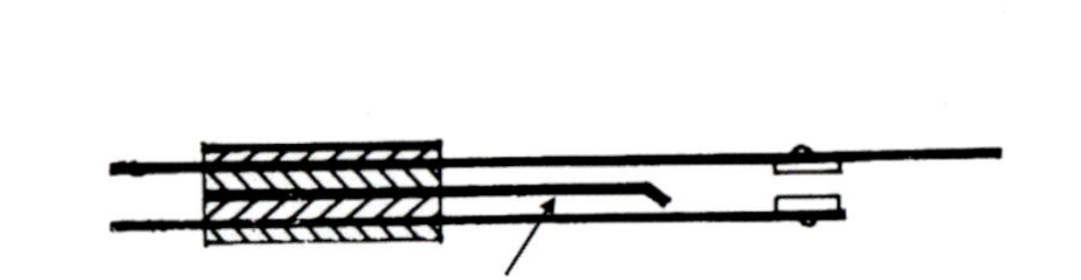

Damping blade in the middle.

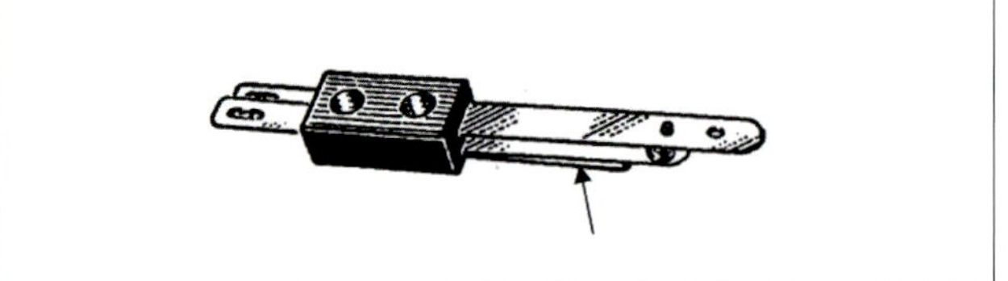

Damping blade on the bottom.

Extra blade: Some switches have an extra blade, usually made of a heavier metal than the regular blades. This "damping" blade, also called a "pre-tensioner" or "back up" blade, is used to stiffen the stationary blade to prevent the switch blades from bouncing too much. The damping blade does not have any wires connected to it. It should only touch one of the switch blades. If the damping blade gets bent, it can touch both blades and short the switch.

Loose or missing contact points: If a contact point (nipple) is loose in the blade, if you can wiggle or spin it, it will cause a poor connection. You can sometimes fix a loose contact point by soldering it to the blade. Put a drop of solder on the back side of the contact point where it connects to the back side of the blade. Do not solder the front side of the contact where it contacts the other blade, because the solder might interfere with the movement of the blades. If that does not work, you will have to replace the blade or the entire switch (see following). If a contact point is missing—the top of the switch blade will have a small hole in it where the contact was originally mounted—you will have to replace the blade or the entire switch.

Broken switch blades: I sometimes find switches where a blade snapped off or the solder lug broke off. You will have to replace the blade or the entire switch.

REPLACING SWITCHES AND SWITCH BLADES

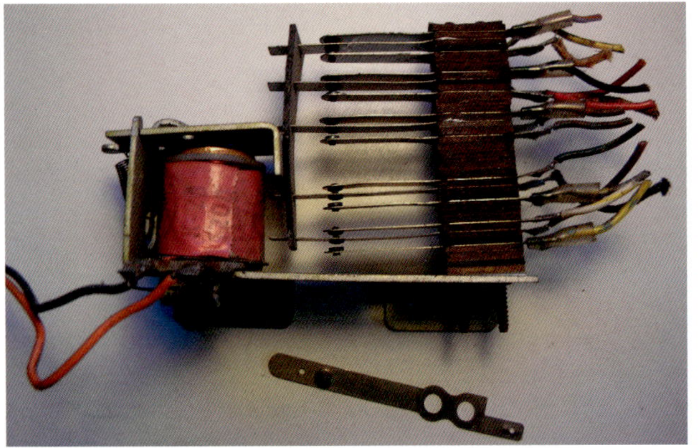

If one switch blade needs replacement, it will be a lot easier to swap out the broken blade than to install and wire an entire set of switches.

If the switch only has two blades, replacing the switch with a new one is fast and easy. If the switch has multiple blades, common on many relays, replacing and rewiring the entire switch could be a big job. You can instead replace a single blade if the switch can be disassembled and if you can get a matching switch blade. Switch blades came in many lengths and configurations and are not always interchangeable.

Replacing one blade is often easier than replacing the entire switch, but it can be a tedious job. Blades are stacked up, separated by spacers, the whole thing held together by two long screws. The screws have to be removed and the blades have to come off and go back on in the same order. Sometimes easy, sometimes not. The little spacers between the switch blades fall out, you drop a screw or a nut, and the whole thing falls apart. #*%&! You just have to control your temper and plug along.

THE SWITCH IS NOT THE PROBLEM

When you first tested the switch by shorting it (bypassing it with a jumper wire or a screwdriver) it either worked, which meant the switch needed cleaning, adjusting, or replacing; or it didn't work, which meant the problem is not in the switch. So, where do you go next?

See if any of the wires soldered to the switch broke off. Solder the wire back on and see if you fixed your problem. You might find that the solder lug end of a switch blade snapped off, and is just hanging there with the wire still soldered to it. The blade or the entire switch needs to be replaced (explained previously).

If the wires are still connected to the switch we have to start looking elsewhere. You can try to follow the wires from the switch to wherever they go in the machine—usually a relay in an electro-mechanical machine or a circuit board in an electronic machine—to see if a wire has broken or come off. If you cannot trace the wires, or if you can't find anything obviously wrong when you do, you might want to read "Schematics" in the "Tracing Circuits" chapter, because that's the next step in locating the problem.

But before you do, try shorting the switch again, just to be sure the switch isn't the problem. Maybe you didn't make good electrical contact the first time. It would be a happy discovery. Fixing a switch is a much easier job than tracing circuits.

 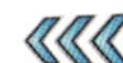

MICRO SWITCHES (ELECTRONIC MACHINES)

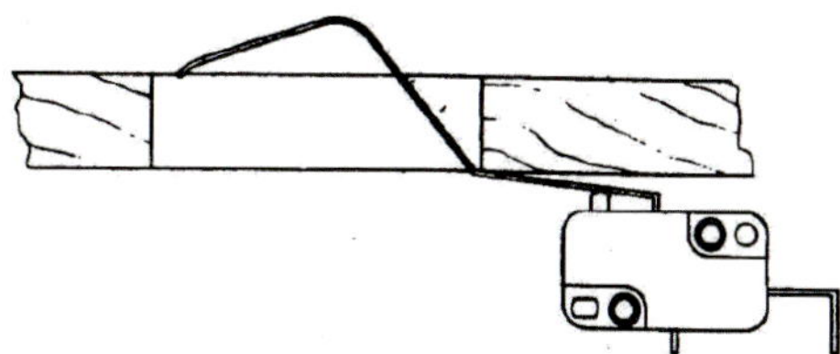

Electronic machines, starting around 1980, use micro-switches, with only one blade instead of the older two-blade switches. Micro-switches do not have contacts that touch each other, and cannot be cleaned or adjusted in the same way. But the switch blade on the micro-switch can get bent and not close the switch when a ball contacts the blade.

You can hear micro-switches click when the switch blade closes the switch. Use your ear and listen to the click to determine if the switch is opening and closing. You can use your multi meter or continuity checker to test the switch, but you should remove the switch from the circuit—unsolder one of the wires connected to it—to get a true reading from your tester. Otherwise the tester could be reading the circuit that connects to the switch instead of the switch itself.

SWITCHES THAT DO NOTHING

It is rare, but pinball machines have been built with switches that do nothing. Some feature was planned for the machine and then not made, or a feature was included on a first production run of the machine that caused problems and was removed.

I had a machine in my shop called Faces, made by Sonic in 1976, that has targets on the playfield labeled "Opens Gate." But there is no gate on the machine, and the targets do nothing. The switches behind the targets are not wired to anything. It didn't take long to discover where the manufacturer had planned to put the gate, and it was obvious that a gate was never installed. And it was interesting that the fellow who owned the machine for fifteen years before selling it to me never noticed the missing gate.

Still, it's about 99.9% likely a switch that isn't working has developed a problem and is not part of a circuit abandoned in mid-production.

POP BUMPERS NOT WORKING

To understand why a pop bumper isn't working, or sometimes working, and how to fix it, it helps to understand the fascinating Rube Goldberg-ish design of the pop bumper mechanism. Pop bumpers are also called Thumper bumpers and Jet bumpers.

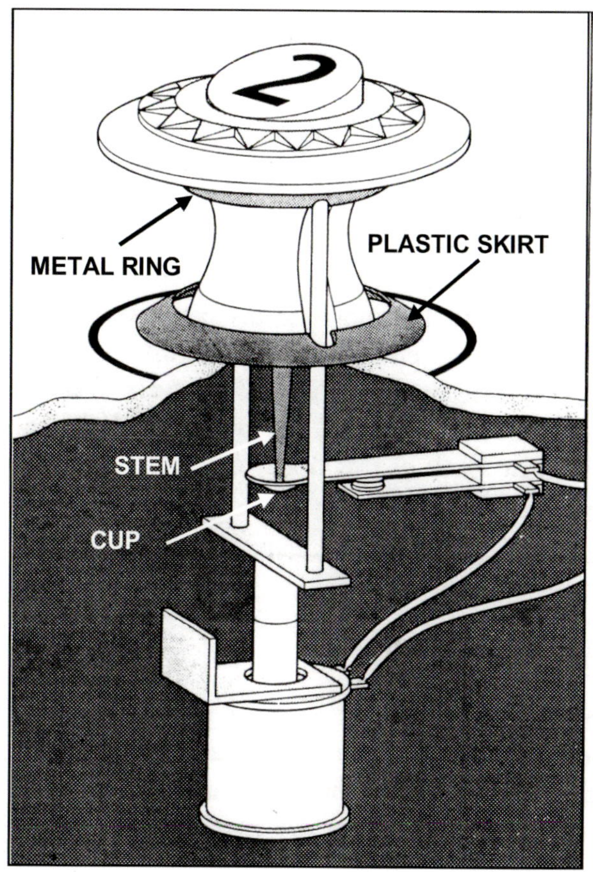

When the ball hits a pop bumper, the ball rolls onto a plastic ring at the base of the bumper called a bumper skirt that tips just a little from the weight of the ball. If you remove the playfield glass and put your finger on the skirt you will see how it moves. Also notice that at the top of the pop bumper, underneath the bumper cap, is a metal ring that moves downward. If you push the metal ring down it should move easily and then spring back up when you let go. Here is how it all works:

1. The bumper skirt (the wide plastic ring) at the bottom of the bumper is attached to a plastic stem below the playfield. When the ball hits the pop bumper, the ball tips the bumper skirt. The skirt tips the stem.
2. That plastic stem is sitting in a small metal cup mounted at the end of a switch blade. When the plastic stem tips, it pushes the metal cup down, which makes contact with another switch blade below it and closes the switch.
3. When the switch closes, the switch activates a solenoid mounted under the pop bumper. The solenoid pulls down the metal ring.
4. The metal ring hits the ball and the ball goes flying off.
5. When the ball is no longer in contact with the pop bumper everything returns to rest: The skirt and shaft return to level, the solenoid is no longer activated and the metal ring springs back up, ready for another go.

And it all takes less than a second. Except when it doesn't. Anything with that many interconnected parts will eventually have problems and need some help. Here's what to do:

THE METAL RING

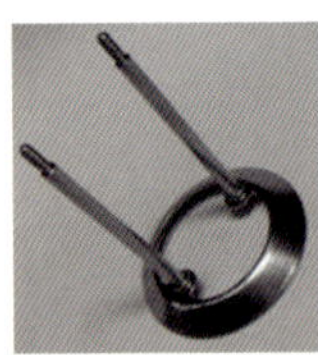

If the metal ring will not move down easily and spring back up, the mechanism is jammed and the pop bumper will not work. Look underneath the pop bumper and work the mechanism by hand. Look for a broken or jammed link or connecting bracket, called a yoke. If the mechanism looks fine, see if the plunger will not move in and out of the solenoid. If the plunger is stuck the solenoid is probably burned and needs to be replaced.

Quite often both problems exist. A link will break or jam, causing the solenoid switch to stick in the closed position and burning out the solenoid.

THE STEM (PLASTIC SKIRT) AND CUP

The stem attached to the plastic skirt, and the metal cup the stem sits in, collect dirt and grime. If they get dirty enough, the stem may no longer slide inside the cup like it should. The stem gets stuck, the cup does not move down, the switch does not close, and the solenoid does not activate. This is a very common problem. Just about every pinball machine that comes into my shop has pop bumpers with gummed-up stems and cups and, even though the pop bumpers may be working, they are often sluggish or intermittent due to the dirt.

I use a Q-Tip soaked in alcohol to clean the cup and the tip of the stem. I then put a small amount of coin lube (available from pinball suppliers) or teflon gel on the end of a toothpick and coat the tip of the stem and the inside of the cup. The stem should slide easily around the cup and return to vertical center when at rest. I should mention that this procedure sounds easy but is often a struggle trying to get your hand in a very tight space. Sometimes using a chopstick or some other longer stick instead of a toothpick may make the job easier.

If the pop bumper is activated when the ball hits it on one side of the bumper, but doesn't activate when the ball hits it from a different side, the plastic stem and cup are not interacting as they should. Most likely the cup needs to be adjusted, bent up a bit or angled a bit, so the stem moves the cup and closes the switch from every direction.

If the skirt is broken or warped, that could cause the activation problem. Replacing the skirt, which requires removing the pop bumper, should solve the problem. Removing a pop bumper is covered below.

THE SWITCHES

There are two switches mounted on the underside of the pop bumper. The top switch closes when the cup moves down and energizes the solenoid. The lower switch closes when the metal ring comes down and energizes the scoring circuit. (The illustration shows only one switch: the switch that is closed by the cup. The second switch, not in the illustration, would be below the horizontal plate in the illustration and is closed when the plate moves down.) These switches get dirty and out of alignment. Clean the switch contacts and adjust the switches so that they close and open when they should.

It is not common, but some pop bumpers have only one switch—the switch that activates the solenoid, not the second switch that activates the scoring. These one-switch pop bumpers are usually wired to a second pop bumper that does have two switches. Activating one activates both.

WEAK POP BUMPER

A weak pop bumper may be caused by the problems listed previously.

The return spring could also be affecting the power of the pop bumper. A different spring with less resistance might make the pop bumper more powerful.

If the machine is an old electro-mechanical, a more powerful coil or a conversion to DC power will make your pop bumpers come alive. See "More powerful coil" and "Converting to DC" in the "Repairs: Flippers" chapter.

DISASSEMBLING AND REASSEMBLING A POP BUMPER

Disassembling and reassembling a pop bumper is not a difficult job, just about anybody can do it, but it can take an hour, at least the first time you do it, and it will require desoldering and re-soldering wires.

Work on only one pop bumper at a time. Take it apart, fix it, put it back together, and test it out before starting another pop bumper. That way, if you screw up, you can examine the other pop bumpers to see what you did wrong. Pop bumpers are very visual. You can easily observe how they work. So having a correctly assembled pop bumper to refer to will be a huge help when you're hopelessly confused. You can also take photos or make diagrams to help you remember how things came apart.

Before you disassemble a pop bumper, examine it top and bottom to see if you can determine what parts might need replacing. Pop bumper parts are inexpensive, and it would be convenient to have all the parts you need before you take the bumpers apart, although you might not discover a broken part until the bumper is disassembled. And now you've

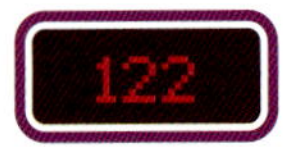

got to place a second order. If you do have a pop bumper apart and need to order parts, be sure to put all of the loose parts in a zip-lock bag or closed container. If you just leave parts sitting on the machine or on the workbench, the universe has a way of guaranteeing missing parts when you return.

Most pop bumpers are similar; their design, operation, and assembly are about the same from one manufacturer to the next. Here's what you do:

1. Remove the bumper cap and the lamp.
2. Raise up the playfield (take the balls out first) and prop the playfield up so you can work comfortably on the underside of the playfield.
3. Make sure the pop bumper you are about to disassemble is the one that needs repair. Sound silly? Some machines have several pop bumpers mounted close together. I've disassembled the wrong pop bumper more than once.

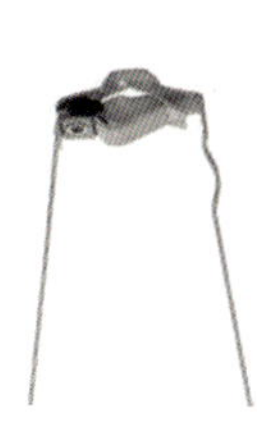

4. The light socket in the pop bumper has two long, thin metal pigtails (legs) that go through holes in the playfield and solder to wires on the underside of the playfield. Unsolder the pigtails from the wires and bend the ends of the pigtails straight so they'll slide through the holes in the playfield when you remove the pop bumper.
5. Unscrew nuts from the two metal shafts of the metal ring.
6. Unscrew the pop bumper body from the top of the playfield.
7. Remove the pop bumper assembly from the top of the playfield.

At this point you can replace any part of the pop bumper assembly:

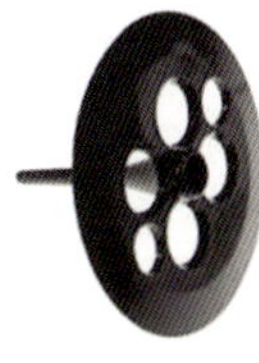

1. The plastic skirt if it is chipped. New skirts not only look much nicer, they might even improve the action because the ball is no longer hitting a chipped edge. Note how many holes are in the skirt (some skirts have four holes, some have six) and how the skirt aligns under the body of the pop bumper.
2. The clear flexible plastic disc, called a trim platter, that goes under the skirt (if your machine has one, many machines don't) and if you want to replace it with a new one (more on this below).
3. The bumper body if it is damaged or discolored.
4. The light socket if it is corroded and not making good contact with the lamp, or if the pigtails are broken off or too bent up.
5. The metal ring if damaged. Check to see that the two legs of the ring are parallel. The metal ring can bend easily but usually can be straightened out without having to replace it.

Next, clean and adjust the switches on the pop bumper assembly underneath the playfield, without removing the under-playfield bracket or parts.

If you need to replace missing or broken parts under the playfield, or if you want to disassemble the solenoid to clean or replace the plunger or coil sleeve, unscrew the pop bumper bracket from the playfield and disassemble all of the parts. Unless the solenoid is burned and needs to be replaced, you can let the solenoid dangle from the wires. There is no need to unsolder the solenoid.

Reverse the process to reassemble the pop bumper.

DISKS UNDER POP BUMPERS

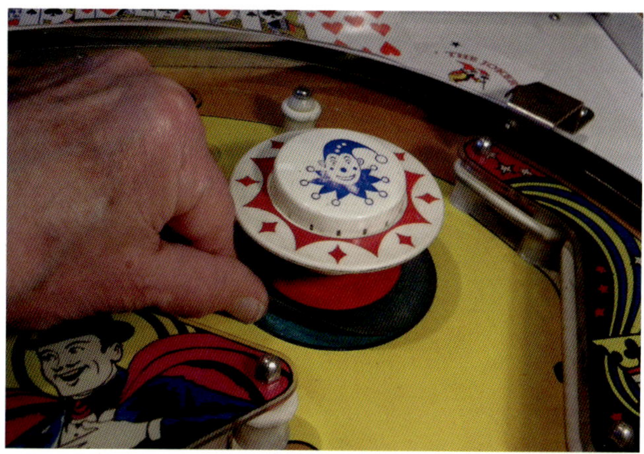

Some pinball machines have clear flexible plastic disks, called trim platters (also called shields), under pop bumpers to protect the playfield paint. These disks are often loose, and they get bent, discolored, and chipped. If you are ambitious you can purchase replacement disks, dismount the pop bumpers, and install the new disks. If you aren't so ambitious, you can carefully cut out the disks with a razor knife—don't scratch the playfield!—and not replace them. Give the playfield around the pop bumpers a good coat of wax every few months and it's not likely the ball will do any damage. I have cut off worn disks on many machines and never had problems with playfield damage.

POP BUMPER REBUILD KITS

Some pinball suppliers sell "rebuild kits" for pop bumpers for $5 or $10. The kits usually include a new plunger and coil sleeve for the solenoid, the yokes (the oblong brackets) that mount on the unit, a replacement spring, and nuts and screws. Rebuild kits do not typically include any of the pop bumper body parts, the mounting brackets, or the solenoid. If you're buying parts for your pop bumper anyway, buy the rebuild kit as well. Old yokes are sometimes broken, old coil sleeves are often worn out, old plungers might be tarnished or pitted, and are easier to replace than to clean. And it never hurts to have a new spring.

CLEANING

Cleaning is a double-edged sword. It can make a pinball machine much nicer looking, but if you are not careful it can damage a machine permanently. *Caveat Emptor*: Pay Attention.

SCORE REELS AND OTHER PAINTED PLASTICS

The paint or printing on plastic parts can come off or smear easily. Plastic parts from different machines and different manufacturers react differently. A cleaner that works great on one machine might damage or remove the paint on a different machine.

Whatever cleaning product you use, try a small or concealed spot first, just to be sure you won't remove paint. And go easy. A light application may be successful, while a heavily applied application might ruin the paint. Constantly watch for any smears, and examine your cleaning cloth for any signs of paint. Use a white or light colored cloth so it is easy to spot any paint you may be removing.

What cleaning agent does work? I've often used plain water on a tissue. I use alcohol: isopropyl 99% if you can find it, 91% otherwise. I use a cleaner I like called Mean Green, though it is toxic and can irritate your skin. Novus #1 is another popular plastic cleaner. I have had success—and disasters—with all of these.

No matter what product anyone recommends, no matter what you hear or read, you won't know if it works until you try it.

Playfield plastics: Playfield plastics often build up a thin coating of dirt on the bottom that might wipe off with a damp tissue or rag. However, the darkening may be due to overheating from the lamp underneath, which is permanent. Switching from #44 lamps to #47 lamps might prevent any new damage. See "Lamps" in the "Components and Features" chapter.

CABINET

The paint on cabinets is usually more durable than the paint on plastic parts, but not always. The same warnings apply: Try a small or concealed spot first. Use a white or light colored rag and be constantly alert for smeared paint coming off on your rag. Cabinets with smooth, undamaged paint usually can handle most any cleaner. Cabinets with rough or crumbling paint are easily damaged.

For cleaning cabinets I have used wood cleaners, furniture cleaners, and my trusty (well, sometimes trusty) Mean Green. I've used playfield cleaners (see "Playfields"). I've even used ammonia, with plenty of ventilation. The different cleaners sometimes work wonders, but sometimes take the paint off.

APRON (SCORE CARD HOLDER)

The apron is the large score card holder mounted at the bottom of the playfield. On older pinball machines, the apron was painted bright white with colorful lettering and the manufacturer's logo. Well, be warned: That lettering smears very easily when you clean the dirt off it. I have not found any sure-fire way to protect the paint. I suggest that you clean around the lettering and let it go at that.

POLISH CHROME PARTS

The chrome on a pinball machine, like the chrome on a 1965 Mustang or a classic Harley Davidson, needs to shine. A little chrome polish on the legs, the coin door if it is metal, and the rails can make your machine look mighty nice. Clean off the polish residue thoroughly so it doesn't wind up on your hands when you play a game.

If any of the metal parts are rusty, surface rust can be removed, or at least reduced, using Brillo or SOS pads (steel wool impregnated with soap). Rub the dry pads on the legs and then wipe off with a damp cloth.

Take off the leg bolts (only one at a time!) and polish the heads with a buffer wheel. The plunger handle can also be polished, or what I do is remove the plunger and use a buffer wheel to shine it up. While it's off, treat yourself to a shiny new barrel spring—the spring mounted between the handle of the plunger and the cabinet.

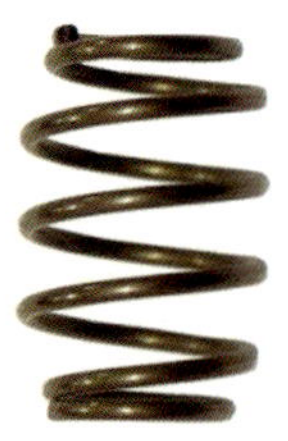

Screw heads on the playfield look really sharp if they are shiny. Buy new screws at a hardware store if the screws are a standard size and shape. If the playfield screws are a special design, you can polish them using a bench grinder. Remove the screws, grab the screw threads with a pair of pliers, and polish the head with a wire wheel. Be careful here: If you grab the screw too tightly you can damage the threads. If you grab the screw too loosely, it will go flying off when it contacts the grinder wheel.

PAINT TOUCHUP

Missing paint detracts from the appearance of your pinball machine. It would just take a little touch up: a dab of paint on a small brush, or even a Sharpie, and what a nice improvement.

Well, as Elvis Presley sang back when rock & roll—and pinball machines—were young, *Don't*. If you are an artist experienced with mixing and applying paints, you might be able to match the old colors and do a passable job of

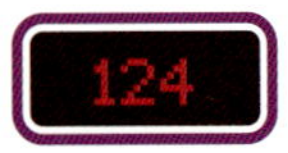

improving the look of your playfield or cabinet or even a peeling backglass. But for most of us amateurs, the cure is too often worse than the disease. The color match is never perfect, the edges are sloppy, and it winds up looking a lot worse than when you started. And touch up jobs will often reduce the value of your machine. Pinball collectors prefer original paint to touch up.

This is especially true of cabinets. I sometimes get pinball machines in my shop that have completely repainted cabinets. People don't like the scraped or faded paint and decide to hide it all with a shiny new coat of paint. Instead of trying to save or recreate the original graphics, the cabinets are repainted a solid color, or painted with some glittery or spiderweb effect—quite often a nice job, but always greatly reducing the collector value of the game.

Instead of repainting, try a full-on cleaning and wax job. If you use the right cleaning materials and take the time—and it will take time—you might transform your machine in just an afternoon.

> Like any machine worth having, an abused or neglected pinball machine can become one of the frustrations of living you play pinball to forget.
>
> —Jim Tolbert, *Tilt*

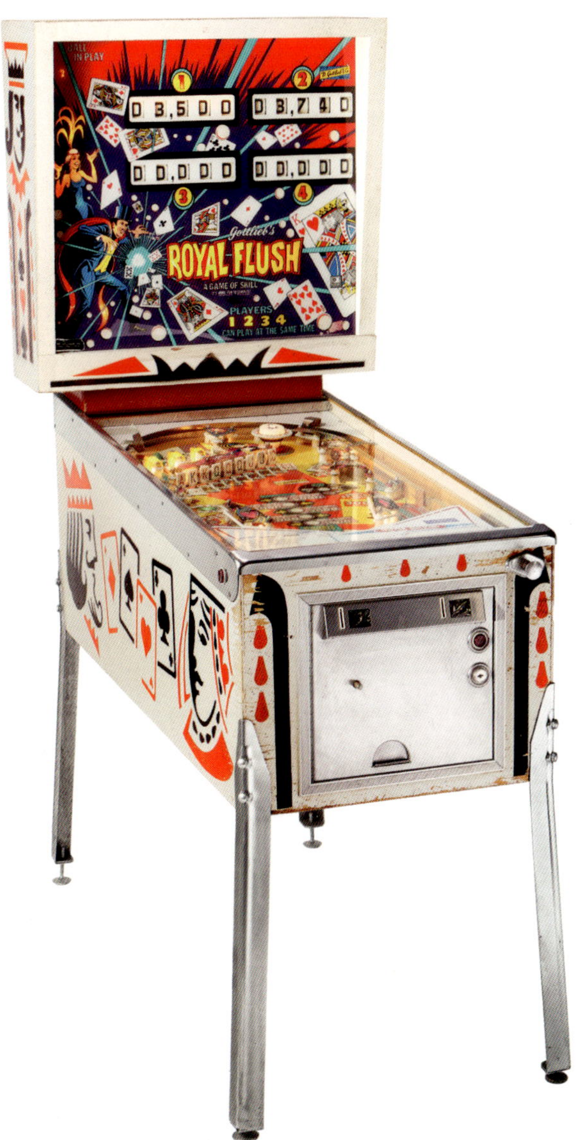

Gottlieb **Royal Flush**, 1976. One of Gottlieb's most popular machines, over 12,000 produced. Gottlieb also made a two-player version called Card Whiz, and an electronic version, Royal Flush Deluxe. Gottlieb had previously built a pinball machine called Royal Flush in 1957.

CHAPTER 11

REPAIRS ELECTRO-MECHANICAL MACHINES

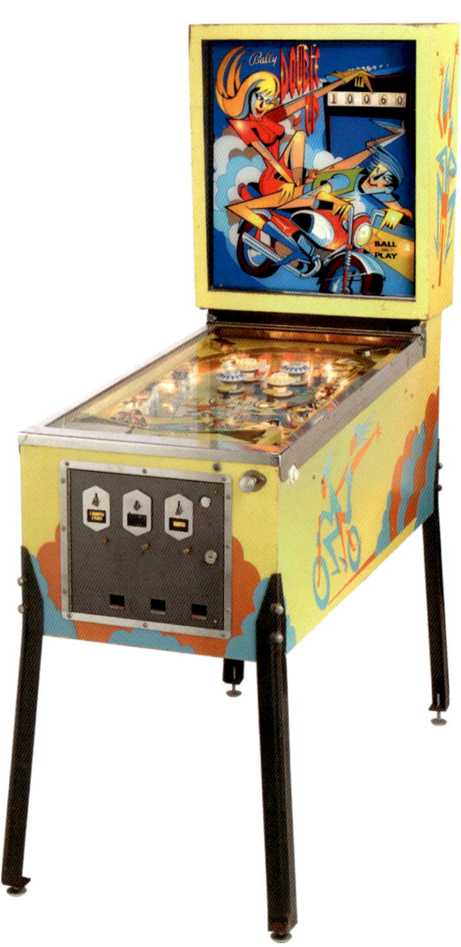

Bally **Double-Up**, 1970. If you've never repaired a pinball machine, single-player machines are much less complicated than two player and four player machines. Usually.

The last electro-mechanical pinball machine was made more than forty years ago, yet there are still thousands of electro-mechanicals out there still being played, still looking great, and still working great. And anyone with any degree of intelligence, perseverance, and some spare time—well, maybe a lot of spare time—can keep an old electro-mechanical machine in good condition.

The repairs in this chapter are in addition to the repairs previously covered in the chapters "Repairs: Flippers," "Repairs: Basics" and "Repairs: All Machines." The repairs in this chapter are only for electro-mechanical machines. Repairs for electronic machines are covered in the next chapter.

SELF-CORRECTING PROBLEMS

Some electro-mechanical problems go away all by themselves. If your electro-mechanical machine has been working and suddenly goes on the fritz, shut it off. If the motor keeps running, or the score reels keep turning, or solenoids keep firing, shut it off. Then turn it back on again.

Very often a switch, a rotating unit, or a relay is stuck, and simply turning the machine off and on again will fix the problem. I was playing a 1975 Gottlieb Pioneer, and right in the middle of a game the motor started running and wouldn't stop. Whirr, click, click, click. I turned the machine off and back on, but the motor kept on running. I tried it a third time. No luck, the motor kept on running. I was too busy that day to open up the game and fix it, but a few days later, I turned the machine on to check it out and repair it and, well, it worked fine. The problem was gone, whatever it was, and it hasn't come back.

If a problem persists, turn off the machine. Don't let a stuck coil stay on, don't let the motor or the score reels run and run, and don't ignore unusual noises. You are likely to burn out something.

TROUBLESHOOTING: GO THROUGH EVERYTHING (WELL, ALMOST EVERYTHING . . .)

This manual is full of suggestions for fixing specific problems with electro-mechanical machines. This is the approach most people take: Isolate one problem and fix it. Then go on to the next. Eventually the machine is working.

There is another, much more thorough approach, and in many ways a much easier approach. Don't struggle with problems you don't understand and problems you need to trace. Don't struggle with schematics that look like a street map of Tokyo. Just go through the entire machine and check everything. Most problems with electro-mechanical machines are out-of-adjustment switches, dirty switch contacts, rotating units that are gummed up, broken wires, and poor connections. If you took a few evenings and went through every component in the machine, you will probably solve most if not all of the machine's problems. In fact, any dedicated professional pinball repair shop will do just that: Go through the entire machine.

You don't need much electrical knowledge and you don't need to understand circuits. Electro-mechanical machines are very "visual." You can look at a switch or a relay and see it opening and closing. You can look at and hand operate rotating units and see if they are moving smoothly or sticking. You can see all of the solder connections and easily spot broken or disconnected wires.

Just take it a little bit at a time. Start with the playfield. Unplug and remove the playfield from the machine, turn it over, and set it carefully on a workbench, or a pair of saw horses, or across the top of the cabinet (be sure it's not resting on a bumper cap or other breakable part) and go through everything. Then unscrew and disconnect the component board from the bottom of the cabinet, lift it out, and do the same. Then remove the back door from the backbox and do the same.

Take your time. When you disassemble a component, make a diagram or take photos of how everything came apart so you can figure how to get it back together. Have a container to hold small parts. Don't put loose screws, clips, and springs on the board you're working on or on the cabinet. They fall off, they roll away, and they vanish. One more suggestion: After you examine a component and make any adjustments or repairs, mark each component with a Sharpie to keep track of what you have and haven't already worked on. Maybe you have a better memory than I do, but I am lost without my "reminders."

This is what you need to do for each component:

1. File all of the switch contacts, and examine every switch to see if the contacts are opening and closing. Adjust if necessary. Most of the switches in the machine will not be out of adjustment. Only one in twenty, or maybe one in thirty, will need adjustment. So if you find yourself adjusting every other switch, stop! You are probably doing something wrong (see "The Warning").
2. On all switches, check for broken or disconnected wires, broken switch blades, and missing or loose contacts on the switch blades. Make sure the screws holding the switches in place are tight. Replace any missing screws.
3. Disassemble and clean every solenoid. Check the coil stop, plunger, linkage, and coil sleeve for wear. Check for broken wires. Tighten loose screws and replace missing screws. If a screw is loose and will not tighten try a fatter or slightly longer screw, but be careful that a new screw is not too long and comes through the top of the playfield!
4. Clean all rotating (stepper) units such as score reels, bonus units, player units, ball count units, match units, and the credit wheel. Make sure rotating contacts are clean and making good contact. Make sure the units rotate smoothly without getting stuck halfway through a step-up or step-down. Don't disassemble the units unless absolutely necessary; you'll about double your work taking apart and putting together these elaborate contraptions. And all you have to do is lose just one tiny part . . .
5. Clean the large connectors (wiring harnesses) that you disconnected to get the machine apart. Use a wire brush on the male pins. Check that no wires are broken off.
6. Put it all back together again and see if the game will play.

All of the above cleaning and adjustments are covered in the Repairs chapters.

AND NOW . . . THE WARNING

There are many seasoned pinball repair people who think my suggestion to "go through everything" is a very bad idea. They think it is too easy to make things worse instead of better, to misadjust switches that did not need adjusting in the first place, or to accidentally knock your elbow on a stack of switches and start a chain reaction of bad-to-worse.

I will leave that decision up to you. Gauge your own ability to work in tight spaces, to have the patience to do a good job and to be very careful in your work. Besides, if you have a non-working machine and there is no one in your area who can fix it, isn't it better to try to repair it than to turn it into a very large room decoration?

One genuine warning regarding motor switches—those dozens of switches mounted around the motor on Gottlieb machines, or on top of the motor on Bally and Williams machines. Motor switches not only open and close, they open and close in a sequence where timing of the switch operation, one switch opening a fraction of a second before or after another switch opens, is critical to the operation of the machine. The switch gap—the distance between switch contacts when the switch is open—can affect this timing. A motor switch with too big a gap may not close fast enough to allow the machine to complete its sequence of actions, making the game unplayable. So unless you see an obvious problem with a motor switch, this is one component you probably want to leave as is.

MOTOR STUCK OR SLUGGISH

If a machine has not been started for a few years, the score motor may have trouble rotating, the moving parts impeded by dirt, or rust, or dried lubricant. Score motors almost never need to be replaced, just a little help. If there is a place to lubricate the motor (see the chapter "Lubrication"), try that first. Sometimes, just attempting to start a game several times will free up the motor parts and gears. Try gently using your finger to help the motor rotate. Just don't force anything, and don't push on any of the switches. Eventually, the motor will respond.

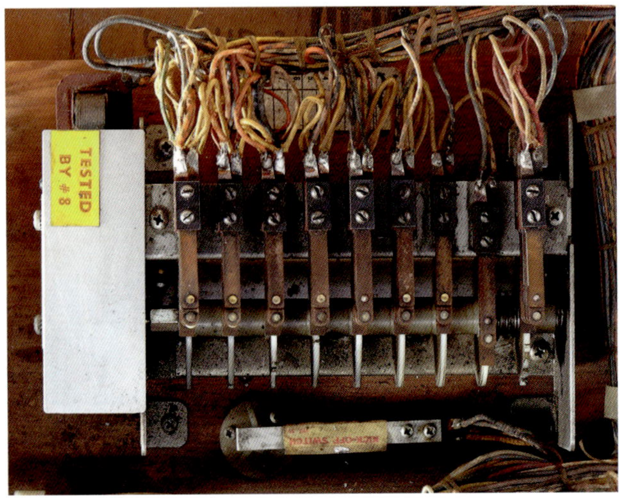

Bally and Williams motors had the switches mounted on top, making them fairly easy to reach and adjust.

MOTOR RUNS CONSTANTLY

If the score motor runs constantly, most likely you have a shorted or stuck switch somewhere in the machine. The problem switch is not likely to be on the motor itself, good news if you just read the previous warning. Check all switches under the playfield, and switches wired to solenoids, targets, rollovers, outholes, and kickout holes. Check the playfield switches that are directly behind rubber rings. You may have to take off the plastic shields covering the switches to see them, but I've often found these switches bent up and shorted (closed). If you find a shorted switch, adjust it and see if your problem goes away.

Stepper (rotating) units that don't reset can sometimes cause the motor to run continually. Stepper units such as bonus units, player units, ball count units, and match units get dirty and gummed up and will stop rotating, getting stuck part way through the reset cycle (see "Rotating Units").

Stuck score reels are often the culprit...

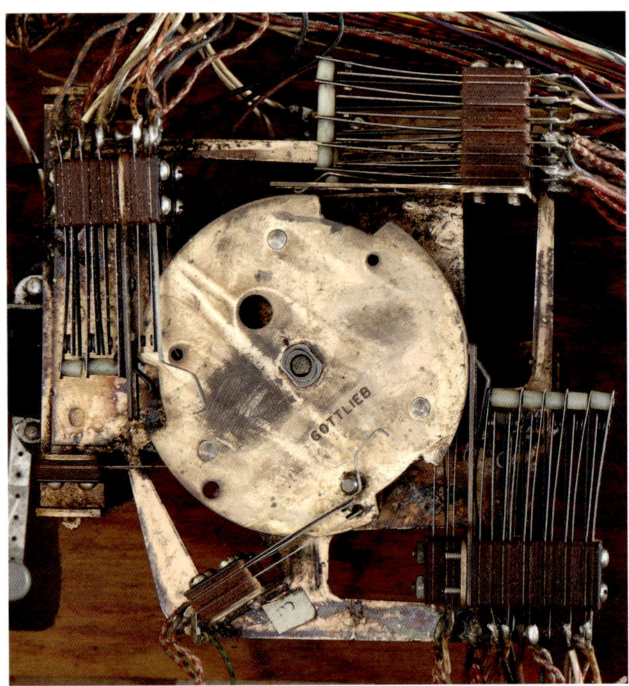

Gottlieb score motors were designed with horizontal cams resembling a carousel, with switches mounted around the motor, and stacked on top of each other, making some of the switches very difficult to get to.

STUCK SCORE REELS

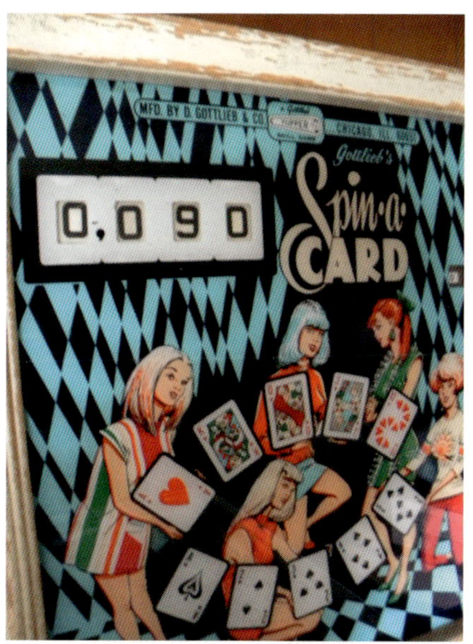

A score reel stuck on 9, a common problem as reels get dirty, and springs don't pull like they should. When a score reel moves from 9 to 0 all six switches open at the same time—a struggle for dirty score reel mechanisms.

Score reels that do not reset to zero will often cause the score motor to run constantly.

Each score reel is a separate unit, called a "drum unit." The drum units snap in and out of their mounting brackets so you can work on them easily.

Remove the non-working score reel drum unit from its mounting bracket. Reset the score reel by hand to zero (hand-operate the solenoid on the score reel). If this stops the motor, you found your problem. The score reel mechanism may be stuck, usually due to dirt or grease, but sometimes due to a broken part on the drum unit. Look for a broken switch or a broken wire or a missing spring.

Score reels usually have to be disassembled to be cleaned. Be careful. These drum units have tiny clips, screws, washers, and springs that are easy to lose and easy to incorrectly reassemble.

By the way, score reels often get stuck on the 9. As a reel turns from 8 to 9, several switches on the unit open and close all at once. Years of grime buildup resists the rotation. A good cleaning, and maybe tightening or shortening the spring, should fix the problem.

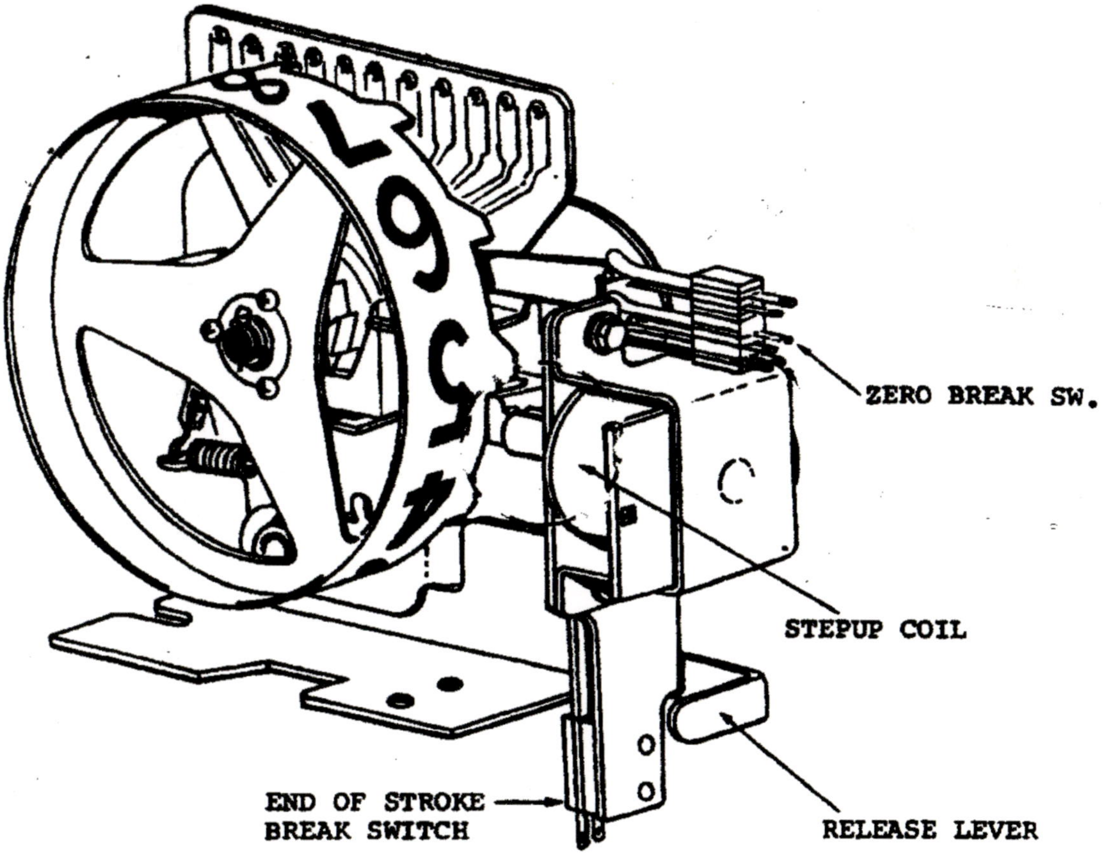

ZERO BREAK SW.
STEPUP COIL
END OF STROKE BREAK SWITCH
RELEASE LEVER

Electro-mechanical drum unit: score reel and assembly. These units snap in and out for cleaning and adjustment.

If the drum unit seems to be okay, one or more switches on the score reel may be out of adjustment—not opening or not closing—which will also cause the score reel to stop moving and cause the motor to run continuously. Operate the score reel solenoid by hand and watch the switches opening and closing. If you are unsure how the switches should operate, take a look at a different score reel, one that did reset to zero, and see how the switches open and close on it. Adjust the switches if they are out of adjustment. Clean the switch contacts.

SCORING PROBLEMS: SWITCHES MOUNTED ON SCORE REELS

Score Reel Switch Chart

"Top" refers to the switch or pair of switches farthest from the coil

"Number" is the number on the score reel showing in front (showing through the window in the backglass)

Switch location on stack →	Top	Middle	Bottom
Williams and Bally			
Reels with three stacks:			
Number 0	open	open	open
Numbers 1-8	closed	closed	open
Number 9	closed	closed	closed
Highest reel (two stacks):			
Number 0	open		open
Numbers 1-8	closed		closed
Number 9	closed		closed
Gottlieb			
Reels with three stacks:			
Number 0	open	open	closed
Numbers 1-8	open	closed	open
Number 9	closed	closed	open
Highest reel (two stacks):			
Number 0	open		closed
Numbers 1-8	closed		open
Number 9	closed		open

As score reels turn, a stack of switches mounted to each score reel opens and closes. Each score reel, except the highest number reel, has a switch stack three high: three pairs of switches (sometimes one switch instead of a pair) mounted one on top of the other. The highest number reel—the reel showing the 100s, or 1,000s, or 10,000s, whatever the highest number is on the machine—has a stack of switches two high. (Have a look at the photograph and then re-read the paragraph.)

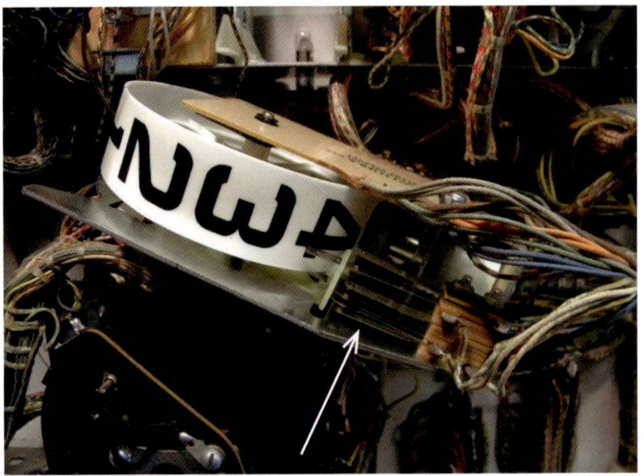

Three sets of switches, one on top of the other, that open and close as the score reel turns.

These switches sometimes get out of adjustment causing problems with the scoring and the operation of the machine. Here is how you can fix the problem:

One score reel at a time, remove the score reel unit from its mounting bracket (it snaps in and out), operate the solenoid by hand, and watch the switches open and close as the reel rotates--that is, if the reel rotates: As I mention elsewhere, score reels often get stuck and need to be cleaned. So first, get the reel to rotate smoothly before looking at the switches.

When the score reel shows 0 the switches are in one position, some open, some closed. The switches are in a different position when the score reel is at 1 through 8. The switches are in a third position when the score reel is at 9. The accompanying chart shows which switches are open and closed for the different numbers (0 through 9) on each score reel. Bally and Williams have the same switch positions. Gottlieb (of course) has a different sequence.

Note that there is also a single switch mounted away from the stack, that opens and closes every time the score reel turns, probably a hundred times every game. This switch, called the end-of-stroke switch (not to be confused with the flipper end-of-stroke switch), can get dirty and out of adjustment, making poor contact. Sometimes one of the switch blades on this switch will snap off and need to be replaced. It only takes a minute to check this switch while you have the score reel out.

If this is too confusing, don't bother with it for now. Don't try to adjust switches if you don't fully understand how they work. It took me a month working on machines before I finally understood the operation, and even then I could never remember which switches should be open and which should be closed. And I still can't. I refer to the chart every time I check the switches. The chart has helped me many times.

SCORE RELAYS

If the score reel switches seem to work correctly, the switch problem may be on the relay for that score reel. The relay is usually mounted in the backbox, hopefully labeled: 10 point relay, 100 point relay, 1,000 point relay. Note that the score relay is not mounted on the score reel (drum) unit. The relay is mounted on the large board that also holds the other components in the backbox.

Operate the relay by hand and make sure the switches on the relay open and close. Adjust and clean the switches. With luck this will solve your score reel reset problem and your running motor problem.

While checking the score relays make sure none of them are activated, stuck in the "on" position. Score relays should only be activated momentarily, when a ball hits a target or rolls over a switch. A constantly activated score relay means there is a shorted or closed switch somewhere on the machine, most likely on the playfield, and this switch is the probable cause of the running motor. Examine all scoring mechanisms on the playfield: targets, rollover switches, pop bumpers, kickout holes, spinners, kickers, and the switches behind rubber rings. Chances are quite good you will find a switch that is out of adjustment and stuck closed, with switch blades touching when they should not be.

POOR CONNECTIONS

The playfield, the cabinet, and the backbox are wired together by means of large quick disconnect wiring harnesses, also called connectors or Jones plugs, each holding as many as twenty wires. These connectors unplug so that the machine can be disassembled for travel and for repair. The male pins on the connectors will get dirty and corroded, causing problems. Very often when something isn't working, the problem is with a poor connection on one of these connectors. Carefully unplug each connector. Grab both ends of the connector and ease or wiggle the connector off. I repeat this elsewhere, but it is important that you don't pull on just one end of the connector, because you could break the connector in half. Clean the pins with a wire brush or steel wool and then reconnect.

Chicago Coin machines: The last electro-mechanical machines made by Chicago Coin (Chicago Dynamics) around 1976 did not have the old-fashioned wiring harnesses. They had plastic snap-together plugs like the plugs in electronic machines.

SOUPING UP AN OLD MACHINE: TRANSFORMER HIGH TAP

Old electro-mechanical machines often lack zip. The flippers, pop bumpers, kickout holes, sling shots, and all the solenoid-powered items on the playfield could use double bypass surgery. Even after cleaning and adjusting all the switches, cleaning the coil plungers, waxing the playfield, and getting a shiny new ball, the game may still be too slow for your enjoyment.

There is a quick and easy wiring change you can make on the machine's transformer that will increase the voltage to the coils by about 3 to 6 volts that will make the coils hotter (figuratively and literally) and the action faster.

On most electro-mechanical machines, two of the transformer's lugs, usually next to each other on the top of the transformer, are marked "Normal" and "High." On some transformers the taps are numbered, left to right, 2, 4, 8, and 10. 2 is the High tap, 4 is the Normal tap. The pinball machines came from the factory with the connecting wire soldered to the Normal tap. The High tap had no connections to it.

The Normal tap was meant to be used for regular 110–120 volt current. But back when arcades had ten or twenty pinball machines in a row, all drawing off the same wiring

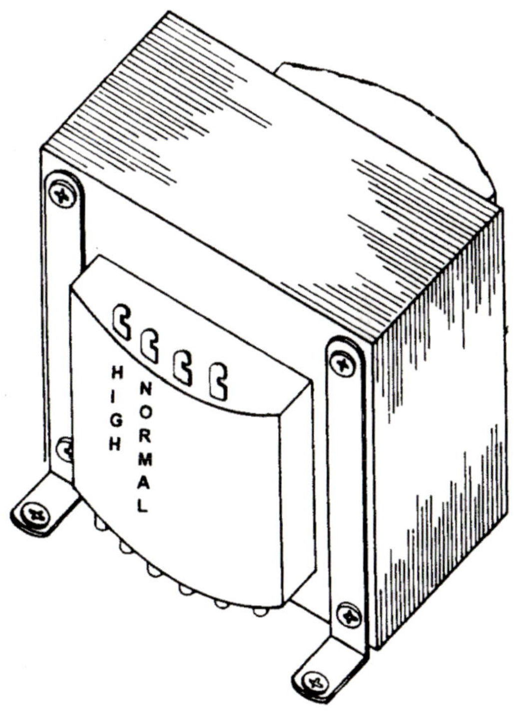

Most electro-mechanical transformers had four lugs at the top and were labeled. On Gottlieb machines left tap was High tap and the second from the left was Normal. Bally had the same configuration, but used the number 2 to designate High tap and the number 4 to designate Normal. Williams transformers were reversed: The right tap was High tap (labeled High Tap) and the second from the right was normal (but not labeled "normal." It was labeled "24 Volts").

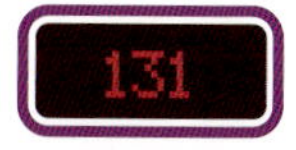

circuit in the building, the machines towards the end of the line sometimes experienced a voltage drop. The High tap on the transformer would compensate for the lower line voltage coming into the machine. You simply unsoldered the wire from the Normal tap and soldered it to the High tap.

The manufacturer's instructions warned people not to move the wire from the Normal tap to the High tap unless the line voltage was low, for fear of damage to the components. But after years of playing, as these old machines get older and slower (or seemed to, compared to today's faster games), switching the wire from the transformer's Normal tap to the High tap can bring new vigor to old machines without damaging the games. The coils do operate hotter, generating more heat, and are more likely to burn out if they get stuck in the on position or if a flipper is held in too long, but many people feel the resulting improvement in play is more than worth the risk.

I high-tap every electro-mechanical machine that comes in my shop. Sometimes I notice a dramatic improvement in performance, but sometimes it is subtle or not noticeable at all. I've yet to have a machine come back due to damage from the High tap. The High tap does not affect the lights, which are on a different transformer tap than the machine's operating components.

This is one of those "it worked for me" suggestions. Use your own judgment. But be sure to unplug the machine from the wall before getting near the transformer. The high tap has 33 volts or 57 volts coming out of it, depending on the machine, enough to give you a shock. The other side of the transformer has 120 volts going into it—enough to kill you.

Important Warning: On some transformers, the lug on the bottom of the transformer, where the power cord attaches, is labeled "normal." This is **NOT** the "Normal" tap you want to unsolder. This is 120 volts—line voltage—coming into the transformer. Do not move this connection. **Do not touch it when the pinball machine is plugged in.**

Chicago Coin transformers: Many Chicago Coin (Chicago Dynamics) machines had a plug mounted on the floor of the cabinet that let you high-tap their transformers without doing any rewiring. The plug had two positions. Sometimes the plug was labeled 30 volts (Normal tap) and 33 volts (High tap). Sometimes the plug was labeled "Playfield Action: Normal" (Normal tap) and "Playfield Action: Stronger" (High tap). Some Chicago Coin machines did not have the plug, but there is an unmarked High tap on the transformer. The top right lug, with a white wire, is 30 volts (Normal tap), while the bottom lug directly below is 33 volts (High tap). But *DO NOT* rewire these transformers unless you know for sure which tap is which. And don't do anything with the machine plugged into the wall. Some of the wires on that transformer are line voltage.

RELAYS: BUZZING, LOUD HUMMING

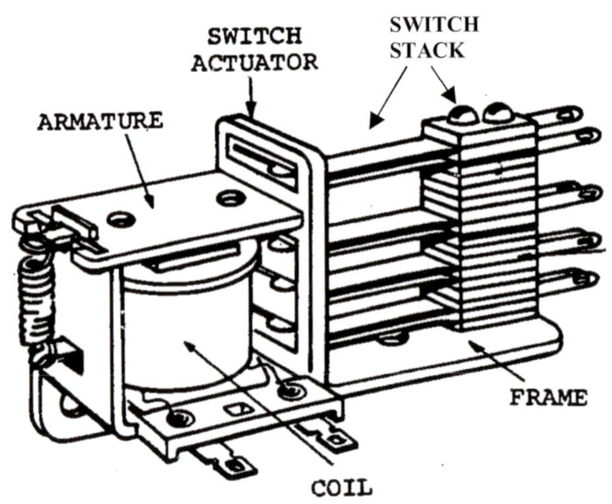

When a relay is activated, it pulls a metal hinged plate called an armature against it. The armature, in turn, opens and closes one or more switches. Remember that a relay is a wire coil with a solid center; a solenoid is a wire coil with a hollow center.

Relays buzz when the armature vibrates against the metal core of the relay. This is not harmful and does not usually affect the operation of the game, but it can be annoying. Some relays are activated only briefly, but some relays may be activated for the entire game or activated when the game is over. Those that are activated for long periods of time, and buzz or hum, can be very annoying. Here are some possible solutions:

1. Some relays buzz because they are not mounted securely to the cabinet. Tighten all hold-down screws and replace missing screws.
2. Examine the relay to see if the armature is seated properly and if the spring is attached properly. If you don't know what to look for, examine other relays in the machine to see how they work. Check for dirt or debris that may be preventing the armature from moving as it should.
3. Look at the armature where it contacts the coil. If the armature shows an indentation, usually resembling a half moon, this will cause buzzing. Remove the armature and smooth out the indentation with a grinding wheel, or replace the armature if you can obtain a replacement.
4. Too much spring tension can also cause buzzing relays. Try replacing the spring with one that is slightly longer or that has less strength.

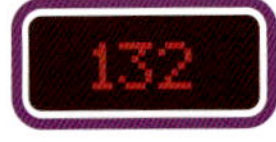

5. If none of the above remedies work, try this: Put a piece of a self-sticking paper label or electrical tape on the armature where it contacts the relay core. The paper or tape will serve as a damper, quieting the noise. The paper or tape is thin enough that it shouldn't affect the magnetic pull on the armature.
6. Another possible solution to buzzing relays: Replace the coil. Relay coils have a brass or copper strike plate where the armature contacts the coil, quite often in the shape of a half moon. If that plate comes loose or wears down it can cause buzzing. A new relay coil may solve the problem.

If you replace a relay, notice that the relays in your machine are mounted to their brackets with a copper or brass screw and a copper or brass washer between the bracket and the relay. The copper or brass magnetically isolates the relay from the bracket, helping prevent a problem called "residual magnetism" where the relay stays magnetized and acts as if it is always activated, causing all kinds of problems in the operation of the machine. Be sure to use the brass or copper screw and washer on the new relay.

COIN DOOR RELAY

One very annoying buzzing relay is on the front coin door. This relay is activated the entire time the machine is turned on, whether a game is playing or not. This is the coin lock-out relay. The purpose of this relay is to open a lever that allows a coin to drop into the coin box inside the machine. When the machine is turned off completely, this coin lock-out relay deactivates. It releases the lever on the coin mechanism so that anyone who drops in a coin will automatically have the coin routed to the coin return slot. The manufacturers did not want people losing their coins when the game was turned off. Angry people make for damaged pinball machines.

Over the years the coin lock-out relay will often start to buzz a lot, and it may not be that easy to quiet. If the machine is for home use you can simply disable the relay by disconnecting or cutting one of the wires to it and silencing it. If you do this, be sure to rig the relay so the armature is held against the relay. A piece of electrical tape securing the armature against the coil usually works. That way the coin mechanism, if there still is a coin mechanism in the machine, permanently accepts coins. If there is no coin mechanism the coin relay is doing nothing anyway. Instead of disabling the relay, you can remove the relay and the linkage parts. If you do remove anything, save the relay and the parts in case a future owner wants to put them back. Some pinball collectors like everything to be original. Tape or insulate the end of any cut wires so they do not accidentally touch part of the metal door and blow a fuse or cause a short.

Coin Door Shock Warning: On many pinball machines from the 1960s and earlier, there was full 120 line voltage to the coin door. Unplug these machines before opening the coin door. Be very careful when cutting and insulating the wires. If you do not know which wires are which, get an electrician or a professional pinball repair technician to do the rewiring. Read more about this important issue in the "Repairs: Basics" chapter.

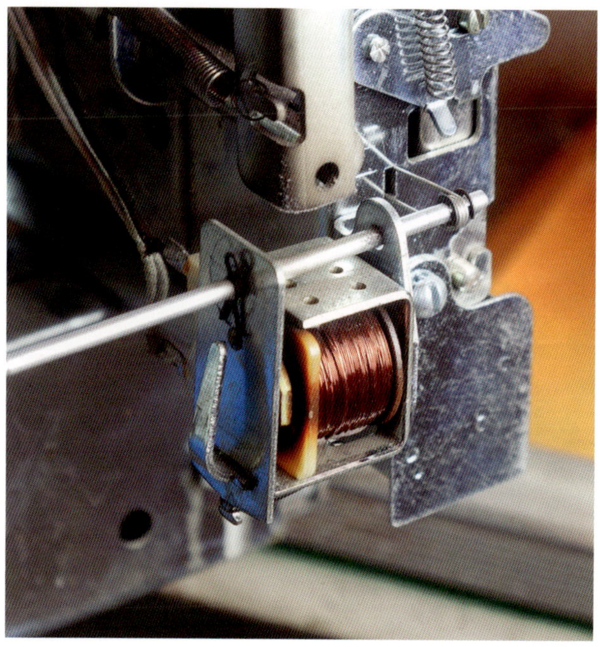

Coin door relays had different ways of operating the coin lock out, but always had some sort of **sliding bar** to move a chute that returned quarters to the player when the machine was turned off.

CONSTANT BUZZING OR LOUD HUMMING

A relay or solenoid making a continuous buzzing or humming sound indicates a coil that is constantly activated. As I mentioned, there are a few relays that are constantly activated, but most relays are not constantly activated. No solenoids are constantly activated (with a few rare exceptions in some 1950s machines).

If a coil is making a noise, check on it. If the coil is activated when it shouldn't be, there is a stuck or shorted switch somewhere and the coil can burn out. Shut the game off until you can determine where the problem lies.

RELAYS WITH LAMPS

If you look at the rows of relays mounted on the floor of the cabinet, you may see one or two relays that have light sockets and miniature light bulbs mounted on the relays. Chances are these relays have the word "delay" in their description.

The lamps on these relays are not for illumination. They are #455 bulbs, flashers. They turn on when current is first applied to them, and turn off one or two seconds later.

The flashers were used on these "delay" relays as part of the operating circuit. When the relay was activated, the bulb lit up. As soon as the bulb turned itself off, it opened the circuit to the relay. The relay stayed energized (closed) for the few seconds the bulb stayed lit. That's what "delay" refers to. The light bulb delayed the circuit from opening by the amount of time the bulb stayed lit.

There are different reasons delay circuits are used in pinball machines. Usually the delay allowed a score motor or a series of scoring sequences to go through a full cycle without being cut off in mid-operation.

Here is the problem. Not all flasher lamps, even brand new ones, operate quite the same. Some flashers do just what they are supposed to do: light up right away and turn off in a couple seconds. But some flashers do not turn on immediately. Some stay on more than a few seconds. Flashers that take time to turn on and flashers that stay on too long will cause trouble in delay circuits. If you have delay circuits in your machine, you may have to try two or three lamps before you find one that works correctly.

These #455 lamps are the same ones that flash on and off behind the backglass. In the delay circuits the bulbs do not flash on and off. They light once and turn off.

UNMARKED RELAYS (MISSING LABELS)

Just about every old electro-mechanical machine has a few unmarked relays where the labels fell off. If you take the time you can identify these relays. Start by making a list of every relay in the machine that still has a label. Make a diagram of the relays, their physical location, and write down each description: name of the relay or code letter.

Let's say you have 25 relays in the machine: 20 are still labeled and 5 are not. Compare your list (the 20 relays you have identified) with the complete list of relays on the schematic. Check off your 20 relays on the list in the schematic, then note which 5 relays are listed on the schematic that no longer have labels.

For each of those five relays, look on the schematic to see what wire colors are connected to each relay. Most likely each relay coil will have different colored wires connected to it. Then look in the machine, match up the wire colors, and you should be able to identify the relay. Sound like a tedious job? It is, but it will be a huge help if you or a repair person is trying to troubleshoot a problem.

You could also operate each unmarked relay by hand and see what they do, but keep in mind that most relays control more than one operation, activating more than one circuit.

MATCH (0-9 UNIT)

When you own your own machine, getting a match and winning a free game is not always important. So when the match unit (also called the 0–9 unit) finally quits working, which they often do after years of constantly rotating, most people ignore it.

On some machines the match unit serves a purpose, activates a circuit, in addition to the match. Alternating bonuses and alternating extra point lights are sometimes controlled by the rotating match unit. If your machine has a so-many-points-when-lit light that is always on, or always off, a stuck match unit may be the culprit.

The match unit is sometimes inside the backbox, sometimes under the playfield. Some match units are large rotating units, 4–5" in diameter, and very reliable. But many match units are a small rotating assembly, 2–3" in diameter, and they often get stuck or jammed.

The small-diameter match units had two different designs. On some of the small match units, on each side of the

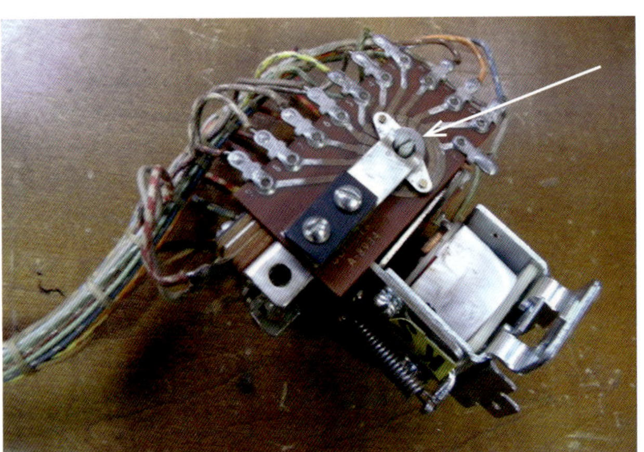

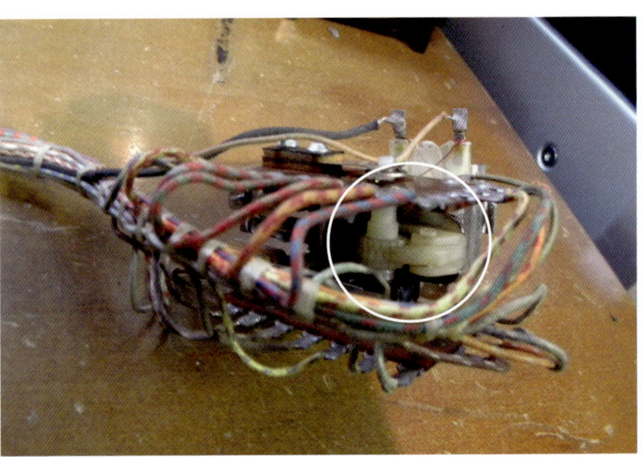

Outside and inside views of a typical Gottlieb match unit. *Left:* Small rotating arms, one on each side of the unit, were held in place by pressure from the flat metal blade. *Right:* The tiny plastic rotating gear often gummed up or jammed or lost its spring.

unit, there is a little metal arm that rotates on a bakelite (fiber) board with ten or more contacts on it. On other small match units, there are rotating plastic discs mounted on the inside of the unit.

There is a gear wheel in the middle of the unit, usually white plastic, and a small plastic lever operated by a relay that moves the gear wheel one notch at a time. Every time the ball hits a 10 point bumper, or maybe a 100 point bumper, the match relay is energized and moves the match wheel one notch. Each notch is wired to a different match number (00, 10, 20, 30, etc.). So during a game as the ball hits scoring switches, the match wheel rotates, until sooner or later it no longer rotates, staying stuck on the same match number game after game.

The rotating gear wheel gets very dirty over time, and sticky, and needs to be cleaned. Try alcohol on a Q-Tip. Don't use WD-40, as it can damage the plastic. (The WD-40 company says that WD-40 is safe on certain plastics but not on others. Since I don't know one plastic from the next, I just try to avoid using anything that might cause problems.)

You may have to remove the match unit (some unplug just for this purpose) and disassemble it to clean it, not fun but not difficult. Make a diagram or take a photo of all the parts you remove, where they go, and in what order.

Quite often the problem is the little plastic lever that turns the gear wheel. It can slip off the gear wheel, it might be dirty and not grabbing the gears, it might be missing the spring that keeps it on track, or it might be worn down and needing replacement. It's one of those simple repairs that can sometimes take an hour of struggling. Are you sure you want the match unit working?

ROTATING (STEP UP, STEP DOWN) UNITS

Every electro-mechanical machine has rotating "stepper" (step up, step down) units: ball count, player advance (first player, second player), bonus increase, and maybe other operations. The terms rotating unit, stepper, stepper unit, and step up/step down unit are used interchangeably.

There are two common designs: mechanisms about 5" or 6" square made of bakelite, with a rotating metal "wiper" unit looking like a brass spider or spread out fingers (left illustration); or a rotating, flat bakelite unit with spring-loaded contacts (right illustration).

Take a look at one of the rotating units in your machine. The unit will have one or two solenoids. One solenoid, the step-up solenoid, moves the metal fingers or the rotating disc one click at a time, one step around a circle of metal contacts. The other solenoid, if there is a second solenoid, is the step down solenoid, and resets the rotating part to start—sometimes one step-down at a time, sometimes all at once, woosh.

ROTATING UNIT WILL NOT ROTATE

Operate each solenoid by hand. Push in the plunger and see how well it operates, how well it moves the unit. A solenoid that does not move at all may be burned (see "Burned coils"). More likely the solenoid is fine, but the mechanism will not respond due to dirt built up over the years. You will need to clean the moving parts of the unit, wherever the moving parts are connected or jointed or making contact.

Sometimes all you need to clean a rotating unit is rubbing alcohol on a tissue or Q-Tip. If the levers and connections are metal you can apply a small amount of WD-40, but be careful not to get it on non-metal parts. Have a tissue ready to catch any excess drips. (If you do not regularly use WD-40, practice a few times on something other than your pinball machine until you get the feel for how much pressure to apply to produce a small spray.)

If the gunk buildup is thick—and on some old units it really is—scrape it off with a knife blade. I have seen rotating mechanisms where the levers were stuck together so tightly that it took repeated cleaning and wiggling the levers back and forth to free them up.

UNIT MAKING POOR CONTACT

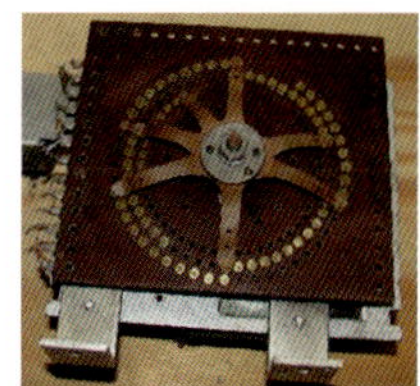

Common rotating "stepper" units. They "step up" and "step down" the ball count, player, bonus, and other features. A "wiper" unit is on the left. A spring-loaded unit is on the right.

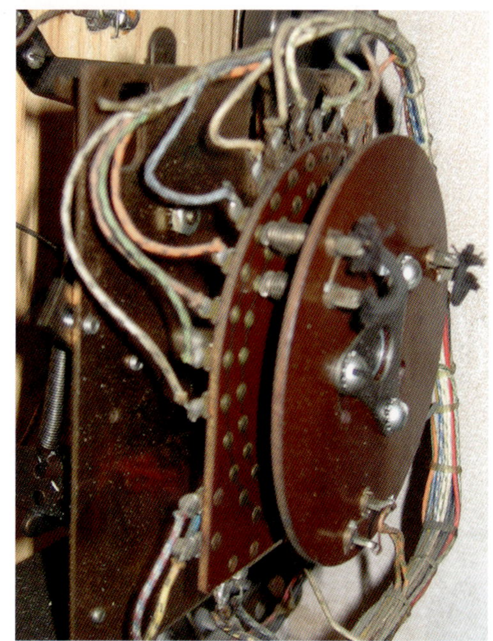

Another common problem with rotating units is poor contact between the rotating contact points and the stationary contact points. Clean the contact points with alcohol or fine emery paper or a pinball flex file until the points shine. Be careful not to bend the metal fingers.

Once you have cleaned the contacts, operate the solenoid by hand and observe the rotating part of the unit as it moves from one stationary contact point to the next. There should be good physical contact. The contact points on the rotating part should line up perfectly with the stationary contact points.

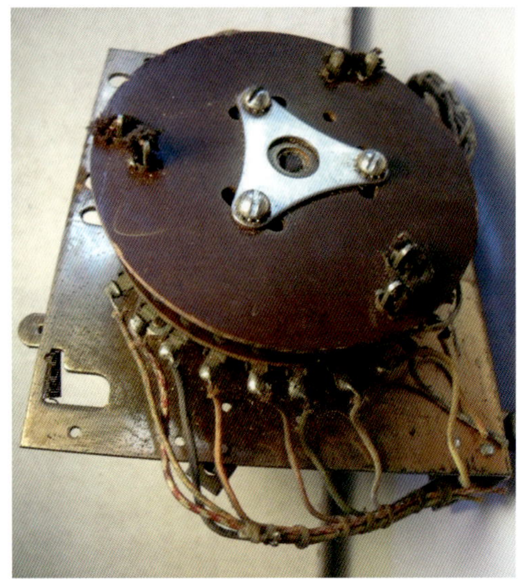

Note the slots under the screws on the triangular plate. If the contacts on the top disk do not align with the contacts on the bottom disk, loosen the three screws, rotate the top disk (clockwise or counterclockwise, it does not matter) just enough to align the contacts, and tighten the screws. Tight.

If the contact points are misaligned—if they are sitting to the right or left of center, or sitting between two stationary points—the alignment is usually adjustable. If the unit has a mounting cover plate, loosen the screws on the cover plate (just a little bit!), adjust the rotating contacts so they park themselves directly on top of the stationary contact points, and retighten the screws.

ADJUSTMENT SCREW

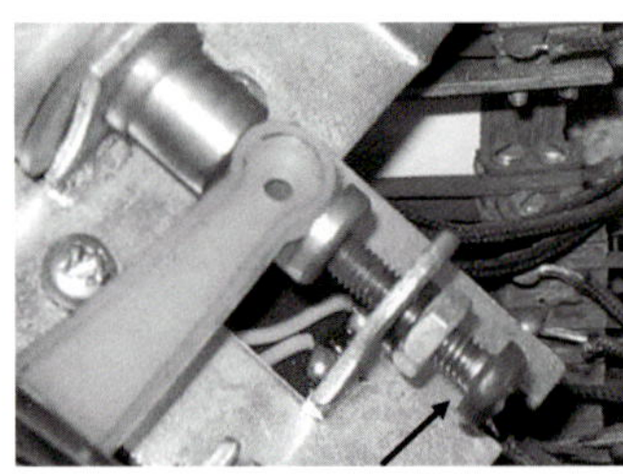

Adjustment screw.

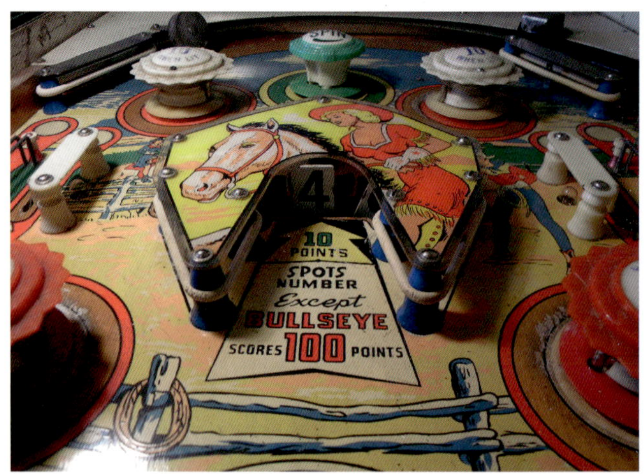

Gottlieb **Corral**, 1961, with roto target mounted vertically. Most of the target is hidden below the playfield.

Some rotating units, instead of having a cover plate, have an adjustment screw that acts as a stop for the plunger that steps the rotating unit up and down, so the rotating contact stops exactly on top of a stationary contact. Adjusting the screw in or out will adjust how far the plunger moves and how far the rotating unit advances with each step. If the rotating unit is not stopping where it should (contacts exactly lining up, one on the other), or if the ratchet is not meshing with a gear, move the adjusting screw in or out just a little bit until everything meshes. It's actually a lot simpler than it sounds.

After you adjust the rotating unit, operate the solenoids by hand to make sure everything moves smoothly and make sure the unit doesn't stick or get sluggish. The manufacturers recommended putting a light coating of pinball lubricant on the stationary contact points. My experience is that you do not need the lubricant. I clean the contacts and then leave them alone. They work fine.

ROTO TARGETS

Some electro-mechanical machines, particularly 1960s Gottlieb games, featured a rotating "roto target." The target wheel had ten or a dozen targets around it, and it was designed so only one or two targets were visible at a time. You would try to hit the visible targets, then try to hit a switch that would spin the roto target, exposing new targets to hit. Sometimes the roto target is mounted horizontally on the playfield, resembling a carousel. Sometimes the roto target is mounted vertically like a Ferris wheel, with most of the targets concealed underneath the playfield.

Roto targets are ratcheted using gears, levers, and springs. At the end of a spin, one of the targets on the roto target wheel stops at a spot exactly in front of a switch.

If you hit the target and nothing happens—no scoring, no chimes—chances are the target is not aligned properly.

That is, the target is not exactly stopped in front of the scoring switch. If the numbers or letters on the target wheel are off center, that's a sure indication that the roto target is not aligned properly. It is not difficult to adjust the target so it works correctly.

Lift the playfield (remove the balls first) and look at the roto target mechanism. There are two places where the roto target can work loose and get out of alignment. The target is mounted on a shaft and held in place by two set screws. Quite often the set screws loosen. Look underneath the rotating part of the mechanism and you will see the set screws.

If the roto target is loose on the shaft, tighten the set screws. This usually requires a hex wrench. Position the rotating targets so they line up with the switch or switches behind the targets. You only have to align one of the targets—it doesn't matter which target—and the rest should automatically line up correctly.

The second place where you may find loose screws causing the target to be out of alignment, is the top bakelite plate of the roto target. The top plate of the roto target is held to the roto target mechanism by three screws, and they screw into slotted holes that allow some adjustment. Loosen the three screws; don't take them out completely. Rotate the top plate so that a target (any target) is exactly over the switch. Tighten the screws. It sounds confusing, but it should become obvious as you look at the unit.

Check your work. Rotate the roto target or operate the solenoid by hand. As the roto target rotates, each target in turn should line up exactly over the switch. Now start up a game and manually hit each target to see if it is scoring and ringing the chimes.

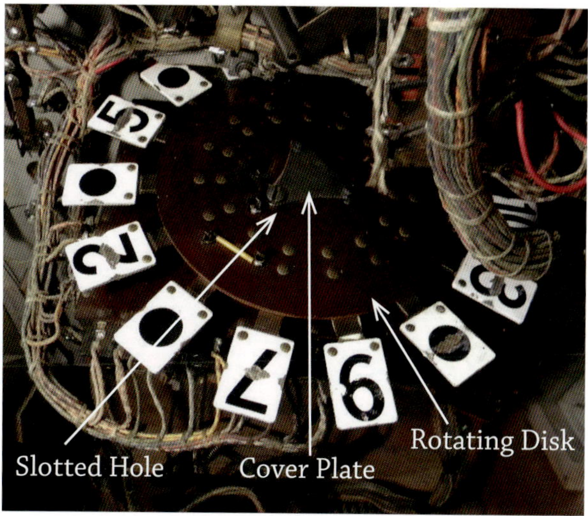

Three screws holding down the triangular shaped cover plate screw into slotted holes in the rotating disc. By loosening the screws the disc can be rotated slightly to align the targets. Underneath the rotating disc are set-screws that hold the disk to the rotating motor shaft. If these screws work loose, the rotating targets will not line up properly. Realign the targets and tighten the set screws.

If this does not solve the problem, see if you can actually see the switch close (or not close) when the target hits it. Just keep adjusting the target until it closes the switch.

> Our industry exists because the hard working, hard playing American public eagerly buys our product, welcomes the relaxation, the release from worry, the low cost amusement that we create. What we really have to sell is America's greatest, most democratic, nationwide, continuous performance show!
>
> —Herb Jones, advertising manager, Bally Pinball Company, 1941

CHAPTER 12

REPAIRS: ELECTRONIC MACHINES

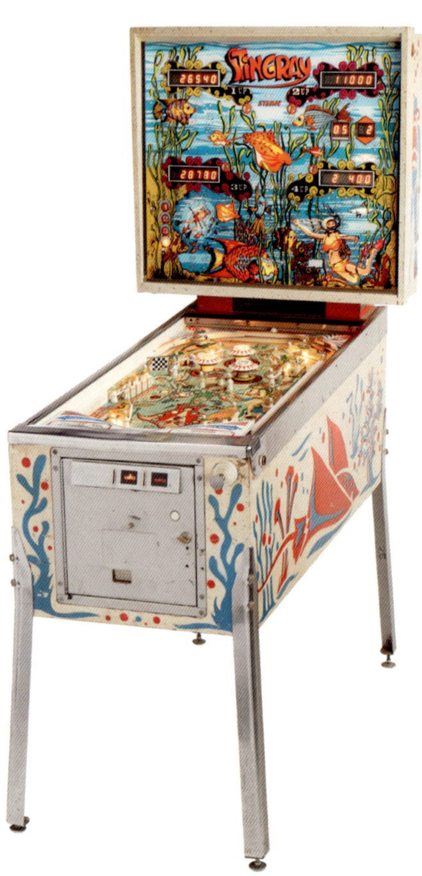

Stern **Stingray**, 1977.

The repairs in this chapter are in addition to the repairs in the chapters "Repairs: Basics," "Repairs: Flippers," and "Repairs: All Machines."

Many repairs and adjustments for electronic machines are the same, or very similar to, the repairs and adjustments for electro-mechanical machines. With electronic machines, however, problems are often caused by bad or shorted resistors, capacitors, diodes, or ICs (integrated circuits, computer chips). If you don't know circuit boards and solid state electronics, you may not be able to figure out what is wrong with your machine. Unlike electro-mechanical machines, where you can observe switches and relays opening and closing, you can stare at a circuit board until the cows come home and not be able to determine anything. And unlike electro-mechanical machines, which are virtually indestructible, it's very easy to permanently damage electronic components.

Circuit board tracing requires a working knowledge of electronic circuits, the ability to read the schematics, and equipment to test the voltages. It would take a full course in solid state electronics to understand the workings of circuit boards, well beyond the capabilities and interests of most pinball owners. The main problem with trying to explain circuit boards without a thorough discussion of electronic circuits is that a little knowledge is dangerous. Pinball owners are likely to cause more damage if they get into the electronic circuits without knowing fully what they're doing.

Still, there is a lot that any pinball owner can do to troubleshoot electronic machines. Even the newest high tech machines have adjustments, settings, and maintenance that you can do at home.

FIRST RULE OF ELECTRONIC REPAIR

Turn off the power before using any metal tools. An electronic machine that is turned on, even in game over mode, has current pulsing through the circuit boards. The slip of a screwdriver or the shorting of a wire can cause a full meltdown.

Static electricity can damage solid state components. See "Static Electricity." in the "Repairs: basics" chapter. This is not a problem for electro-mechanical machines, only electronic machines.

Be careful when touching capacitors, particularly the large cylindrical ones found in power supplies. Capacitors can hold a high voltage charge even after the machine is turned off and unplugged.

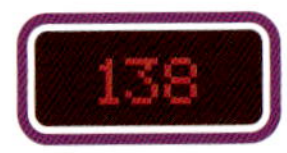

CIRCUIT BOARDS

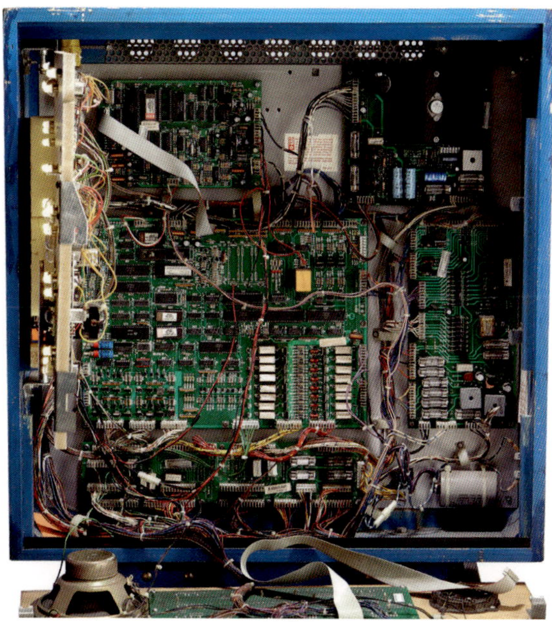

Circuit boards are often the cause of problems with electronic machines, but troubleshooting the boards requires a knowledge of solid state electronics that most pinball owners don't have. Still, there are some things you can try.

REPLACEMENT CIRCUIT BOARDS

The quickest, easiest way to fix a defective circuit board is not to fix it at all. Replace it. New replacement boards are available for many older electronic machines. You just unplug the old board, pull it out, mount the new board, and plug in the connectors. A ten-minute job, no prior experience required. For some machines, new all-in-one circuit boards are being manufactured, a single circuit board that replaces all of the old circuit boards. Most new replacements, however, are specific circuit boards, such as a CPU or a lamp driver or a power supply, to replace specific defective boards. You need to know which board in the machine is the defective one, which may be obvious—maybe it burned, maybe it controlled a specific game function that stopped working—or may not be obvious at all. And you may discover that more than one board is defective.

Replacement circuit boards cost anywhere from about $30 or $40 for a small board to as much as $600 for some CPU boards and the all-in-one boards, which some people may consider a bargain for a machine that sells for thousands of dollars, or may seem outrageously expensive for a machine that you can buy, already working perfectly, for the same amount of money.

Keep in mind that a non-working circuit board may have a very minor problem. We recently repaired a board that had one defective diode, a 95¢ part. But then we knew where to look, which most owners do not.

There are pinball repair businesses that specialize in circuit board repair. Assuming you know which board is defective, you can ship the board to the repair shop. They'll fix it and send it back. In the case of the board we repaired by replacing a diode, the repair bill was $60, a mighty inexpensive fix. "Professional Repairs" are covered later in the manual.

SWAPPING CIRCUIT BOARDS

People with no knowledge of solid state circuits but who have several pinball machines often swap circuit boards, replacing burned or corroded or defective boards with boards taken from parts machines and junkers, machines made about the same time by the same manufacturer. The replacement boards fit, they have the same plug-ins, and they even work. Except when they don't work, or don't work quite the way they should. Two circuit boards from two similar machines may appear to be identical but often aren't. A substitute circuit board might be missing some of the playing action or scoring. And it is possible you could damage your machine. I do not recommend it.

A different problem, for me anyway, is when we get a machine in the shop that doesn't have the original circuit boards and the owner doesn't tell us (maybe the owner doesn't even know). We once spent more than an hour trying to figure out why a circuit board wasn't working correctly until discovering that we were working on a board from another game.

Still, a long-time pinball friend of mine who has several electronic machines and a box full of old scavenged circuit boards, is always swapping boards, and keeps his machines running. He loves it. I do not recommend it.

ONE MORE CIRCUIT BOARD TIP

One thing you can do if your machine isn't working is examine the circuit boards for any obviously burned components. If you find a diode or resistor or other electronic component that burned, you've located a part that needs to be replaced, and maybe solved the problem. If the component was defective, simply replacing it will fix your machine. But if the component burned because of a problem elsewhere in the circuit, the replacement component will also probably burn out. But it may not burn out, it may work just fine, or it may take weeks or months to burn out a second time. Many electronic components are very inexpensive, a few dollars. Replace the component and see what happens. Lady Luck may shine on you.

TEST MODE (DIAGNOSTICS)

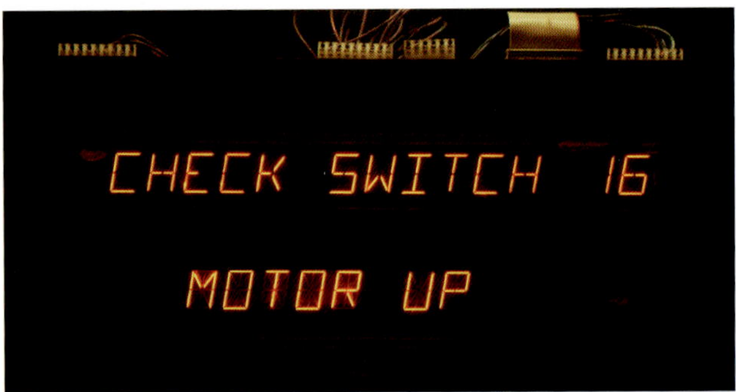

All modern electronic machines allow you to perform a series of diagnostic tests where the machine will automatically operate each component, one at a time. You can run a test of the switches, the solenoids, the lamps, the digital displays, the sounds, and other functions. You will be able to observe what is and isn't working. Modern electronic machines will also display problems as they occur, and even tell you what is not working.

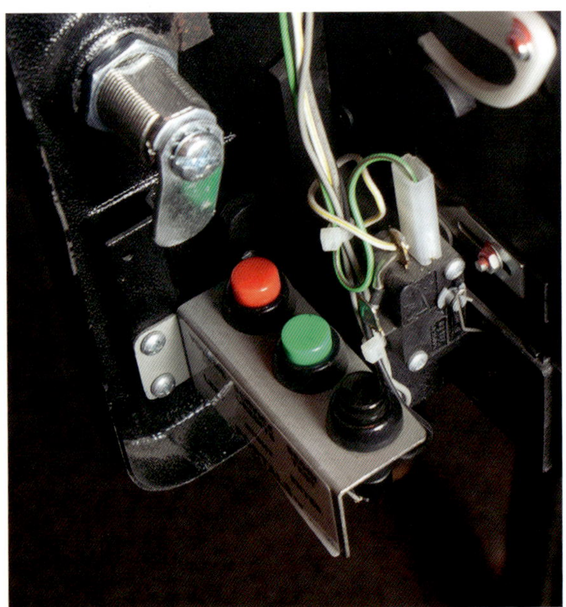

The diagnostic switches—a set of two, three or maybe four push buttons—are inside the coin door, sometimes on the door, sometimes on the cabinet next to the door's hinges. These are the same switches that control the game adjustments covered previously under "Game Adjustments."

Your machine's instruction manual includes detailed step-by-step instructions, which can be confusing if you've never run tests before. But if you take your time, and repeat tests you don't understand the first time, you'll figure it out.

Having that information may, or may not, help you locate the cause of the problem, depending on your knowledge of circuits. If nothing else, it will help you explain the problem to a repair person over the phone, who may immediately know the problem just from your description. Quite often the problem is on one circuit board that can then be removed and taken or mailed to a circuit board repair shop, or replaced with a new circuit board.

POOR CONNECTIONS

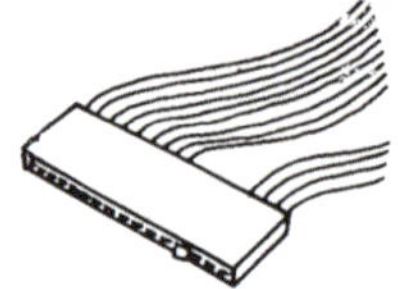

All kinds of problems with electronic pinball machines are due to poor wire connections.

The plastic wire connectors that plug into the circuit boards, sometimes called Molex plugs, and the pins the connectors plug into, also called "headers," are often the culprits. Before you touch any connector or circuit board, ground yourself to prevent static electricity (see "Static Electricity"). Be sure the machine is turned off. Do not plug and unplug solid state connectors with the power turned on; you can damage the components.

Check each connector. Make sure it isn't burned and that all the wires are attached firmly to it. Simply reseating the connectors can often get a game working again.

When you remove and reinsert a connector, pull it straight out and push it straight back in. Do not wiggle it, as this could break the solder connection on the pin. Do not pull a connector out by the wires, as you can pull the wires right out of the connector. When reseating the connector, make sure all the pins line up correctly and that you haven't reversed (flip-flopped) the connector. Most connectors have a key—a blocked pin hole—to prevent plugging the connector backwards or in the wrong socket.

Electronic problems are also caused by the pins on the circuit board that the connectors plug into. The pins can be burned, corroded, or dirty, making for poor electrical contact. The pins sometimes have broken solder joints. Look at the

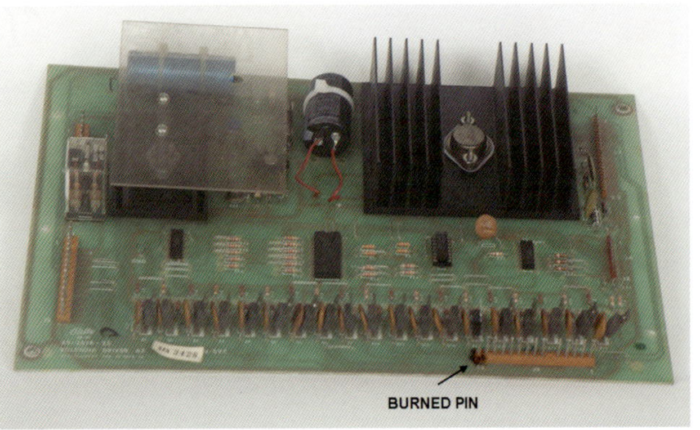

base of the pins for a crack in the solder. Resoldering the pins will solve the problem. After soldering the pins, clean the solder with alcohol on a Q-Tip. This will remove any flux residue, which may contain tiny particles of solder, enough to short the connection.

The pinball machine's lamps are often a heavy current draw that can overheat and burn the plastic connector plugs. By switching to lower power lamps you may save your connector (see "Lamps").

DIGITAL DISPLAYS

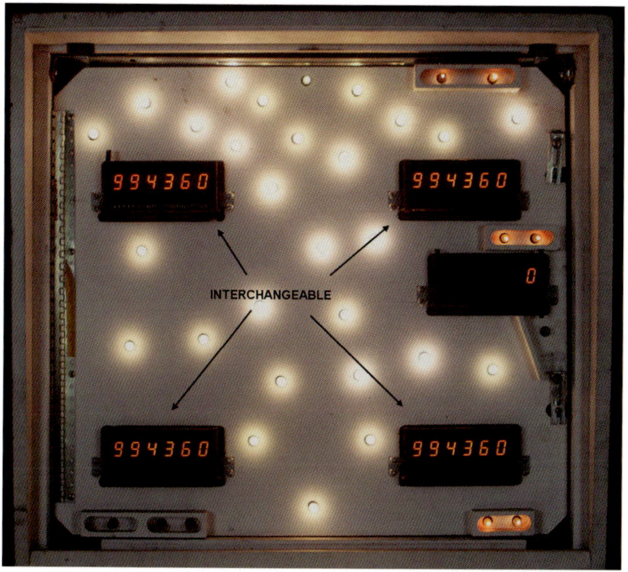

If a digital display is not working, or partially working, the problem could be something as simple as a loose or disconnected cable, it could be a defective display, or it could be a problem in the circuit. So the first thing to do is check the cables and connectors, making sure they are not disconnected or loose.

Digital displays sometimes burn out. Many displays are plug-in modules and can be easily swapped for replacement modules, if you can find a replacement. Most pinball suppliers have new modules, though the prices are often expensive. Some displays are soldered to circuit boards with dozens of connector "feet" soldered to the board. If you don't replace the entire circuit board, you may have a tedious and delicate job unsoldering the old display and soldering in the new display.

If you have an older electronic machine, with a separate digital display module for each player, you might consider swapping modules within your machine. If you are willing to sacrifice the four-player option, you can remove the fourth player's scoring module and use it to replace the burned out module. You'll now have a three-player machine, at least until you can locate a replacement or find someone to repair the damaged module. This, of course, is not an option for newer machines that display everything on one or two digital modules.

Ground yourself before touching the modules. Don't let static electricity damage your machine (see "Static Electricity").

Quite often a problem with the display has nothing to do with the display module itself. One of the solid state components controlling the display could be defective. The defective component could be on the display circuit board or on the CPU board. This will require a knowledge of solid state circuitry.

IC (COMPUTER CHIP)

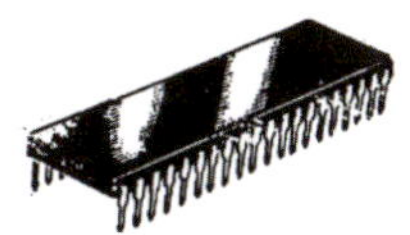

Poorly connected ICs (also known as integrated circuits, computer chips, or logic chips) sometimes cause electronic problems. ICs are the rectangular black plastic components mounted flat on the circuit boards. Some ICs are soldered directly to the circuit boards, but some ICs are plug-in units mounted onto a socket. Sometimes the connection, IC to socket, is not good. Removing and reseating the IC might solve your problem.

First, ground yourself to prevent static electricity. Carefully pull off the ICs that are socket mounted and reseat them again, being careful not to bend or break the tiny pins. And be sure not to reverse the IC when putting it back in, or you'll do all kinds of damage to the circuit.

If you have an IC that was removed from the circuit board and you don't know which side is which, here's how to tell: One of the connectors on the circuit board, where the IC plugs into the circuit board, is called the "1" connection. It is always in one of the corners, usually the bottom left-hand corner, and sometimes designated by a 1, but usually designated by a printed dot.

The IC that you removed also has a 1 pin that connects to the 1 connection on the circuit board. The 1 pin on the IC, usually the bottom left corner pin, may have a dot, or a 1, or a circular indentation next to the pin. More often the IC will have a stripe or a notch, or both, at one end of the chip. That end of the chip connects to the same end on the circuit board where the 1 connection is. If this is confusing—it sure sounds confusing, with so many different configurations—look at another IC on the circuit board and you will see how it is attached.

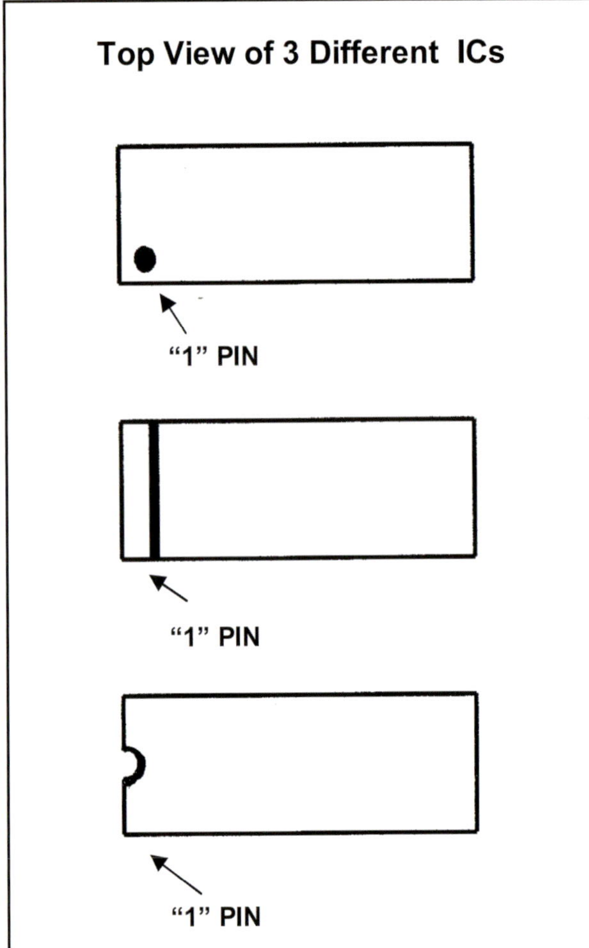

Top view of three different ICs.

"ADJUST FAILURE"

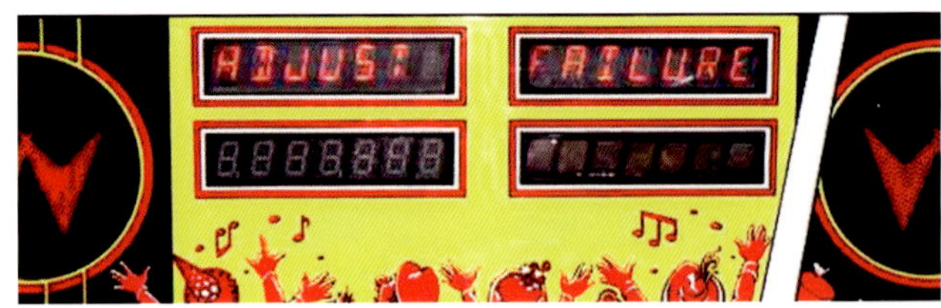

Unlike computers, pinball machines do not often display cryptic, indecipherable messages. But one I've seen several times on Williams machines and some Bally machines when you turn on the game is "Adjust Failure." The game will not start. This indicates a problem in the memory protection circuit and is often caused by weak batteries or poor connection between the batteries and the battery holder.

Open the coin door. Inside the door are three push buttons mounted next to each other. The button closest to the coin door's hinge is usually the Advance button (it will be labeled). Push the Advance button once. Very often this will restore the original factory settings and clear the Adjust Failure. Close the door, turn the machine off, wait a few seconds, and turn the machine back on. You should be able to start a game.

By doing this you will lose the high score memory and any custom settings in the game, such as free play. You have to go to the game adjustments again and reset your custom settings. You might want to replace your batteries first.

> Life and pinball machines have a lot in common. They are both guided by a combination of skill and luck.
>
> —**Keith Temple**, *Pinball Art*

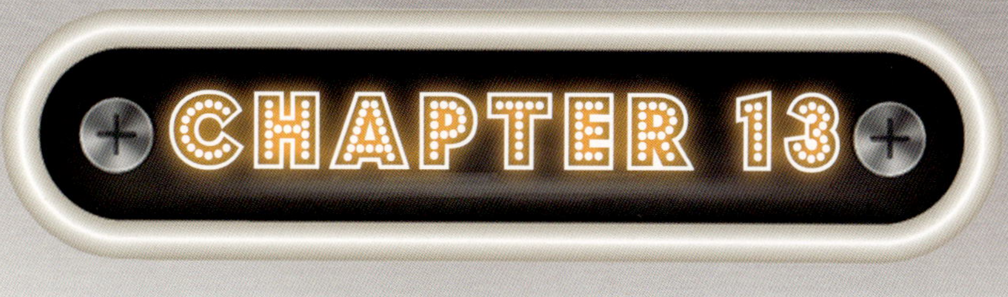

Chapter 13

Tracing Circuits

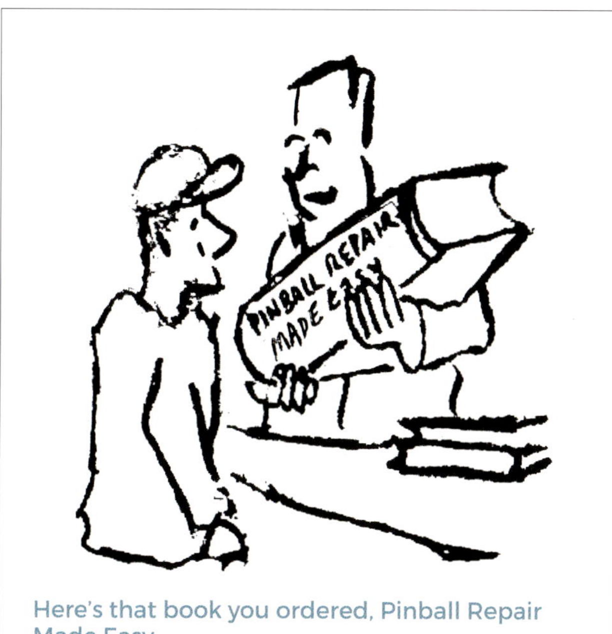

Here's that book you ordered, Pinball Repair Made Easy.

I don't mean this to be a manual for doing complex troubleshooting. There are many repairs, easy repairs, you can make without struggling with the schematic and all those components that you have no idea what they're for. But if you have the time and inclination, it is not difficult to trace some of the simpler circuits, and in the process learn how electrical circuits function.

Looking inside any pinball machine, the mass of wires and electrical components can be an intimidating sight. What you are looking at are many different individual circuits. Some of those circuits can get quite complicated. But many of the circuits are simple, and locating a problem is often easy to do.

First, you need to know how to use a wiring schematic. If you know how to read pinball schematics you can skip to "Tracing Circuits with the Schematic."

PINBALL MACHINE SCHEMATICS

Much of this information about schematics applies to both electro-mechanical and electronic machines. However, the specifics are really for electro-mechanical machines, which have much more rudimentary circuits: easy to explain, easy to understand.

Unfold your schematic, take a deep breath, and have a look. *Cowabunga Buffalo Bob!* If you just want to fold the schematic back up, that's okay too. You can skip to the "Professional Repairs" chapter. But if you're willing to give the schematic a few minutes of your time, its mysteries—at least some of them—will be revealed.

The schematic does not show the physical location of switches, or coils, or wires, or any components in the machine. Components that are shown next to each other on the schematic could be at opposite ends of the pinball machine. Switches shown next to coils on the schematic are often nowhere near the coil they are wired to. Wires shown on the schematic as crossing each other may not even be near each other inside the pinball machine.

The schematic uses symbols for the components in the pinball machine. Most of these symbols are universal, the same for all electrical applications, not just pinball. Each schematic symbol also has a label which is sometimes self-explanatory and sometimes requires a code chart.

SYMBOLS AND LABELS

Coil Symbols

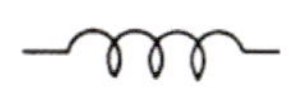

There are maybe forty or fifty coils in a pinball machine. The symbol resembles a looped wire. Solenoids and relays use the same symbol. You cannot distinguish between them from the symbol.

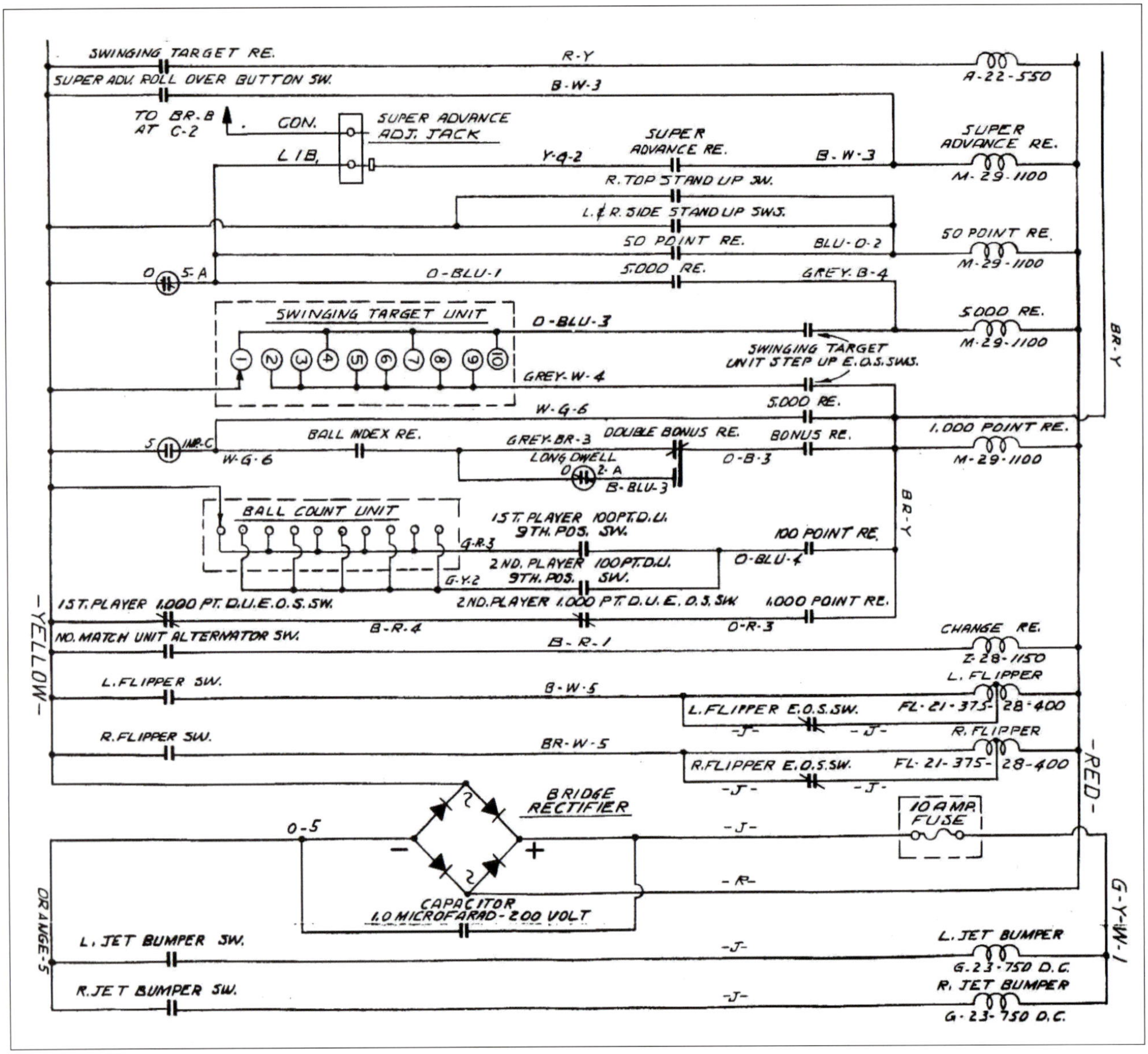

A small part of an electro-mechanical wiring schematic from a 1970s Williams machine. Rotating units (in this schematic the Swinging Target Unit and the Ball Count Unit) are indicated by a dotted line around the unit. Note the diamond-shaped bridge rectifier symbol indicating that the pop bumpers ("jet bumpers") are DC powered.

Coil Labels

On most schematics, the label next to a coil symbol will tell you which coil it is, such as "100-point relay" or "right flipper," and often will tell you the coil part number. On some schematics, particularly Gottlieb schematics, coils are labeled with a code instead of a description: a letter, or letters, or a number-letter combination, requiring you to look up the code on a chart on the schematic. I'm sure the Gottlieb Company had a good reason to label a coil, say, M instead of 10-point relay or 100-point relay, but I don't know what it was.

All schematics include a list of coils. The list is sometimes in two sections: "relays" and "other coils" (solenoids). The list of coils on the schematic also tells you where each coil is on the schematic. Like old road maps, schematics have letter and number coordinates along the edges of the schematic. For example, a coil on the schematic at 10-E will be in the area of the schematic where 10 intersects with E.

Do not confuse these coordinates with coils having similar identifying code numbers. There may be a coil numbered 10-E, and there may be a coil on the schematic at location 10-E. The two 10-Es are not related.

Switch Symbols

There are a hundred or more switches in an electro-mechanical machine and they use three different symbols:

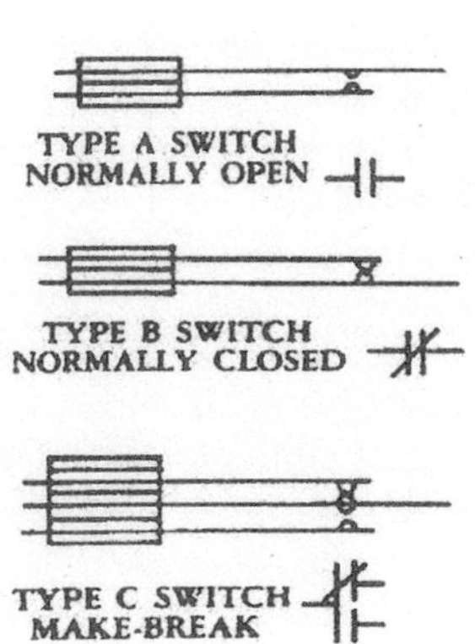

Open: The switch is open, with the switch blades not making contact. Activating the switch closes the switch.
Closed: The switch is closed, the switch blades are making contact. Activating the switch opens the switch.
Make-break: A two-way switch where one side opens as one side closes.
The switch position shown on the schematic, open or closed, is the position of the switch when the circuit is not activated, when the machine is turned off.

Switch Labels

Most switch labels on schematics are self-explanatory, except for motor switches, and except for some Gottlieb switches.

Gottlieb Switches

On Gottlieb schematics, relays are labeled with a letter code or number-letter code instead of what the relay actually does. Switches that are mounted on and activated by relays are labeled with the same code as the relay that activates the switches. For example, a switch labeled M will be on the M relay and activated by the M relay. If there are several switches on the relay (there almost always are), all of the switches will be labeled M. And what is the M relay? You need to go to the list of relays near the bottom left hand corner of the schematic and find out what it is. (The M relay is Gottlieb's code for either the 10 Point Relay or 100 Point Relay. Gottlieb could have put "10 PT. RELAY" or "100 PT. RELAY" next to the switch symbol, but of course they didn't.)

Motor Switches

Switch symbols that have a circle around them are motor switches. The switches are on the score motor. On Gottlieb pinball machine schematics, motor switches do not have a circle around them, but they are labeled "MOTOR."

On motor switches, the number-letter designation next to the switch refers to the physical location of the switch on the motor. The number refers to the switch stack (there are anywhere from four to ten switch stacks on a score motor), while the letter refers to the position of the switch on the stack: A is the lowest, B second from the bottom, etc. For example, "3B" means motor switch stack #3, second switch up from the bottom. The schematic usually includes a motor diagram showing the switch positions. Sometimes the motor diagram is on a card stapled to the side of the cabinet, sometimes the motor diagram is in the instruction manual.

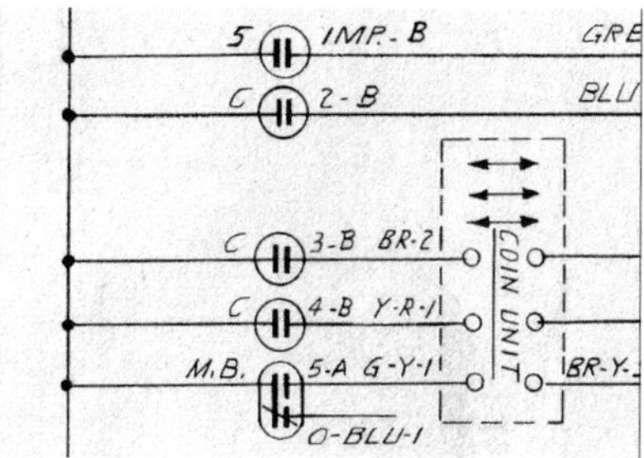

Wiring Symbols

All those horizontal and vertical lines on the schematic are wires. Be careful where you see wires crossing each other. Some of those crossed wires are connected together and some are not. Different schematics and different manufacturers used slightly different—and contradictory—symbols to indicate which wires are connected to each other and which wires are not.

Wiring Symbols Type #1

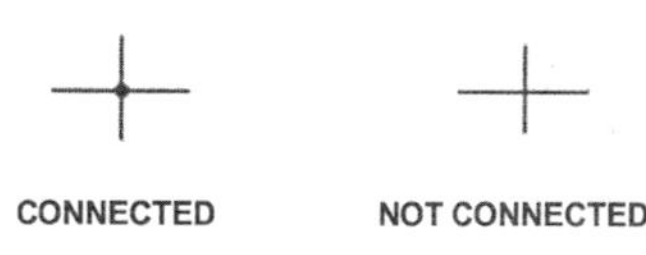

If the schematic has a black dot at some crossed wires and no black dot at others, the crossed wires with the black dot are connected. The crossed wires where there is no black dot are not connected. They have nothing to do with each other, and if you are not paying attention, the illustration can throw you off the trail. More than once I've followed the wrong wire on a schematic.

Wiring Symbols Type #2

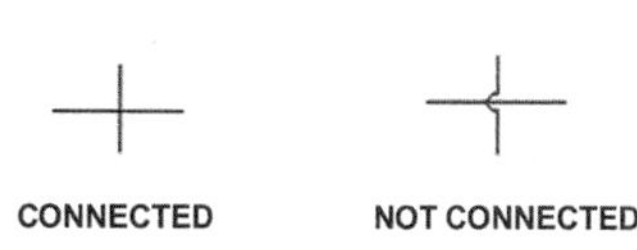

If the schematic has crossed wires (without the black dot) at some intersections, and a little loop where the wires cross at other intersections, the looped wires are not connected. The crossed wires, even though they don't have a black dot, are connected. This is a major flaw in what should have been universal symbols, but pretty soon you get used to recognizing how the wires are, or aren't, connected.

Wiring Colors and Labels

Most wires in a pinball machine are color coded. Some wires are a solid color. Some wires are two-color and sometimes three-color. Some wires are a predominant color with a second "tracer" color.

On the schematic the wire colors are sometimes spelled out, and sometimes indicated by letters: "R" for red; "R-Y" for red and yellow; and "R+Y" for red with a yellow tracer.

Most of the letter codes on the schematic are obvious: R for red, Y for yellow, O for orange, W for white, and MAR for maroon. Green wires are labeled G or GR and gray wires are usually spelled out. Some gray wires are labeled SL for slate. There are three colors starting in B, and the codes usually are B or BLK for black, BL or BLU for blue, and BR or BRN for brown. But you don't have to guess. There is a wire color code chart on most schematics.

The color designations are often followed by a number, such as R-2. When a certain color or color combination is used in more than one place in a pinball machine, the number indicates that this is the first, or second, or however many times the color is used. "R-2" means that this is the second use of a red wire in the machine.

On some schematics, particularly Bally, each color is given a number as well as a letter designation, such as 1 for Red, 2 for Blue, and 3 for Yellow. So using this example, a wire labeled "12" will be a red and blue wire. A wire labeled "12-2" will be the second red-and-blue wire in the machine. These color codes are not as confusing as they sound. They are actually easy to figure out, just a bit hard to explain.

Some wires on the schematic are labeled "J," which stands for "jumper." Jumpers are short lengths of wire that usually connect several relays together when the relays are mounted next to each other and share a common wire. Jumper wires can be any color, and the letter J is of no help in determining the color. However, it is easy to locate jumper wires in the machine. Interconnected jumper wires will all be the same color.

Some wires on the schematic have no color designation at all. This indicates wires without insulation. These are bare metal stranded wires, usually stapled to the underside of the playfield and inside the backbox. These uninsulated wires are usually common wires to lamp sockets.

Finally, here is something you will discover quickly: Wire colors fade. Very often you can't tell what color the wire was. Reds turn to a dull brown. Yellows turn to a dull brown. Whites turn to a dull brown. Sometimes, if you follow a wire from its faded end up into the bundled wires where the wire has been hidden away, you might see the original color. More likely you'll have to make some educated (or uneducated) guesses.

Lamp Symbols and Labels

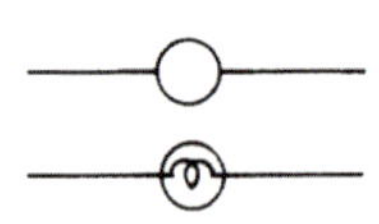

There are fifty or more lamps in most pinball machines. The symbol for a lamp is either an empty circle or a circle with a looped wire inside it. Lamp symbols are almost always grouped together on the schematic, on a circuit separate from the coils.

There are more lamps in a machine than shown on the schematic. The lamps labeled "illumination" or "playfield" or "name and scene" are the decorative lamps that are always lit when the machine is turned on. The schematic may show five or six lamp symbols wired together, but there could actually be twenty or thirty lamps in the circuit. Most lamp labels are self-explanatory.

Fuse Symbols and Labels

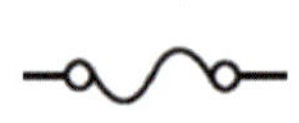

There are anywhere from three to ten fuses in an electro-mechanical machine, and sometimes twenty or more fuses in an electronic machine. The fuse symbol is a wiggly line looking like a sine wave or a sideways S. There is always a fuse between the wall plug and the transformer, and between the transformer and the operating and lamp circuits.

Fuse labels will indicate the amperage (5 AMP, 10 AMP, etc.) and if the fuse is slow-blow (SLO-BLO, time delay).

Score Motor (Electro-mechanical Machines)

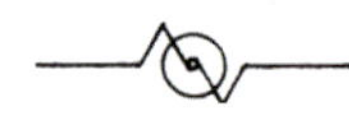

There is only one score motor in an electro-mechanical pinball machine, usually simply called the motor. The symbol varies with different manufacturers, but always has some sort of circle in the center of it. The label may just be an M, or it may say SCORE MOTOR. Electronic machines do not have score motors.

Transformer

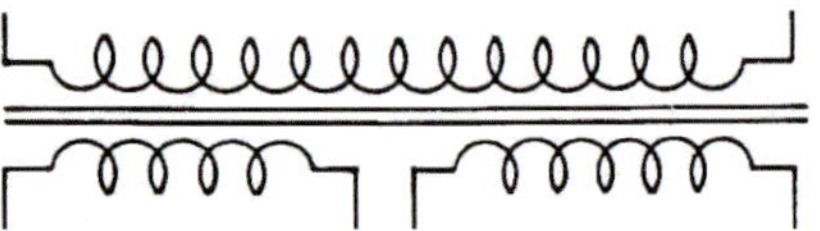

There is only one transformer in a pinball machine. The symbol is usually back-to-back coils, separated by two or three straight lines. The label will show the voltages.

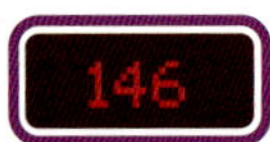

Step Up/ Step Down Units

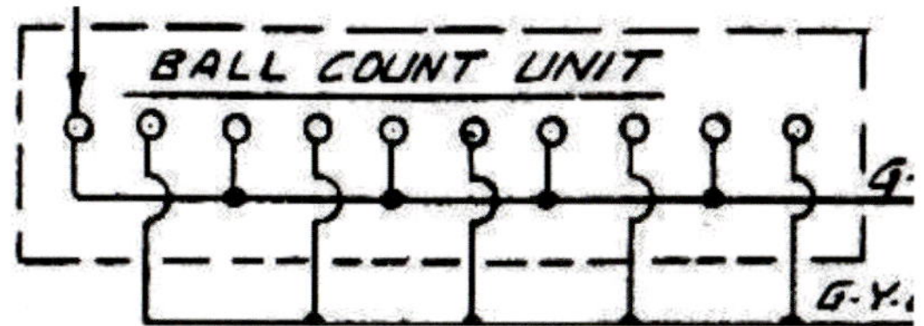

Electro-mechanical machines have several rotating (stepper) units: player, ball count, bonus, match, etc. The schematic symbols look like large and often confusing wiring diagrams, with a dash line outlining the unit. The label always includes the word UNIT.

Other Symbols

Below are some additional symbols for components typically found on both electro-mechanical and electronic machines. On schematics, the symbols sometimes include specifications (ratings) for the components.

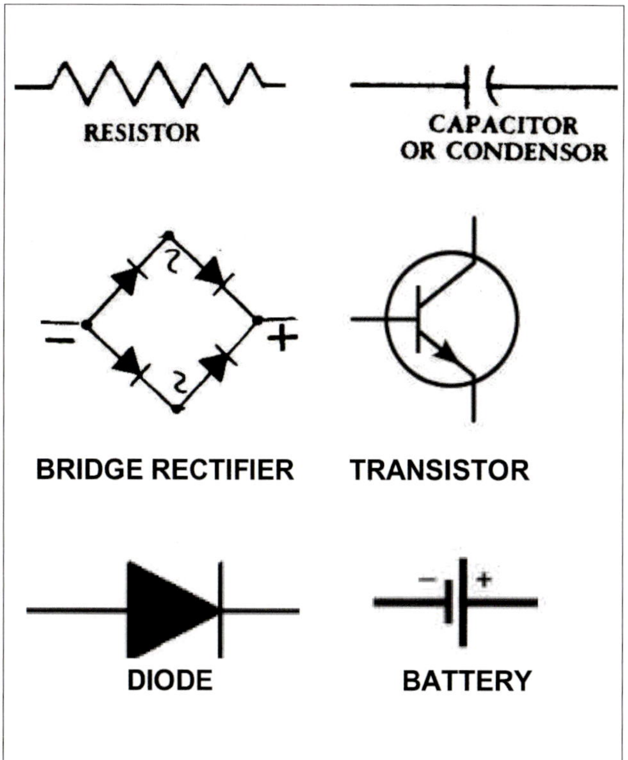

Schematic Label Abbreviations

Given the tiny amount of space on schematics for labels, abbreviations are necessary. Some abbreviations are obvious, but some are not. Some common schematic abbreviations are:

ADJ	Adjustment (usually a plug-in option)
BK	Break
BOT	Bottom
DU	Drum unit (score reel assembly)
EOS	End of stroke
IMP	Impulse (part of the score motor)
IND	Index (part of the score motor)
L	Left
PL	Player
POS	Position
PT	Point
R	Right
RE	Relay
RO	Roll over
SCM	Score motor
SLO-BLO	Slow blow fuse
SU	Step up
SW	Switch

USING THE SCHEMATIC TO TROUBLESHOOT

Let's say you've got something in your machine that isn't working or isn't working correctly. Maybe a switch or target that does not do anything or only does part of its job. Maybe it scores but does not ring the chime. Maybe you have a bonus that does not advance when it should, or a ball that won't kick out, or a game that won't end.

Almost all problems in electro-mechanical circuits can be traced to a switch that is shorted or making poor contact, a rotating unit that is gummed up and won't move, or a wire that has broken off a component.

Before you start studying the schematic, examine the switch or unit that isn't operating. Are all the wires connected? Are the contact points clean and making a good connection? If it's a switch or target, does the switch open and close? If it's a rotating unit, does the rotating unit rotate smoothly?

If the problem is not obvious during examination, you'll need to locate the non-working component on the schematic and see what switches, relays, and other components are part of that circuit. Where do you start? Here's how I would proceed:

Step 1. If you know that a certain solenoid or relay is part of the non-working circuit, that's the best place to start on the schematic. All coils are listed on a chart on the schematic, including their location on the schematic. So it is easy to locate a coil on a schematic and then trace the wires coming off of the coil to switches and other components in that circuit.

Step 2. If all you know is that a switch or target isn't working, you have more work locating the switch on the schematic. Switches can be anywhere on the schematic, and unlike coils, switches on the schematic have no chart to guide you. You may have to hunt through fifty or more switches to find the one you're looking for. It's not as bad as a needle in a haystack, but not as easy as locating a coil. Once you do find the switch on the schematic, follow the wires from the switch to the relay or components that the switch operates. The wires may also lead you to other switches in the same circuit.

Step 3. Locate those components inside the pinball machine, examine them, clean and adjust them, make sure the wires aren't broken off, and see if your problem goes away.

Let's try an easy example: The ball hits a 10-point target but nothing happens. You've already examined the non-working 10-point target switch and found no problems. It opens and closes, it makes good contact, and it has no disconnected wires. So:

Step 1. You locate that 10-point target switch on the schematic and find out it activates the 10-point relay.

Step 2. Locate the 10-point relay in the machine. Don't forget there are relays under the playfield, on the bottom of the cabinet, and in the backbox. A relay could be located most anywhere. But I'll help you out: The 10-point relay will be in the backbox.

Step 3. Start a new game and see if the relay is activated, pulling in its armature and probably making a buzzing noise. If it is not activated skip to Step 4. If it *is* activated, you have found the reason the target switch is not working: Another switch that also activates the 10-point relay is shorted: closed when it should be open. Locate the shorted switch using the schematic: What switches on the schematic activate the 10-point relay? Then find those switches in the machine and see if you locate the shorted switch.

Step 4. If the 10-point relay is *not* activated, check that all of the wires are attached and see if the relay energizes (pulls in the hinged armature attached to the relay) when the playfield switch is closed. If the relay does energize go to Step 6. If the relay does not energize, there is an open circuit somewhere between the playfield switch and the relay. Look for a poor connection where the large disconnect plugs attach the playfield to the component board on the floor of the cabinet, or to the backbox. If the connections are good, go back to the schematic and see if there are any additional switches between the playfield switch and the relay. Chances are one of those switches is not closed when it should be, or is closed but not making good contact. All of the switches are labeled on the schematic. Locate each switch in the machine one at a time and make sure it is wired, closed, and making good contact.

Step 5. Once you get the relay to energize see if that solved the problem.

Step 6. If the relay does energize--closes when the target switch is closed--but the circuit is not working, clean and adjust the switches on that relay. Each of those switches operates a different circuit in the machine. One of those relay switches operates the 10-point score reel. Another of those relay switches rings the 10-point chime. And on it goes, each switch leading to another circuit. The switches open and close and look fine, but at least one of them is not making good contact. Chances are good that you will find your problem here.

If you did not fix the problem at this point we need to expand our search. (Maybe we first need a break, especially if it's lunchtime.) So far we've been testing a very simple circuit: A playfield switch energizes a relay, the relay opens and closes switches, and those switches energize other circuits.

One or more of the switches on the relay are not accomplishing what they are supposed to do. Each of those switches is somewhere on the schematic—and the switches can be anywhere on the schematic—identified by the relay we were just looking at and distinguished from the other switches on the relay by the wire colors. In other words, there will be several switches with the same label, which refers to the relay that opens and closes the switches, but each switch will have different wire colors. (*"I thought you said this is an 'easy' example."*)

So, one switch at a time, locate each switch on the schematic, see what components are activated by the switch, and then go through the previous steps. These relay switches are not part of the relay circuit itself: They are not wired to the relay, they are only opened and closed by the relay. These switches are somewhere else on that schematic, probably spread all over the schematic. It's a little like tracing a family tree as it branches out into cousins and nephews you never knew existed. The trick is to keep each branch (each circuit) separate and take on one circuit at a time until you find the problem.

Anyway, I'm going to stop here. The more you work with schematics they will get faster and easier to use. As you learn more about pinball circuits, which will happen just by experimenting and trial and error, the source of problems will be more obvious, requiring less searching on the schematic and fewer places to have to look in the machine.

ELECTRONIC SCHEMATICS

In the instruction manual for your machine electronic schematics are spread out over several pages. Each page of the schematic shows a different circuit, with wires ending on the side of the page that pick up again on another page. Instruction manuals also include diagrams of each circuit board, showing the actual physical location of the components on each board. Unfortunately, an untrained person can make only limited use of these schematics. You are much more likely to have success troubleshooting an electronic machine by going through the diagnostic procedures (the "test mode") while following the step-by-step instructions in the instruction manual.

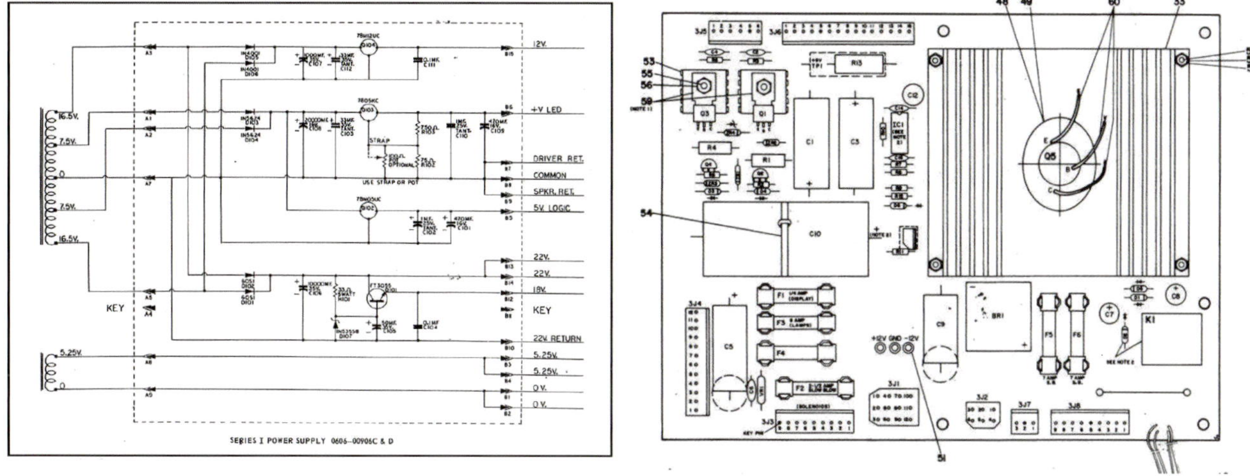

Schematic of a power supply board (left) and physical layout illustration of a power supply board (right). Instruction manuals will include both.

For a while, people had been in the virtual reality thing with video. Maybe people started to realize that pinball is reality. The flippers smack, the ball hits the bumper, it's very physical. It's not virtual reality.

—Michael Houk, special effects producer

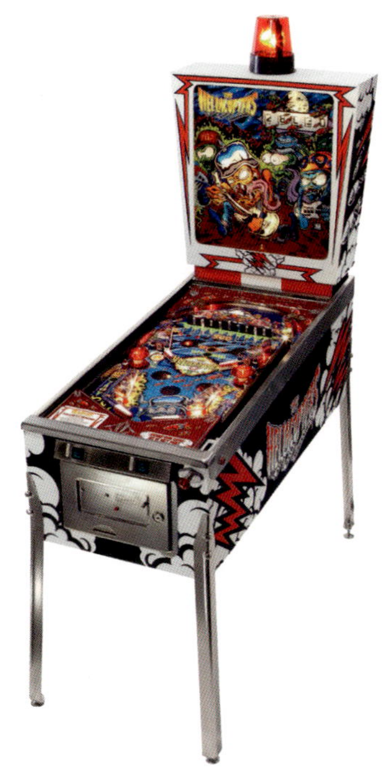

Hellacopters, originally a 1973 Gottlieb King Pin, "re-themed" with new artwork by pinball designer Donny Gillies, for the Swedish band The Hellacopters.

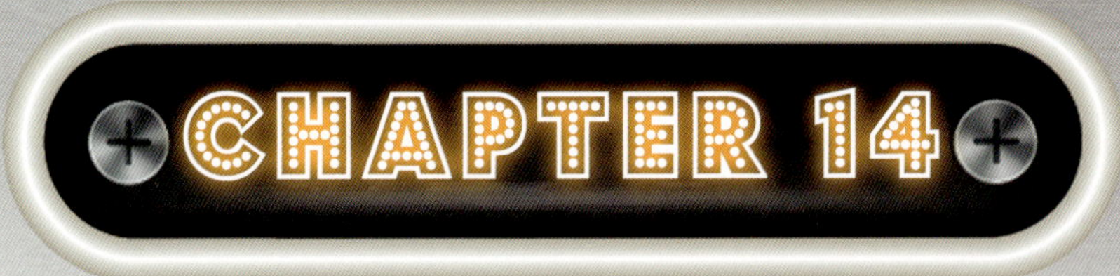

CHAPTER 14

LUBRICATION

Most moving parts of pinball machines should not be lubricated. Lubrication too often gums up operation and attracts dirt. But some parts will perform much better when lubricated. Where lubrication is recommended, use a lubricant specially formulated for pinballs or coin machines, sometimes called coin lubricant, available from some pinball suppliers. Some pinball suppliers recommend a product called Lubriplate. You can use a Teflon gel lubricant instead, available from hardware stores and electronic suppliers. A small tube or jar will last you forever.

When lubricating, use a very small amount. Wear gloves, or use a tool to apply the lubricant. Petroleum and Teflon based products may be toxic. Do not use Vaseline as a lubricant. It leaves a dirty, gummy residue.

COIL PLUNGERS

Do not lubricate coil (solenoid) plungers. If coil plungers are dirty, clean with rubbing alcohol. If plungers are rusty, use steel wool or a wire grinding wheel to remove the rust.

Some manufacturers recommended putting flaked or powdered graphite on solenoid plungers. This can get messy, and in my opinion is not needed.

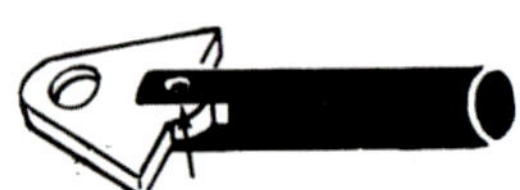

Pinball manufacturers recommended lubricating the pivot points where the plunger connects to the link. Personally I have never lubricated these joints and have never had a problem.

DROP TARGETS

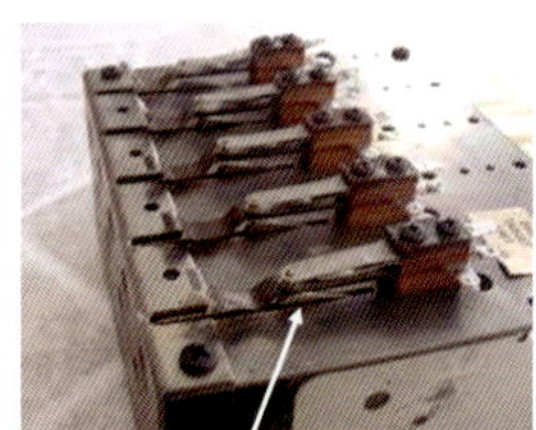

Rear view of a drop target bank (targets not visible). The arrow points to where the back of the drop target slides up and down, closing the switch behind the target. A little lubrication at this juncture will help the targets drop smoothly.

Some manufacturers suggest lubricating the backs of drop targets, where the targets contact the switch mechanisms, so that the targets drop and reset smoothly without getting stuck halfway up or down. Stuck drop targets are likely to be caused by problems with the springs or warping of the targets. A little lubrication might, or might not, help.

POP BUMPERS

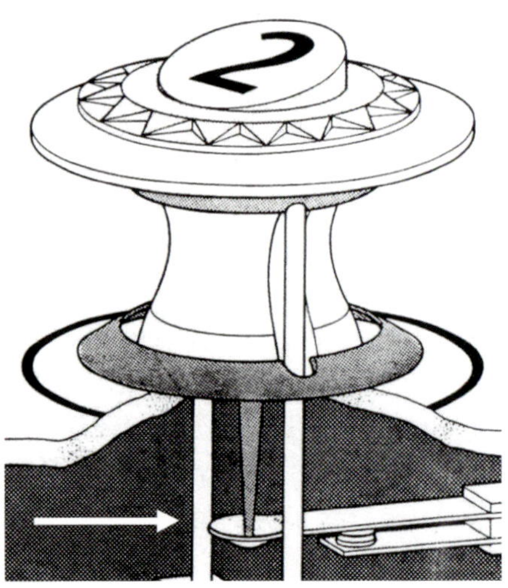

The arrow points to a cup under the playfield with a stem sitting in the cup. Clean the cup and the tip of the stem with a Q-Tip and alcohol, then put lube in the cup. Try using a toothpick to get in there.

Put a small amount of lubricant in the little cup under the playfield, where the plastic stem rubs against the cup. Use a toothpick to get the lubricant into the cup. Some pinball people recommend against this lubrication, claiming that it gums up the pop bumper operation. My experience is that the lubrication keeps the pop bumper working much more smoothly. Lubricating the cup was also recommended by the manufacturers.

BALL SHOOTER

Some people lubricate the shooter, and some manufacturers recommended it, but I don't recommend it. Not only is it unnecessary, but players can wind up with lubricant on their fingers, and then all over the machine, if they grab the shooter rod. It's a better idea to just clean the shooter and the bushing the shooter goes through, with alcohol. If the ball shooter is rusty use steel wool or a wire grinding wheel. If you're still getting resistance, replacing the bushing may solve the problem.

CONTACT POINTS ON ROTATING UNITS

I do not lubricate the contacts on rotating stepper units (ball count units, player units, bonus units, match units, etc.). I clean the points with alcohol and polish them lightly with fine emery cloth or crocus cloth. I've never had a problem with not lubricating the points. However, many pinball people and the manufacturers recommend lubricating the points. They suggest a very light coating of lubricant.

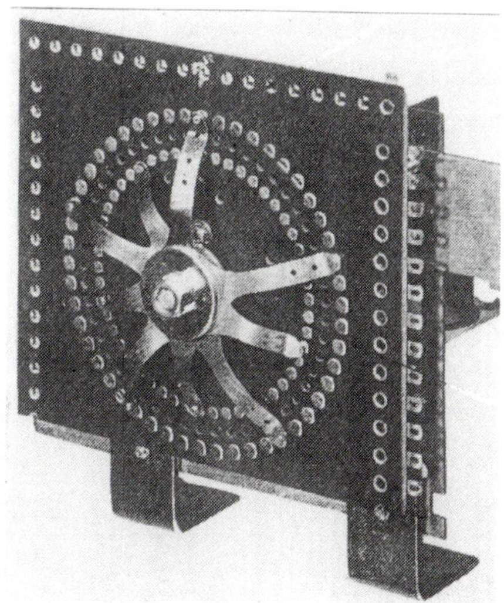

The arrow points to the outer circle of contacts. The manufacturers recommended a very light coating of lubricant on both the inner and outer circles of contacts.

SCORE MOTOR

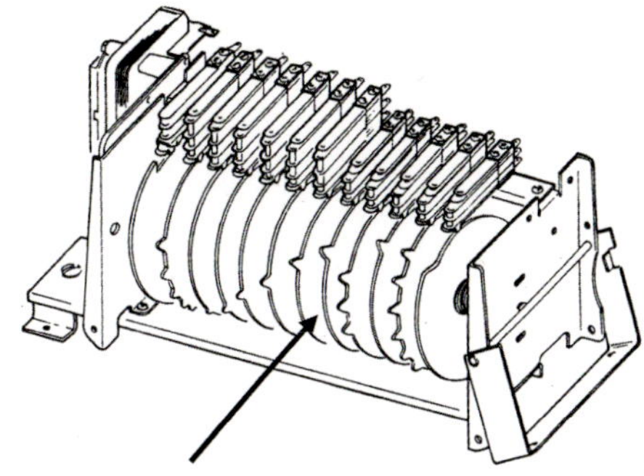

Lubricate edges of cams

Some manufacturers recommended lubricating the edges of cams.

Some electro-mechanical score motors have a place to put a few drops of oil. Motors requiring oil will have a hole, sometimes with a sponge-like substance visible inside the hole or other spot specifically marked for oil. Chances are the motor hasn't been lubricated in years. Use a light machine oil, such as 3-in-1 or oil designed for sewing machines. You only need a few drops, maybe once a year.

Many electro-mechanical score motors do not have an oil hole. These motors should not be lubricated until you can determine if they should be lubricated at all, and if so, where to put the lubricant. The instruction manual or a card stapled to the cabinet will have lubrication instructions. If there are exposed gears, put a drop or two of oil on the gears. Do not oil the rotating motor armature.

The manufacturers recommended putting lubrication on the outside edges of the score motor cams (the large rotating discs, also called biscuits) where they contact the switches. I don't lubricate motor cams and have never found a need for it. Besides, some score motors have ten or more cams, and that is a lot of lubricating.

SCORE REELS

Most score reel mechanisms (drum units) have a rotating plastic cam underneath the score reel. The cam opens and closes switches when the score number comes around to 9 and 0. On some score reels, the cam is a molded part of the reel itself that the switches rub against (see 152). I lubricate the rim of the cam.

On other score units, the cam is a separate plastic component mounted underneath the score reel that has a slot that a plastic pin rides in (see page 152). I lubricate the slot and the tip of the pin.

This lubrication was not recommended by the manufacturers, but I find that it sure helps those old score reels rotate from the 9 to the 0 without getting stuck.

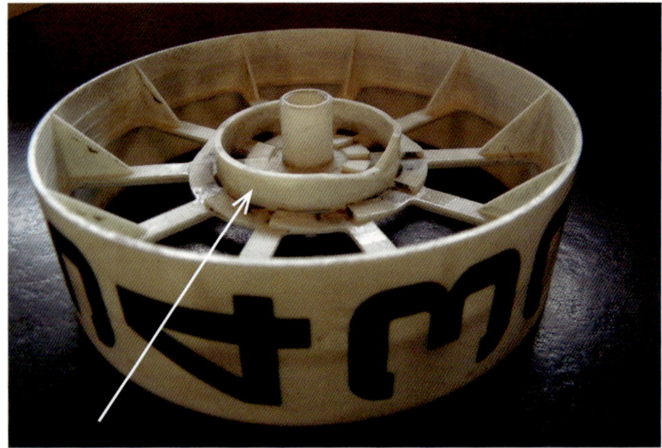

Score reel with cam molded onto the reel. Lubricate the edge of the cam (where the arrow is pointing).

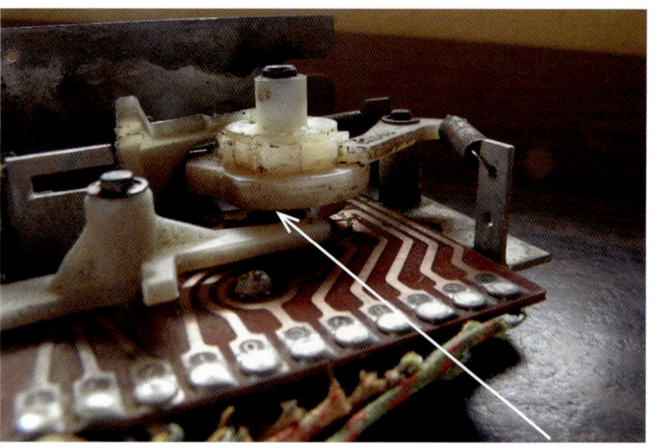

Score unit with the score reel removed to show the rotating cam and pin in the slot under the cam. Lubricate the pin and slot.

> Pinball people know. Playing the game can increase power of concentration, improve mind-eye-hand coordination, expand the imagination, further self-confidence, introduce kindred spirits, promote healthy competition, and provide a very special sense of relaxation. At the drop of a coin.
>
> —Jim Tolbert, *Tilt*

Two examples of early woodrails with scoring indicated by painted numbers on the backglass:

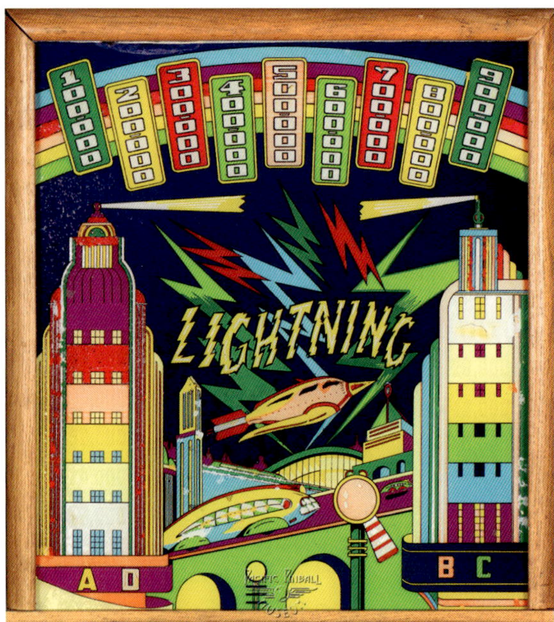

Marvel Manufacturing Company **Lightning**, 1947. 5 balls for five cents. Marvel made pinball machines from 1944 to 1948. Donated to the PPM by Pat and Gordon A. Hasse, Jr. Photography by RobPerica.com

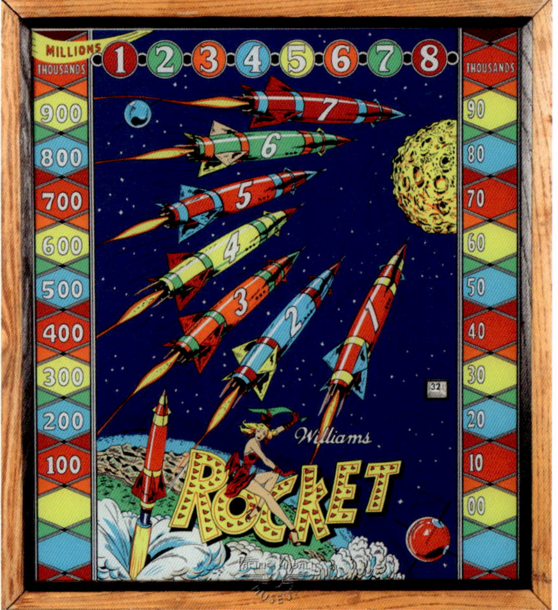

Williams **Rocket**, 1959. Harry Williams, the founder of Williams Pinball, designed this machine himself. A game was still 5 balls for 5 cents. The next year, 1960, cost of a game jumped to a dime. Donated to the PPM by Larry Zartarian. Photography by RobPerica.com

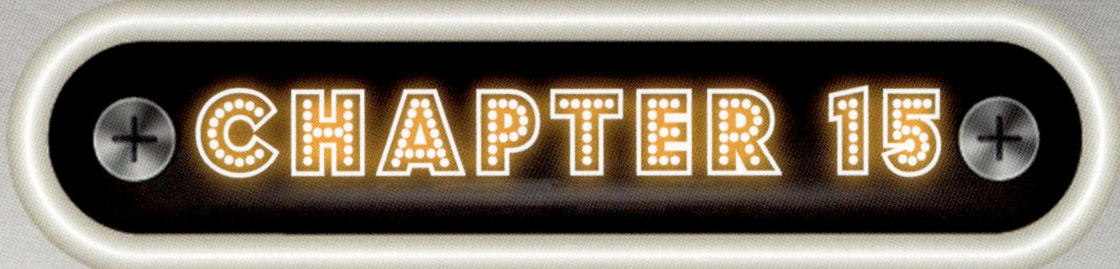

SOURCES AND RESOURCES

PINBALL PARTS, SUPPLIES, SCHEMATICS, MANUALS, AND BOOKS

All of the old pinball manufacturers are long gone, but new parts for most pinball machines are readily available. There are many independent pinball businesses that sell parts, instruction manuals, and schematics for new and old machines, as well as supplies, tools, and books. Many of these businesses have retail stores if you live nearby, and all have mail order sales through the internet or over the phone. Type in "pinball machine parts" or "pinball machine supplies" in a search engine and take your pick.

Many of these pinball businesses are very friendly and are willing to take time to answer your questions, make suggestions, and even help you troubleshoot problems. You might get better prices if you comparison shop, but if you find a helpful, reliable supplier, pay a little more if necessary.

You can buy pinball parts and supplies on eBay and Amazon, but the prices are often higher than those at regular pinball suppliers. Don't think you're getting a good deal just because someone lists it on eBay and a bunch of people bid on it. You also might wind up buying the wrong parts, or inferior parts, when you deal with unknown internet sellers. Regular pinball suppliers, people who are in the pinball business, know their business. Tell the supplier what machine you have and what you need, and let the supplier help you determine what you should purchase.

PARTS FOR OLD MACHINES

Some pinball parts are universal, or just about universal, that can fit machines of all vintages, and are readily available: balls, lamps, rubber rings, solenoids, relays, switches, and many types of flippers, bumper caps, and other interchangeable parts.

For old machines, some parts may be hard to locate, and if found are expensive. Always keep your eye out for someone getting rid of a broken, worn out pinball machine that you can get for free or for $10 or so. The "junker" is probably not worth fixing, but there are many valuable parts in it, often in excellent condition, that you can strip out and save. I've often repaired machines using parts I rescued from old graveyard machines. The machine might even have a usable playfield glass or even a still-attractive backglass.

You can find used parts on the Mr. Pinball website (type in "Mr. Pinball" in a search engine), which has a free classified ad section.

NEED MORE HELP?

I've piled about everything I can think of into this manual, everything I've encountered in more than twenty years fixing pinball machines. But as sure as Monday follows a rainy weekend, sooner or later you will have a problem I didn't cover or a question I didn't answer.

Over the years several pinball repair guidebooks have been published. Every one of them was a bit different, covered different problems, and had different suggestions. Most of these manuals are long out of print, but all of them occasionally show up on Amazon and eBay. Keep your eye out, and unless someone is asking an exorbitant price, grab the guidebooks up when you find them.

Of course, every answer to every pinball problem is on the internet. Often helpful information, but often misinformation: inaccurate, worthless, and possibly dangerous. It's amazing how many people are willing to give advice without knowing what they're talking about. Just in the last week (as I write this), I found wrong information about a circuit board problem on a Gottlieb/Premier game, and wrong information on a problem with a Williams rotating unit. I found a strong admonishment to never lubricate anything on a pinball machine, which is just nonsense. As the Old Professor said, "Know your source."

Get to know people who own and love pinball machines, and develop a network of friends who share information and help with troubleshooting. I had a good pinball buddy who I got together with about once a month, solving each other's pinball problems (and drinking each other's beer). I remember one time I was struggling to find a short in the wiring of an old machine. I had gone through the circuits over and over and could not find the problem. Just about the time I was ready to take a sledge hammer to the #*&!! machine, my friend just happened to stop by. He looked at the machine for about one minute, spotted a broken wire, and said "Is that your problem?"

> Continue experiencing the joy and frustration of the game, introduce others where able, and ignore the critics.
>
> —**Marco Rossignoli**, *Pinball Perspectives*

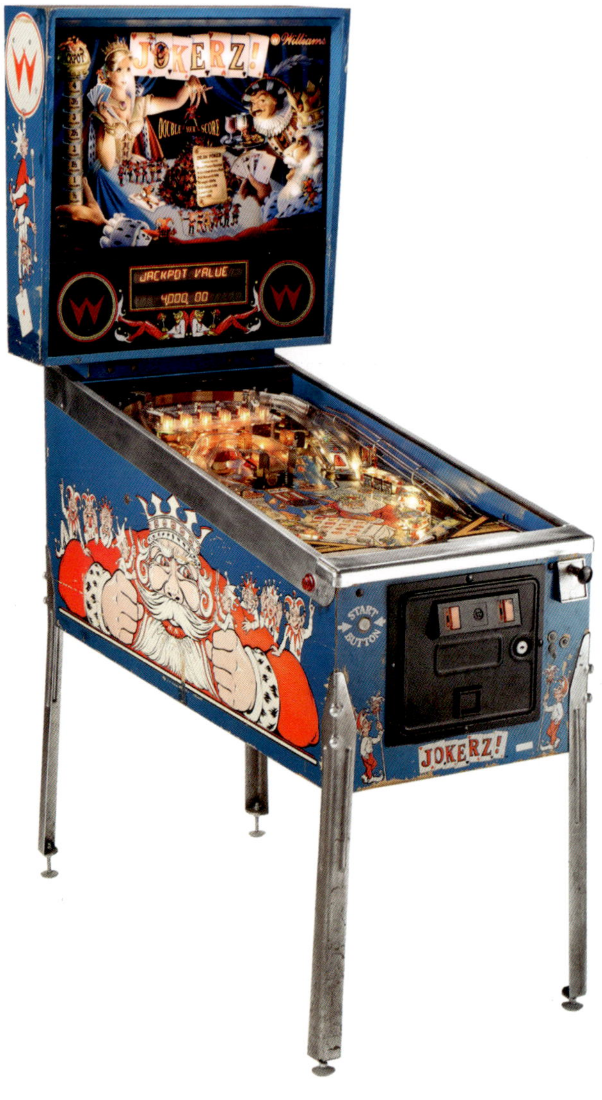

Williams **Jokerz**, 1988. A hand of cards in the backbox spins when you hit a target, giving you anything from a pair of 2s to a royal flush.

CHAPTER 16

PROFESSIONAL REPAIRS

If your machine needs repair beyond what you can accomplish yourself, there are many people who repair pinball machines. Some are professionals with full time businesses. Some are hobbyists who do occasional repairs for others. Some charge by the hour, some charge by the machine, and some will just help you out for free.

Many repair people make house calls, but some will require you to bring your machine to their shop. Finding a repair person who will make house calls is obviously a major benefit to you, but be sure to ask what the charge is for travel time.

If you know anyone who owns a pinball machine, that person may know someone who does repairs. If there is a nearby pinball show, you will find someone there who repairs machines.

Try the internet. Type in "pinball repair" and your city and state in a search engine and see if you get any local results.

Do you still have a Yellow Pages? Look under Amusements, Amusement Devices, Arcades, Coin Operated, Novelties, Vending, Music, Jukeboxes, and Video Games. Amusement businesses (route operators) are usually listed under one of the above categories and they sometimes do repairs for home machines, or they are likely to know someone who does repairs. Quite often there is someone at the amusement business who moonlights after work, repairing pinballs.

"CAVEAT EMPTOR" (PINBALL OWNER BEWARE)

Before you hand over your machine to a repair person, find out how much experience the person has and how knowledgeable the person is about your particular type of machine. Electro-mechanical machines and electronic machines are different technologies from different eras. Early electronic machines had totally different and often rudimentary circuits compared to new machines.

For electronic machines, the problem is quite often on one circuit board. If you can isolate the problem, or if you can explain the problem over the phone to a repair person, all you may need to do is remove the one circuit board and send it out for repair.

Get a firm price from the repair person before agreeing to the work. Repair charges vary dramatically from business to business and from city to city. There is no "industry standard," no typical fees. If it is an hourly rate, find out how many hours the machine will require. Don't wind up spending more money on a repair than what it might cost to buy another machine or maybe just a replacement circuit board. Ask about the cost of parts. Some repair people mark up parts considerably.

Find out what kind of guarantee comes with the work. Many repair shops guarantee their repairs for at least 30 days.

I always give my customers a detailed list of what repairs I made to their machines, so if the machine breaks again, both I and my customer can see if it's something I already tried to fix, or if it is something unrelated to my previous work.

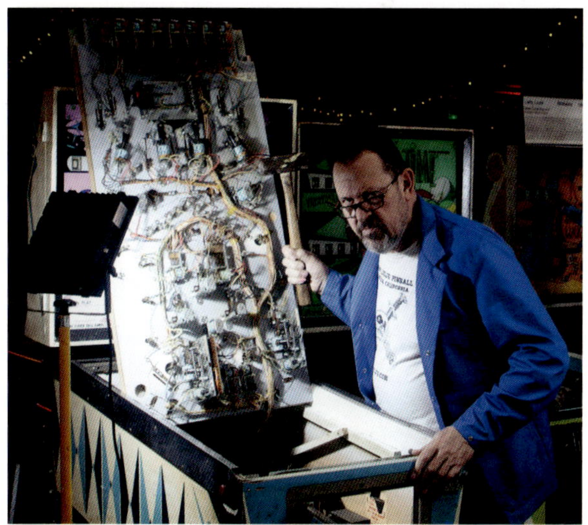

A professional pinball repairman will approach every repair in a conscientious, business-like manner, and will always have the right tools for the job.

CHAPTER 17

START A PINBALL BUSINESS

Why not? Turn your hobby into a business. Have fun and get paid for it, too. Best of all, it can be done one step at a time.

PINBALL ROUTE

Start a small pinball route. We all know that there are almost no pinball machines in bars, bowling alleys, and pizza places anymore. Full time amusement operators have switched to arcade games or just jukeboxes and coin-operated pool tables. But there is still a market for pinball machines. If no one in your area is serving that market, why not you?

This is the kind of business you can operate in your spare time. It's the perfect sideline business for a pinball person. You can start with two or three, or even just one pinball machine. The machines don't have to be shiny new models either. People love to play old pinball machines as much as the newest behemoths.

Where do you begin? First, find out if anyone else is supplying pinball machines to local businesses. You probably don't want to step on anyone's toes or damage their livelihood (or find out the Mafia is still around). You're looking for a need to fill, not a rival to fight.

Decide if you're up to the work involved. You'll be moving heavy machines regularly. You'll have to get good at on-site repairs, especially the flippers. People may not mind if a target isn't working, but they will not play a machine with broken flippers. You'll have to keep an eye on your machines. Machines get beat up in bars and other rowdy places. Coin mechanisms get jammed, glass gets scratched. The machines have to look nice. They will need regular cleaning. Lamps will burn out and need to be replaced.

Talk to local business owners who might have room for a pinball machine and ask if they'd be willing to have a machine in their establishment. You can make whatever percentage split you and the owner agree on. Typically operators get 75% of the income. The business owners get 25%, not much, but a business owner may just want the machine because it pulls more people into the establishment.

Make inquiries to your city, county, and state governments to learn if you need a special operator's license or amusement license for machines that are out on route, and if so what the requirements are. Years ago, pinball machines were licensed the way slot machines in gambling establishments are licensed. A few jurisdictions still license pinball machine route operators, but most don't have any requirements other than those that apply to all businesses. Arcades, by the way, are often heavily regulated, so be sure some government bureaucrat doesn't mistakenly think you are starting an arcade.

PARTY RENTAL BUSINESS

If dealing with bars and rowdy customers is not to your liking, consider starting a pinball party rental business. Operators easily get $100 or $200 for a one-day rental, setting the machines for free play for party guests. The machines won't get beat up, and you won't have to be constantly cleaning and repairing machines and counting quarters. You won't have to share the income with anyone. And no special amusement licenses are required for private party rentals.

SALES AND SERVICE BUSINESS

Eventually someone will want you to repair a machine or want to buy a machine from you. That's how I got into the business. Someone asked, I said sure—even though I wasn't sure at all what I was doing—and Bell Springs Pinball Alley was born.

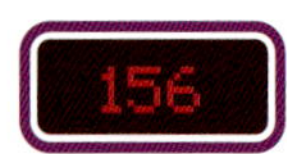

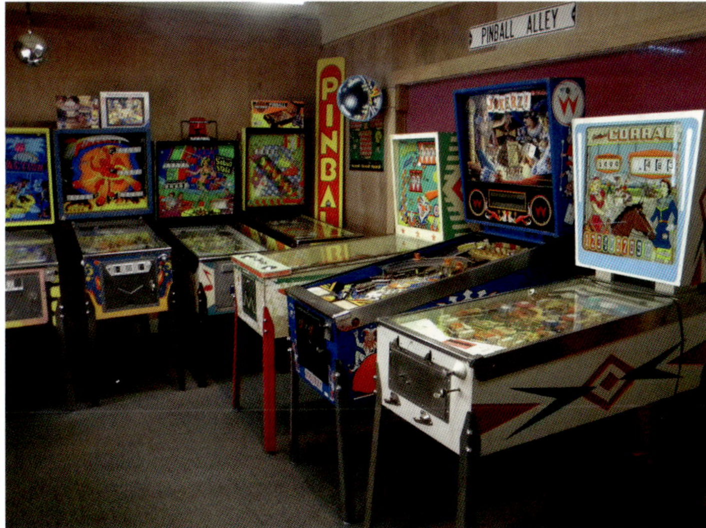

Showroom floor of the first location of Bell Springs Pinball Alley: Half of a two-car garage.

THE "BUSINESS END" OF YOUR BUSINESS

If it looks like it's worth a try, treat it like a real business, even if it's just one pinball machine. It's easy to set up a small business. You don't need to incorporate or get too complicated, and you can run it out of your home. According the US Small Business Administration, there are 19 million home-based businesses in the United States, most of them one-person, part time ventures. Why not you?

Without getting into too much detail, starting a pinball business will have the same requirements as every other small business in your community.

SETTING UP A SMALL BUSINESS

Most small businesses in the United States are sole proprietorships—one person businesses. They are not incorporated and they are not LLCs (limited liability companies). Sole proprietorships do not need to register with the IRS or any other federal agency.

Some cities and counties require small businesses to get a business license and some don't. Some states require you to have a sales tax permit and some don't. Some localities require you to file for a fictitious name (a DBA or "doing business as") and, again, some don't. Contact your city and state, or search their websites, to find out their requirements.

Keeping records: You'll need to keep financial records. They can be quite simple: a record of your income and a record of your expenses. You can do it with pencil and paper, which despite the omnipresence of computer platforms, is still how thousands of small businesses keep their books; or use spreadsheets, or accounting software such as QuickBooks, or some smartphone app. They all produce the same records.

Insurance: Very little insurance is required by law. You may need to get special vehicle insurance coverage. Most vehicle insurance policies do not cover commercial operations. Liability insurance, which covers you if a customer is injured, is optional in most states.

Income taxes: Sole proprietors file a business Schedule C with their 1040 return. If you make a profit, that profit is combined on the 1040 with any other taxable income you might have, to figure your income taxes. Sole proprietors also pay self-employment tax, which is combined Social Security and Medicare for self-employed people. If your state has an income tax, you also have to report your business income on your state tax return.

If you want to know more about all of these business issues, get a copy of my book *Small Time Operator: How to Start Your Own Business, Keep Your Books, Pay Your Taxes, and Stay Out of Trouble*, which is currently in its fifteenth revised edition. It has everything you need to know.

I'm sure you know that a pinball business is definitely labor intensive. But for all of us who are in the pinball business, it's truly a labor of love. And a great way to pay for your favorite hobby. Small time operators can, and do, make good money.

> You can start out easily and simply. You don't have to make the Big Plunge, selling everything you own and going into debt. More than two-thirds of all new businesses are started as part time or weekend ventures, started by people still holding on to a job while they experiment with their new business. Start slowly, try it out, and learn as you go. You'll get there.
>
> —*Small Time Operator*

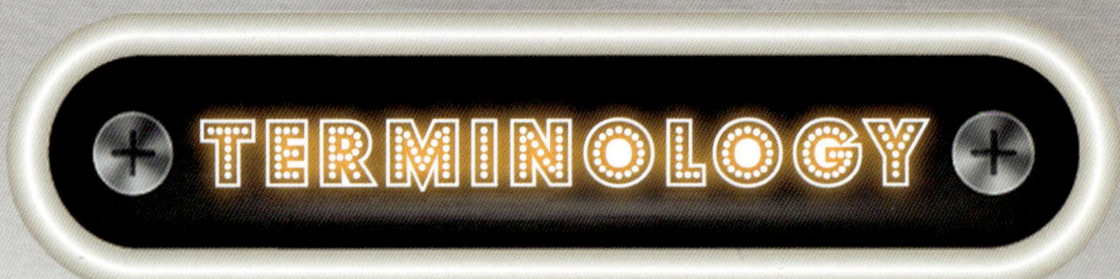

You will not understand this manual, and you won't be able to describe your problems or the parts you need, unless you understand the terminology. I have also discovered that different manufacturers use different terms for the same thing. Ah yes, just to keep us on our toes. So here, alphabetically, is terminology used in pinball.

AC: Stands for alternating current. The power coming into a pinball machine from the wall outlet in your home is AC. Some pinball machines run on AC and some convert the power to DC (direct current).

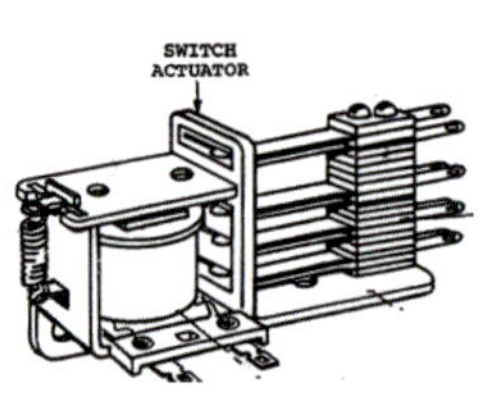

Actuator: The part of a mechanism that operates a switch. In pinball, "actuator" is most commonly used to describe the part of a relay mechanism that attaches to the switches. The switch blades that are moved by the actuator are called the actuator blades.

Add-a-ball: A game where high scores and specials earn extra balls instead of free games. Don't confuse add-a-ball with multi-ball; they mean different things. Multi-ball games can have two or more balls in play at the same time.

Animation: Same as Backglass animation.

Anti-cheat switch. If you bang the cabinet or the front coin door very hard, the anti-cheat switch vibrates and closes, or opens, and shuts down the machine, ending the game. Also called a Slam Switch, Bounce Switch, or Kick-off Switch. A few schematics call it a tilt switch, although it is not a tilt switch and is not part of the tilt circuit.

Apron: Large score card holder mounted at the bottom of the playfield. Also called a Score Card Holder.

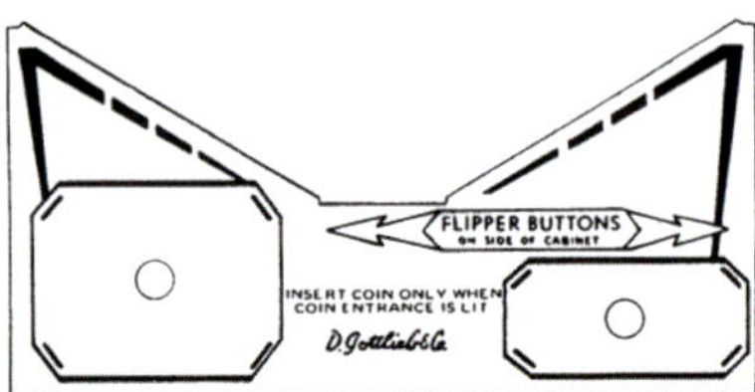

Armature: Armature has several meanings. The metal plate attached to a relay, which is pulled towards the relay when the relay is activated, is called an armature. Some solenoid plungers are also called armatures. The rotating part of a score motor is called an armature.

Attract mode: Newer electronic machines, in game over mode, start flashing lights and making sounds to attract players. "*Play me.*"

Backbox: The upright part of the pinball cabinet, mounted at the back of the machine, holding the backglass and showing scores, ball in play, etc. In electro-mechanical machines, the scoring mechanisms and many of the game adjustments are in the backbox. In electronic machines most of the circuit boards are in the backbox. Also called a Light Box or simply the Head.

Backflash: Same as Backglass.

Backglass: The painted, decorated glass that sits upright in the backbox, displaying scores, ball in play, tilt, game over, match, etc. Starting in the 1980s, backglasses were eliminated and replaced with Translites, though some people still call translites "backglasses."

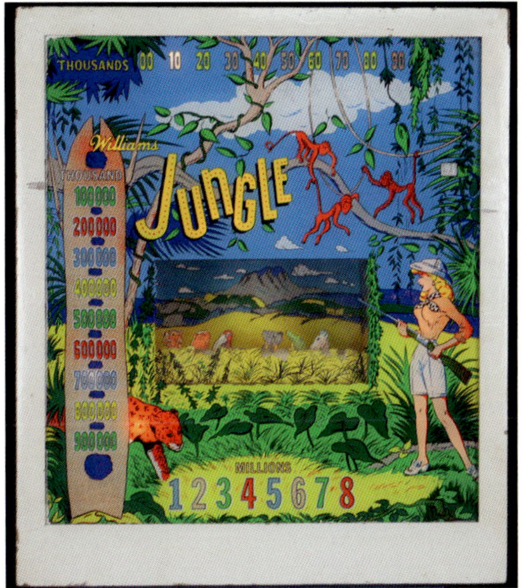

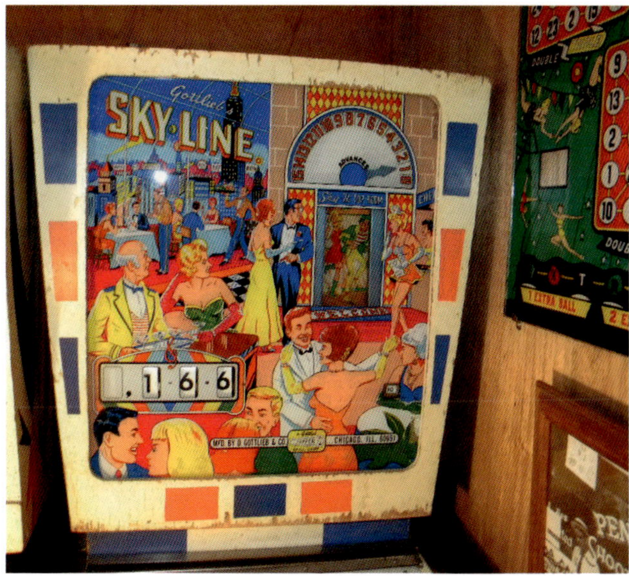

LEFT: Williams **Jungle**, 1960. A shooting gallery game (without the gun) has eight wild animal heads in the backbox. When the pinball hits a target, one animal head drops down. Seven more to go.

RIGHT: Gottlieb **Sky Line**, 1965. With the Chicago skyline in the background, the elevator to this penthouse restaurant has a rotating pointer that moves up one floor every time you hit the advance target, and the elevator doors open to welcome the delightfully drunk couple.

Backglass animation: Actual physical (mechanical) activity taking place inside the backbox and visible behind the backglass. Sometimes the animation was primarily for entertainment, and sometimes the animation was an integral part of the game.

Backup blade: Same as Damping Blade.

Bagatelle: Table game invented in the 1700s where a ball is hit by a cue stick or a plunger and could land in different holes on the playfield. Precursor to the pinball machine. Antique table model pinball machines were often called bagatelles. Some toy pinball machines today are still called bagatelles.

Bakelite: A hard, fiber-like material, about 1/16" thick, and usually reddish brown in color. Used in electro-mechanical machines for wiring connectors and rotating (stepper) units such as bonus, match, player, and ball advance units. Also used as links connecting solenoid plungers to the mechanisms they operate.

Ball count switch / Ball advance switch: Same as Trough Switch.

Ball lifter: Manually operated mechanism that lifts a ball from under the playfield to the playfield surface. Used on all pinball machines until the 1960s, when the automatic ball return was invented.

Ball return solenoid: Same as Outhole Solenoid.

Ball saver: Same as Down Post. The term Ball Saver also refers to a feature in electronic pinball machines where a ball that drained very quickly is returned to the shooter lane to give you another try.

Ball shooter: A plunger that shoots the ball from its resting place at the bottom of the playfield into play. Most ball shooters are hand operated: pull it back and let it go. Some ball shooters are activated by a push button or sometimes a flipper button or some other mechanism (a trigger on a gun). Some electronic machines activate the ball shooter automatically as part of the game play.

Bank: Same as Relay Bank.

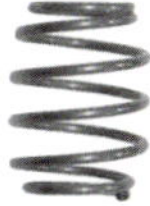

Barrel spring. Barrel-shaped spring about 3/4" long mounted on the ball shooter between the handle of the shooter and the outside of the cabinet.

Bat: See Flipper Bat. Also see Pitch and Bat.

Bingo: A pinball game that was primarily designed for gambling. It had no flippers. The playfield had holes where the ball landed and stayed.

Biscuits: Same as Cam.

Bounce switch: Same as Anti-Cheat Switch.

Bottom panel: In electro-mechanical machines, the board on the floor of the cabinet that holds the transformer, relays, and other components. The bottom panel can be unbolted and lifted up onto the top of the cabinet to make it easier to work on the components.

Break: See Make-Break.

Bridge rectifier: Electronic component that coverts AC current to DC current.

Buffer: Same as Rebound Rubber.

Bumper: See Pop Bumper. Also see Mushroom Bumper. Pinball machines made before 1960 had bumpers that did not kick back like pop bumpers and were just called "bumpers."

Bumper skirt: The plastic ring at the bottom of a pop bumper that tilts when the ball rolls onto it.

Bushings: Plastic sleeves that mount to the cabinet or playfield to hold and guide moving parts. Flipper shafts go through bushings mounted underneath the playfield. The ball shooter goes through a bushing mounted inside the front of the cabinet.

Cabinet: When most people refer to the cabinet, they are talking about the long box that sits on the legs and holds the playfield. The upright part of the cabinet is usually called the backbox, light box, or head.

Cam: A rotating notched disc found on score motors and some score reels. Also called a biscuit.

Captive ball: A ball permanently trapped in a small area of the playfield, usually a short lane or horseshoe loop. The ball in play hits the captive ball, moving it and scoring. Captive balls in straight lanes are also called Messenger Balls.

Card holder: Same as Apron.

Channel: Same as Runway.

Circuit: A network of wires, switches, electronic components, and coils that control an operation or a feature in a pinball machine.

Circuit boards: In electronic machines, the flat boards-mounted in the backbox and sometimes in the cabinet that hold the electronic components: ICs, transistors, diodes, capacitors, resistors, etc. There are different circuit boards for different functions: power supply board, sound board, display board, and main (CPU–central processing unit) board. Circuit boards are also called printed circuit boards or PCBs.

Closed flipper action: Same as Zipper Flipper.

Coil. A coil is a spool of wire wound around a solid or hollow core. Activating a coil creates electro-magnetism. Some coils are called solenoids and some are called relays, depending on their design and function.

Solenoids have hollow cores with moving plungers, and are used in all pinball machines. When the solenoid is activated, electromagnetism forces the plunger out (or in, depending on how it is wired). The plunger may hit something, as in the case of the knocker or a chime; or it may be attached to a mechanism that moves a flipper, or kicks out a ball, or resets a target bank. When a solenoid is deactivated, a spring (or gravity) returns the metal plunger to its resting position.

Relays have solid metal cores and are mounted with an armature, which is a hinged metal plate. Activating the relay pulls in the armature. The armature is attached to a switch, or several switches, and opens and closes the switches when it moves. Relays are used extensively in electro-mechanical machines and occasionally in electronic machines.

Some pinball manufacturers use the word coil only for solenoids. Their schematics and manuals will make a distinction between "coils" (solenoids) and "relays." For the purposes of this manual, coil refers to both solenoids and relays.

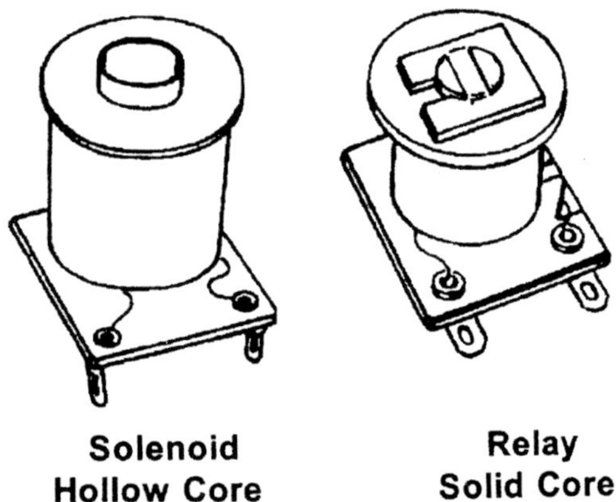

Coil plunger: The moving metal cylinder in a solenoid coil. Also called a Piston.

Coil sleeve: Replaceable plastic or metal liner inside a solenoid.

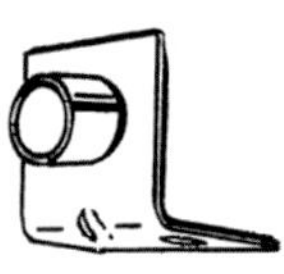

Coil stop: A metal bracket with a short, circular metal plug about $\frac{1}{4}$" to $\frac{1}{2}$" long mounted on it. The coil stop attaches to the end of a solenoid coil. When the plunger enters the solenoid it hits the coil stop and, well, stops. (It's really a plunger stop, but no one calls it a plunger stop.) Also called a Core Plug.

Coin mechanisms / Coin acceptors / Coin rejectors: Three different terms for the same thing: Snap-in brass or plastic units that the coin passes through that starts a new game. Years ago the coin mechanism was also known as a Slug Rejector.

Coin switch: Located on the coin door and activated when a coin hits the switch. Starts a game or adds a credit or an additional player.
Computer chips: Same as Integrated Circuits (ICs). Also called Logic Chips.

Contact point: Same as Switch Contact.

Core: The center of a coil. The coil wire is wound around the core.

Core plug: Same as Coil Stop.

CPU: Short for Central Processing Unit. The CPU board holds the main game ICs and the batteries. Also called MPU for main (or micro) processing unit.

Credits: The number of games on the machine: free games won or quarters already put in.

Credit button: The start button that you push to start a new game. Also called a Replay Button.

Credit switch: In electro-mechanical machines, one of the switches attached to the credit wheel is sometimes called a Credit Switch. Also see Zero Credit Switch.

Credit wheel: On electro-mechanical machines the wheel in the backbox with the number of games showing.

Credit unit: The credit wheel, its brackets, and switches. Also called a Replay Unit.

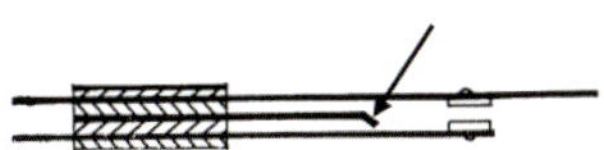

Damping blade: On some switches, the damping blade is mounted next to one of the switch blades and is used to stiffen the switch blade to prevent the blade from bouncing too much. The damping blade is made of a stiffer, thicker metal and does not have any wires connected to it. Also called a pre-tensioner blade or back-up blade.

DC: Stands for direct current. All electronic and some electro-mechanical pinball machines convert the AC power coming from your home wall outlet to DC.

Digital machines: Same as Electronic Machines.

Diodes: Solid state electronic components that block current from flowing in both directions and protect circuits from voltage spikes. Diodes are wired across most coils in electronic machines.

Disappearing post: Same as Down Post.

Displays: Digital scoring devices found on all electronic machines.

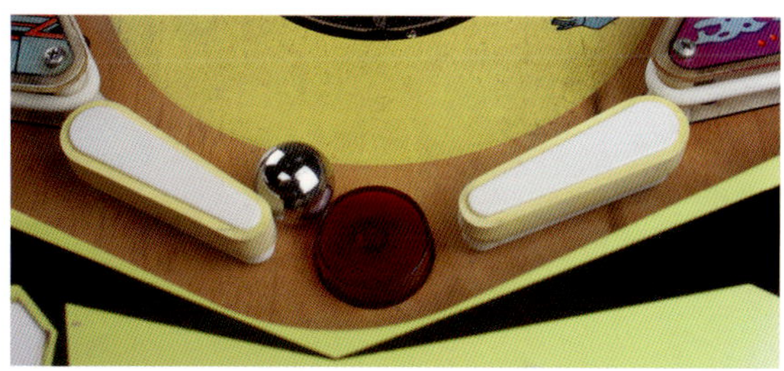

Down post: Some pinball machines have a large round post between the flippers that rises up above the playfield when you hit a certain target, and blocks the ball from draining between the flippers. Hitting another target drops the down post again. Also called an Up Post, a Disappearing Post, or a Ball Saver.

Drain: When the ball leaves the playfield, getting by your flippers or rolling down a side exit lane (outlane), some people call it "drained." Machines are often described as how hard or easy they are to drain the ball. A "real drainer" is not usually a well-liked game.

Drop targets: Spring-loaded targets that drop below the playfield when hit by the ball. Drop targets are often clustered in groups, or banks.

Drum unit (DU): Score reel assembly: the reel, solenoid, switches, components, and mounting bracket.

Eject hole: Same as Kickout Hole. Not to be confused with the Outhole.

Electro-mechanical: All pinball machines manufactured before the invention of solid state, digital scoring machines are electro-mechanical, also known as EM. Electro-mechanical machines have rotating scoring reels. Very old electro-mechanical machines before 1960 had lights on the backglass that showed the scores. Electro-mechanical operation is accomplished by an elaborate network of relays and switches. Machines that have digital scoring and solid state circuit boards are called electronic, solid state, or digital. There is a tremendous difference between the operation and the repair of electro-mechanical machines and electronic machines.

Electronic machines: Modern pinball machines first made about 1976, with digital scoring and IC controlled operation. Also called Digital or Solid State machines. Older pinball machines, pre-electronic, were called Electro-mechanical.
EM: Stands for Electro-mechanical.

End-of-stroke switch: When a flipper is operated, a lever on the flipper mechanism underneath the playfield opens a switch that reduces the power to the flipper coil. The switch is activated at the end of the flipper's stroke, so it is called, quite logically, an End-of-Stroke Switch. Until the last few years every pinball machine flipper had an end-of-stroke switch. Some of the newest machines have eliminated the end-of stroke-switch. End-of-stroke switches are also on score reels and some other mechanisms.

EOS: Stands for "End-of-Stroke." See End-of-Stroke Switch.

Exit lane: Same as Outlane.

Exit hole: Same as Outhole.

Firing chute: Same as Runway.

Flasher: A lamp that flashes on and off.

Flip flags: A playfield feature of four hinged blocks, the size and shape of dominos, that flip back and forth when hit by the ball. Found on only three Bally machines: Wizard, Flip Flop, and Slap Stick.

Flipper: The rotating bats on the playfield that, when you push the buttons on the sides of the cabinet, swing—flip—with speed and force (hopefully) and hit the ball (hopefully). Come on, you know what a flipper is.

Flipper bat: A flipper has two parts: the rotating arm that contacts the ball, and the metal shaft that the arm mounts to. The rotating arm is the flipper bat. Most people just call it the flipper.

Free play: Adjustment on a pinball machine that allows you to play games without inserting coins.

Game motor: Same as Score Motor.

Gap: Space between two switch contacts.

Gate: A hinged metal wire or flapper that the ball rolls under, allowing the ball to exit, but not re-enter, an area of the playfield. Some pinball machines have a gate at the top of an outlane, that swings open to reroute the ball back to the shooter or to a ball-saver that sends the ball back into play.

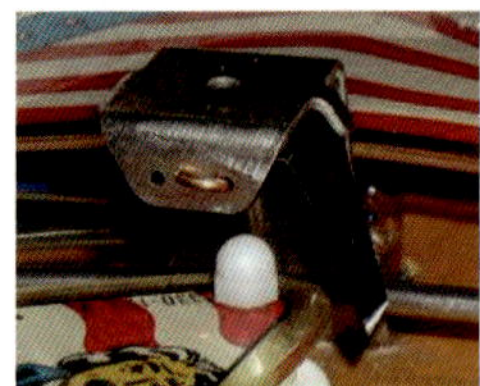

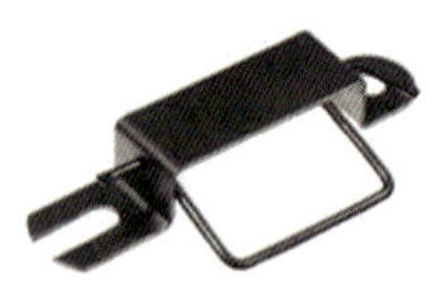

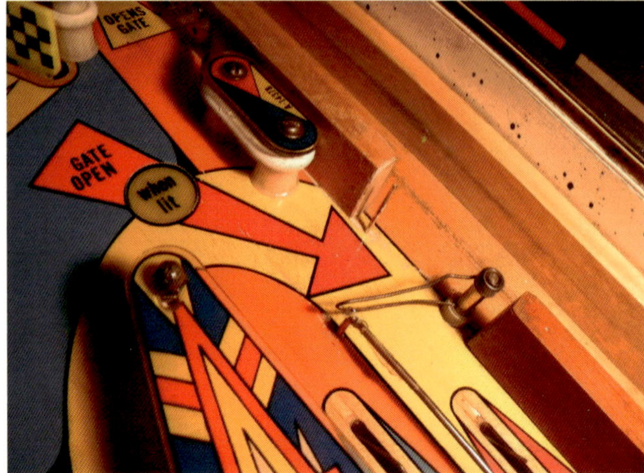

A gate is also a solenoid powered arm that, when open, directs the ball to the shooter lane to be shot again. Gottlieb **Out of Sight**, 1975.

Gobble hole: Hole in the playfield where the ball drops under the playfield and is either routed to the outhole and out of play, or routed to another spot on the playfield where it re-emerges into play.
Half moon: A type of credit unit manufactured by Gottlieb in the mid-1970s that looked like an arch, or half moon, with the number of credits showing behind a rotating plastic disc.

Hand feed: Refers to the operation of old pinball machines from 1960 and earlier where the balls were lifted onto the playfield using a manually operated ball lift mechanism.

Head: Same as Backbox.

High tap: A setting on the transformer that increases the voltage to the coils, increasing their power.

Hold Down bar: Same as Lock Down Bar.

Home models: Some pinball machines were manufactured as "home models" or "professional home models" (the same thing). They were made just for the home market, produced without coin mechanisms, without the match feature, and often without some other features found on commercial machines.

ICs: Integrated circuits. Small electronic components that program electronic pinball machines. Typically anywhere from ½" to 2" long, rectangular, and black, with metal prongs or "feet" that are mounted to circuit boards. Also known as Computer Chips or Logic Chips.

Impulse: One of the rotating cams (discs) on score motors.

Index: One of the rotating cams (discs) on score motors.

Integrated circuits: Same as ICs.

Jet Bumper: Williams's name for Pop Bumper. See Pop Bumper.

Jones plugs: Term for the detachable wiring connectors in electro-mechanical machines. Named after Howard P. Jones, who invented and first manufactured the plugs in the 1930s.

Jumper: Jumper has two different meanings. (1) A jumper is a short length of wire, often with alligator clips on both ends, used to make temporary electrical connections, usually to test circuits and components. (2) Jumper is also a term for a series of short pieces of wire that connect several adjacent relays together.

Kickback: A solenoid-operated plunger found on some machines at the bottom of the left outlane (exit lane). When activated, it kicks the ball back into play.

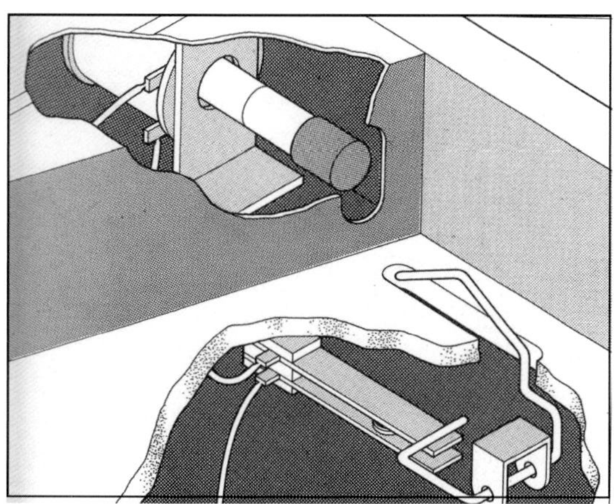

Kicker: Same as Slingshot.

Kicking rubber: Same as Slingshot.

Kick-off switch: Same as Anti-cheat switch.

Kickout hole: A hole in the playfield where the ball lands and is then kicked out by a solenoid. Also called an Eject Hole. Not to be confused with the Outhole, where the ball lands after leaving the playfield.

Knocker: When you win a free game, that great whack or bang—*knock*—is made by a solenoid called the Knocker.

Leaf switch: Switch with metal blades that, when pushed together, make contact to complete a circuit. Used in all electro-mechanical machines and early electronic machines.

Leg adjuster: Same as Leg leveler.

Leg levelers: Screw-on feet that adjust the height of the machine.

Light box: Same as Backbox.

Light shield: Same as Plastics.

Line plug: The plug on the machine's AC outlet cord that plugs into the wall, providing power to your machine. Also called a Wall Plug.

Link: A part that connects a solenoid plunger to the mechanism being operated by the solenoid.

Lock down bar: The long end piece at the front of the cabinet. You unlatch and remove the lock down bar before you can remove the playfield glass and open the machine. Also called a Hold Down Bar.

Lock in: A mechanism or circuit that keeps a relay coil energized or held in the "on" position. Found in most "game over" and "index" relays on electro-mechanical machines.

Locked ball: Sometimes used to describe Captive balls or Trapped balls.

Logic chips: Same as Integrated Circuits (ICs).

Make-break switch: A switch that closes one circuit (make) and at the same time opens another circuit (break).

Match: At the end of a game, the pinball machine lights a random number on the backglass or the display: a 10, 20, 30, 40, 50, 60, 70, 80, 90, or 00. If the last two digits on your score (or the last single digit on games from the 1960s and earlier) are the same as the "match" number—the numbers match—you get a free game.

Messenger balls: Captive balls in short straight lanes that, when hit by a ball in play, are shoved up the lane to hit targets.

Micro switch: Commonly used in electronic pinball machines. See Switch.

Module: In electronic machines, circuit boards are often called modules. Any electronic part that can plug and unplug is sometimes called a module.

Motherboard: Same as the CPU circuit board.

Motor: For the purposes of this manual, "motor" refers only to the Score Motor in electro-mechanical machines (see Score Motor). A few pinball machines use additional motors to reset a relay bank or operate some feature in the machine, though these are very uncommon.

Motor cam: A rotating notched disc that the score motor rotates and that opens and closes the motor switches.

Motor switches: The switches mounted on, and opened and closed by, the score motor.

MPU: Main (or micro) processing unit. Same as CPU.

Multi-ball. A pinball game where more than one ball can be on the playfield at the same time.

Mushroom bumper: Small bumper, shaped like a mushroom, that has a spring-loaded cap that lifts up when hit by a ball, closing a switch. Found on many Bally electro-mechanical machines.

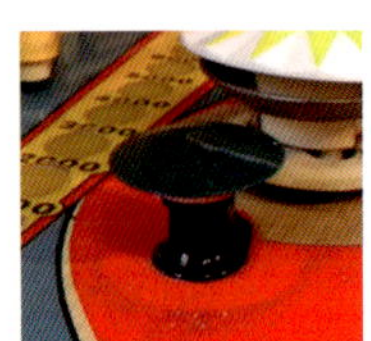

New old stock (NOS): Original pinball machine parts, especially playfields, backglasses, and plastics that were never installed on a machine and are in brand new condition.

Operator: A business that owns pinball machines, putting them out on "routes" such as bowling alleys and bars.

Outhole: That's where the ball lands after it leaves the playfield after it got by your flippers, or drained out a side lane. The switch the ball sits on when it is in the outhole is called the Outhole Switch. The solenoid that kicks the ball out of the outhole is the Outhole Solenoid (also called a Ball Return Solenoid).

Outlane: A lane on the side of the playfield where the ball rolls behind the flippers and leaves the playfield, landing in the outhole.

Over the top: Same as Turn Over.

Pachinko/ Pachinco/ Pacinko: A gambling pinball game from Japan. It is small, about 2 feet by 3 feet, and mounts on a wall, or upright in a cabinet. People shoot tiny balls up, and the balls work their way down the vertical playfield and around little pins sticking out of the playfield, trying to land in a scoring hole. Pronounced "pa-chink-o" and spelled several different ways.

PCB: Stands for Printed Circuit Board. Same as Circuit Board.

Pin game: A pinball machine.

Piston: Same as Coil Plunger.

Pitch and bat. A baseball themed arcade machine resembling a pinball machine where the ball comes out of a hole in the center of the playfield (the pitcher's mound) and is hit by a mechanical bat.

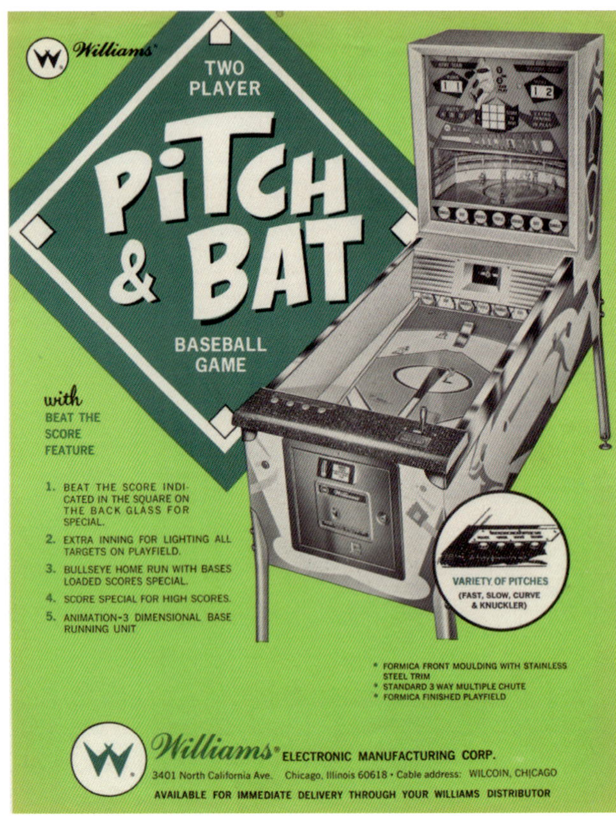

Plastics: The decorated flat plastic shields that are mounted on top of the playfield, usually covering lamps and concealing kickers and rubber rings. Also called Light Shields.

Playboard: Same as Playfield.

Playfield: That's the big board with the bumpers, targets, holes, and flippers that the ball rolls around on. Also called a Playboard or Table.

Plunger: A cylindrical metal rod that moves in and out of a solenoid coil. Also called a Coil plunger or a Piston. The ball shooter is sometimes called a plunger.

Plunger stop: Same as Coil Stop.

Pop bumper: Action bumper on the playfield that, when hit by the ball, hits the ball right back and sends it flying off. Gottlieb named them Pop Bumpers. Bally and Chicago Coin called them Thumper Bumpers. Williams called them Jet Bumpers.

Pre-tensioner blade: Same as Damping Blade.

Printed circuit board (PCB): Same as Circuit Board.

Professional Home Model: Same as Home Model.

Rebound rubber: Round rubber bumper resembling a small wheel, on the top left side of the playfield on some pinball machines. Also called a Buffer.

Rectifier: Same as Bridge Rectifier.

Relay: A relay is a wire coil wound around a solid core. See Coil.

Relay bank: A number of relays mounted next to each other on a common assembly. A relay bank usually has one reset mechanism that resets all of the relays at the same time. Found on many electro-mechanical machines.

Replay: When you light the special or reach the high score and win a free game, it's called a replay.

Replay button: The start button to start a game. Also called a Credit Button.

Replay machine: The manufacturers referred to machines that award a free game as replay machines, to distinguish them from add-a-ball machines. The term was originally created to tell the authorities that the machine was not a gambling game, which many early pinball machines were.

Replay unit: Same as Credit Unit.

Return lane: A ball lane on the side of the machine that returns the ball to the flipper to be hit back into play.

Rollover: A wire switch or button the ball rolls over, scoring points or activating circuits.

Rotating target: Same as Roto Target.

Rotating unit: Same as Stepper Unit.

Roto target: A rotating target feature found on some electro-mechanical machines. The rotating wheel had ten or a dozen targets around it, with only one or two targets visible at a time. You try to hit the visible targets, then try to hit a switch that would spin the roto target, exposing new targets to hit.

Rubbers: European term for rubber rings.

Runway: The narrow alley the ball shoots up when hit by the ball shooter. Also called the Channel, Shooter lane, or Firing chute.

Safety glass: Same as Tempered Glass.

Schematic: The wiring diagram.

Score cards: Printed cards, usually two on each machine, one showing game playing instructions, the other showing scores that win free games.

Score card holder: Same as Apron.

Score motor: All electro-mechanical machines have a score motor, a large rotating unit with cams and dozens of switches that control many of the game functions. The score motor is always mounted on the bottom of the cabinet. Also called a Game Motor or just Motor. Electronic machines do not have a score motor.

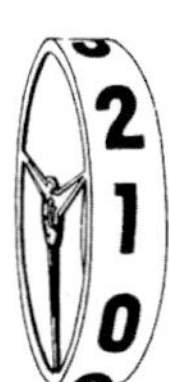

Score reels. On electro-mechanical machines, rotating reels with numbers on them mounted behind the backglass showing the score.

Score reel motor: On a very few old electro-mechanical machines, the score reels were operated by a motor. Very rare. Not to be confused with the Score Motor.

Shooter: Same as Ball Shooter.

Shooter lane: Same as Runway.

Shopped: The term usually means that a machine has been thoroughly repaired, adjusted, and cleaned. Everything works the way it should. There are new rubber rings, no burned out lamps, and shiny balls.

Short / Short circuit: A short circuit—a "short"—is a faulty wiring situation where electrical current bypasses part of the circuit (shortens the circuit), causing components to malfunction and generating more heat than it should: a dangerous situation. The term "short" also refers to intentionally bypassing a component to troubleshoot problems in a machine.

Shut off switch: Same as Anti-Cheat Switch. The on-off power switch under the cabinet is also sometimes called a Shut Off Switch.

Skirt: Same as Bumper Skirt.

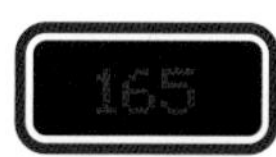

Slam switch: Same as Anti-Cheat switch. Also known as a Slam Tilt.

Sleeve: Same as Coil Sleeve.

Slingshot: A solenoid powered lever mounted behind a rubber ring on the playfield, usually just above the flippers. When the ball hits the rubber ring, the ring pushes against and closes a switch, the slingshot is activated and kicks the ball back into play. Also called a Kicker or Kicking Rubber.

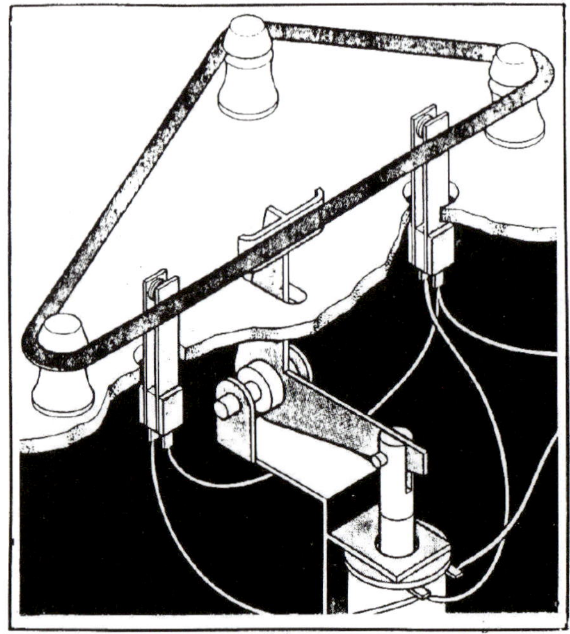

Slow blow (SLO-BLO) fuse: A delayed action fuse that can withstand a momentary surge of high current. Also called Time Delay.

Slug rejector: Same as Coin Mechanism.

Solenoid: A solenoid is a wire coil wound around a hollow core. See Coil.

Solid state machines: Same as Electronic Machines.

Spacer: See Switch spacer.

Special: Special almost always means a free game. When the Special light is lit and you hit the Special target or roll over the Special switch, you are awarded a free game. Some machines allow you to switch the special award from a free game to an extra ball.

Spinner: Spinning target. The ball rolls under the spinner and gets it spinning, racking up points with each rotation.

Stepper (step-up, step down) unit: In electro-mechanical machines, a rotating mechanism that is used as a ball count unit, player unit, or bonus advance unit, as well as other functions. Also called a Rotating Unit.

Stool pigeon: Original name for the tilt. Invented by Harry Williams, founder of Williams Pinball.

Strike plate: The metal ring that the tilt pendulum goes through and hits if you shake the machine too hard.

Switch: On electro-mechanical machines, switches are comprised of two or three switch blades separated by insulating switch spacers. Each blade has a contact, a little rivet or nipple that touches the contact on the adjacent blade to make electrical contact. Also called leaf switches. On electronic machines, switches on the playfield may be the same "leaf switch" type used on electro-mechanical machines, or they may be micro switches, with only one blade.

Switch contact: The riveted point or nipple on the end of leaf switch blades.

Switch spacer: Small bakelite piece separating and insulating the switch blades.

Switch stack: Two or more switches mounted together and all moving together when activated. Often found mounted on relays.

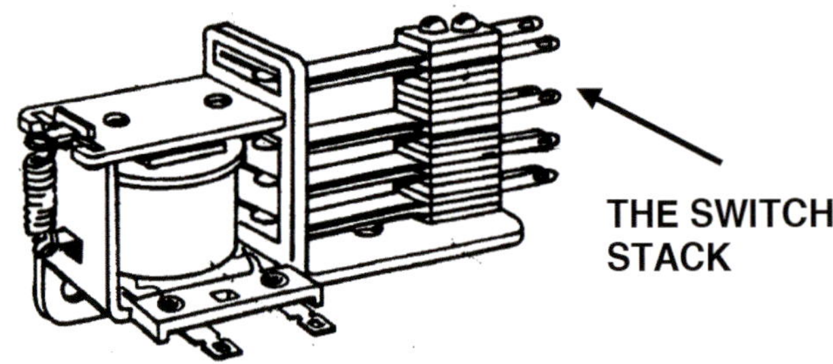

Table: Same as Playfield.

Tap: See High Tap.

Target: A bulls eye or similar upright object that, when hit by the ball, closes a switch behind the target.

Tempered glass: A heat-treated glass that, when it breaks, shatters into tiny pieces not likely to cut you as badly as regular window glass. All manufacturers used tempered glass to cover playfields. Also called safety glass.

Thumper bumper: Bally's name for a pop bumper. See Pop Bumper.

Tilt: Tilt is what happens when you shake a machine too hard. On machines made before 1970, a tilt ended the game. On later machines the tilt cost you the ball in play. Many electronic machines let you program the machine to give a warning or two when you tilt before you are actually penalized.

Tilt switch: A switch that activates the tilt circuit when the machine is shaken too hard. Most pinball machines have two tilt switches: a swinging pendulum and a rolling ball. In a few machines the anti-cheat switch is sometimes called a tilt switch, but it is not really a tilt switch.

Time delay fuse: Same as a Slow Blow.

Tommy: A deaf, dumb, and blind kid who sure plays a mean pinball.

Topper: A headpiece or crest mounted on top of the backbox.

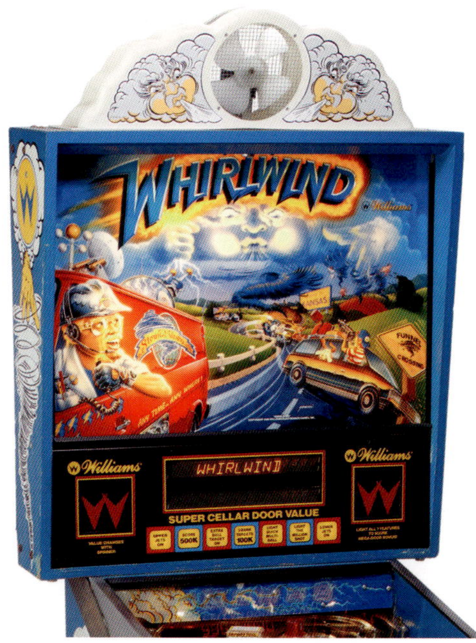

Williams **Whirlwind**, 1990. The "topper" fan blows in your face when the storm is coming.

Transformer: A large, heavy metal unit, about three inches by five inches, that converts (transforms) 120 volt line current into lower voltages to operate the pinball machine. The transformer is mounted on the bottom of the cabinet or, in some electronic machines, in the backbox.

Translite: A flexible, plastic decorated sheet mounted in the upright backbox behind clear glass, that first appeared on pinball machines starting in the 1980s. Before translites were invented, pinball machines used painted backglasses. Some people still refer to translites as "backglasses."

Trapped ball: A ball that lands in a kickout hole on the playfield and stays "trapped" in the hole until released by a game maneuver, such as hitting a target or launching multi-ball.

Trigger lever/arm: The part of a drop target mechanism that lifts the targets.

Trim platters: Clear, thin, flexible plastic or Mylar disks about 3"–4" in diameter mounted under pop bumper skirts on some machines, to protect the playfield paint. Some trim platters are mounted loosely, while some have adhesives that stick to the playfield.

Trough: Guide rails mounted on the playfield and hidden by the apron. When the ball is kicked out of the outhole and back into play, landing in front of the ball shooter, it travels on the trough.

Trough switch: After a ball lands in the outhole and drops below the playfield, the ball rolls on the trough and over the trough switch on its way to the ball shooter. Sometimes called a Ball Count Switch or Ball Advance Switch. Don't confuse the trough switch with the Outhole Switch. They are different.

Turn over: On electro-mechanical machines, after you reach the highest score the score reels can record—usually 99,999—the reels all reset to 0 and keep scoring. You turned the game over, you hit 100,000. Also called Over The Top.

Unit: A group of components assembled and mounted together to perform a specific operation, such as a flipper unit, or a match unit, or a drum unit (score reels).

Up post: Same as Down Post.

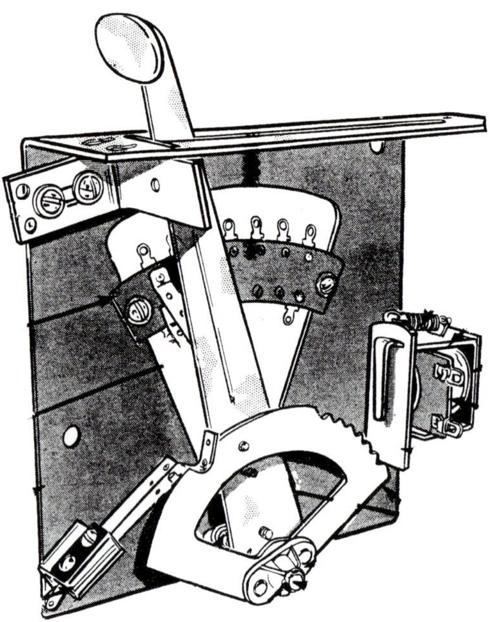

Vari-target: A spring-loaded, ratcheted target that slides back when hit by the ball. The harder the target is hit, the farther it moves and the more it scores.

Vintage: Some people use the word vintage to describe electro-mechanical pinball machines. Some people consider early electronic machines from the 1970s to be vintage.

Wall plug: Same as a Line Plug, mounted at the end of the AC power cord on the machine. Not to be confused with the wall outlet in your home where the wall plug plugs in.

Wedge head: Term for a pinball machine whose upright backbox was trapezoidal: the top of the backbox is wider than the bottom of the backbox. A feature of many Gottlieb machines made during the 1960s. Some pinball machines were "reverse wedge heads": the top of the backbox was narrower than the bottom.

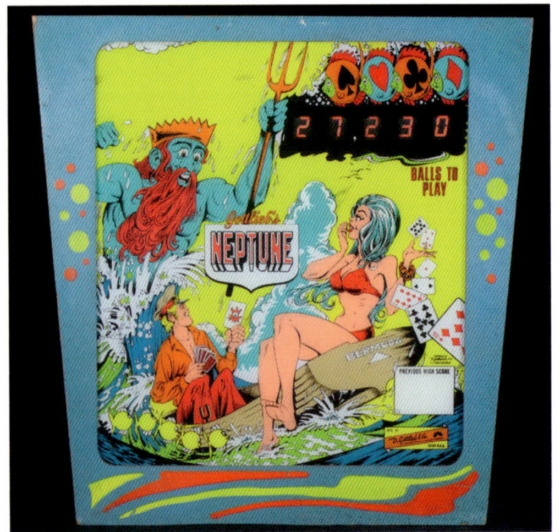

Gottlieb **Neptune**, 1978, an unusual electronic wedgehead machine. Almost all wedgeheads were EMs.

Whitewood: Prototype pinball playfield when designing a new pinball machine.

Wide body: Term for pinball machines that have wider playfields than the standard size. Very few widebody machines were made, most of them in the 1980s and 1990s.

Wiper: The rotating blades on a Stepper or Rotating Unit.

Wiring diagram: Same as Schematic. However, before 1950, the term wiring diagram was occasionally used to describe the physical layout of the wiring in a machine and was a separate, and very different diagram than the schematic.

Wood rail: Term for pinball machines made before 1960, where the trim (railing) around the playfield glass was made of wood.

Wow: In some add-a-ball games when you win an extra ball, a lamp behind the backglass comes on and lights up the word WOW.

Zero-nine (0–9) unit: Same as Match Unit.

Zero switch / Zero position switch / Zero credit switch: On electro-mechanical machines, when the credit wheel shows no games left, the zero switch opens and no more games can be played until another coin is dropped in the slot.

Zipper flippers: Flippers that would slide (zip) together when you hit a certain target, blocking the ball from draining. Hitting another target would return the flippers to their original, normal position. Found on some Bally and Williams machines from the 1970s. Williams called them Closed Flipper Action.

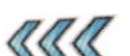

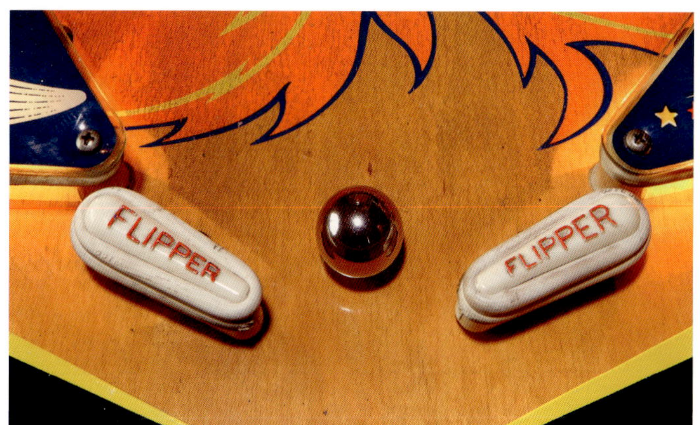

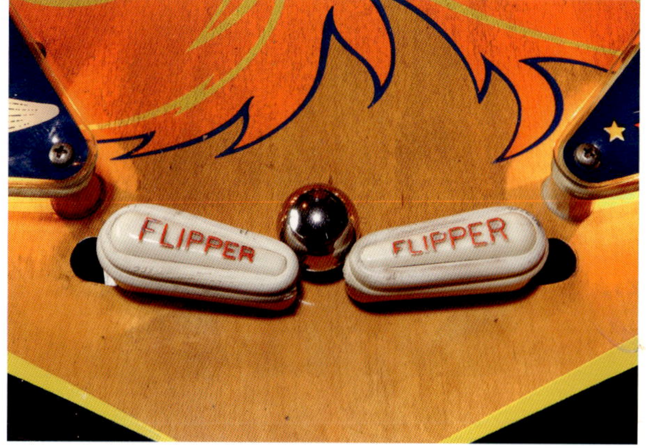

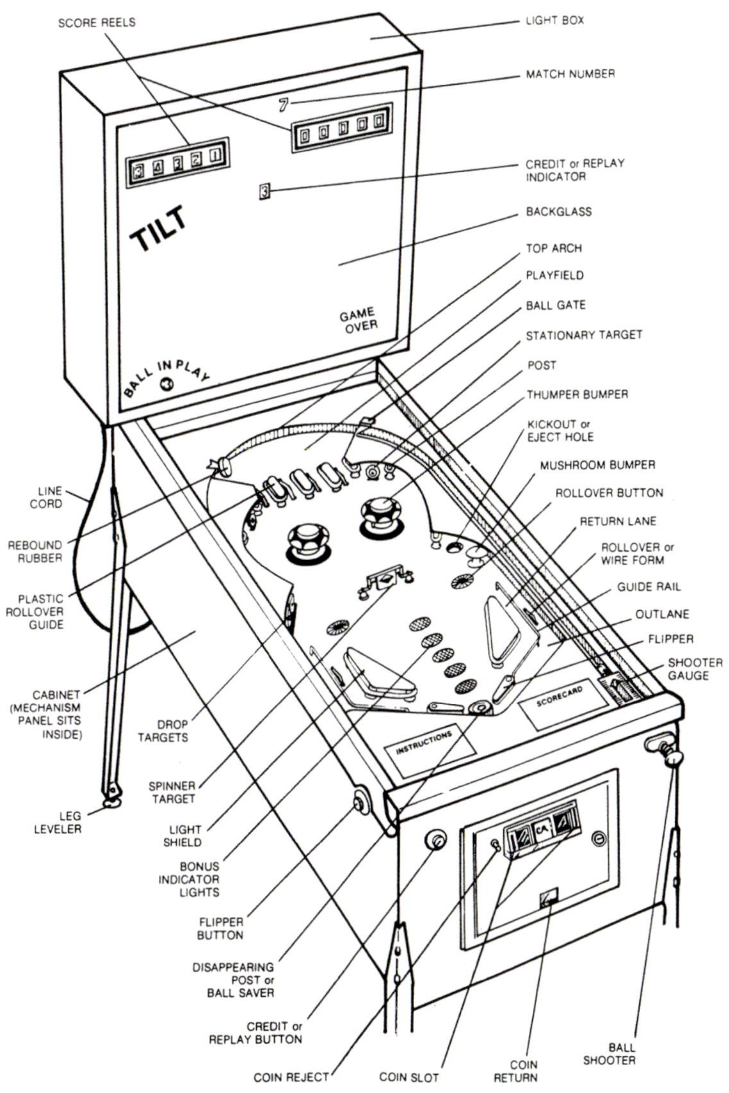

ELECTRO-MECHANICAL MACHINE

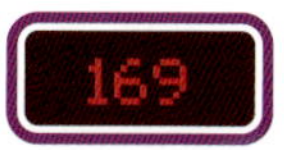

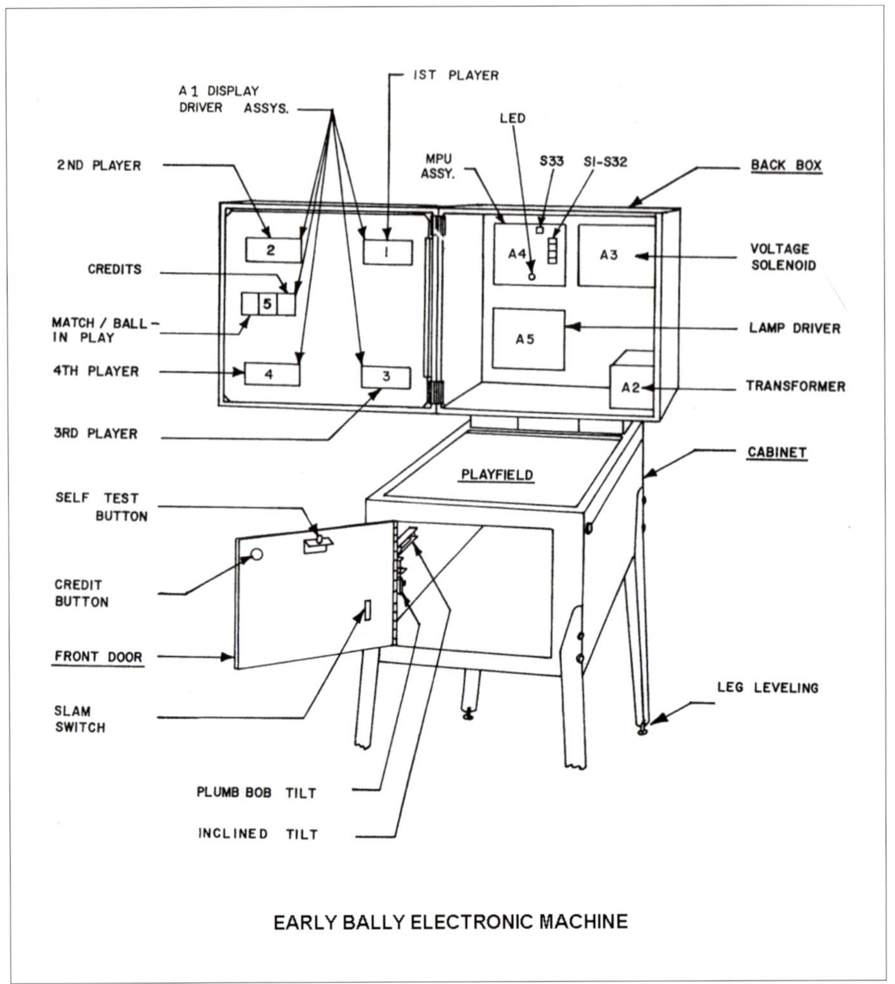

EARLY BALLY ELECTRONIC MACHINE

A pinball machine is a wonderful, crazy contraption with its steel ball flying like a streak of silver. Outside, it is bright and lurid. Inside, it is complicated and intricate.

—Bobbye Claire Natkin and Steve Kirk, *All About Pinball*

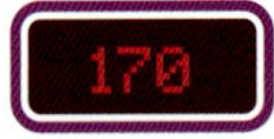

THE PACIFIC PINBALL MUSEUM

The pinball machines in this book are from the collection of the Pacific Pinball Museum.

Located in Alameda, California, across the Bay from San Francisco, the Pacific Pinball Museum has one of the largest collections of pinball machines, and their bagatelle ancestors, in the world: more than 1,700 machines dating from 1879 until today.

The museum's arcade and showroom has a rotating display of more than ninety of their machines that anyone can come in and play. Throughout the museum are murals, jukeboxes, historical artifacts, and hands-on working exhibits.

Founded by Michael Schiess in 2004, and operated by a crew of volunteers who restore the machines and staff the museum, the Pacific Pinball Museum is a 501(c)(3) nonprofit foundation. Donations to the museum are tax deductible.

The museum is at 1510 Webster Street, Alameda, California, 94501. Phone 510-769-1349. Email info@pacificpinball.org. Visit the museum online at PacificPinball.org, or better yet, visit the museum in person and enjoy an amazing day of pinball and fun.

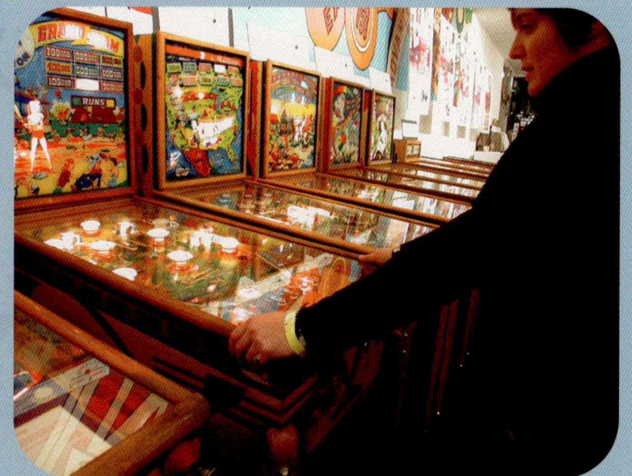

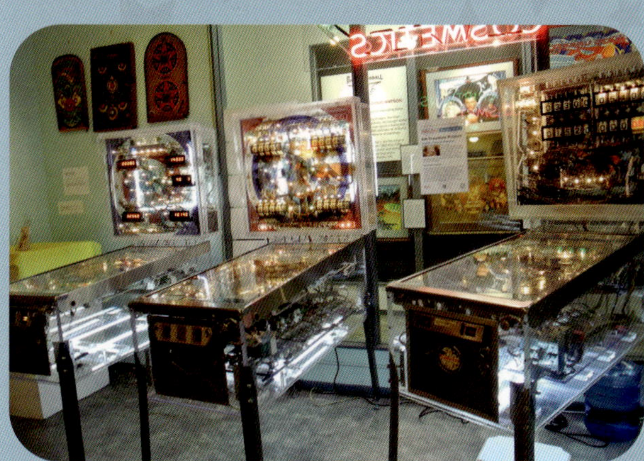

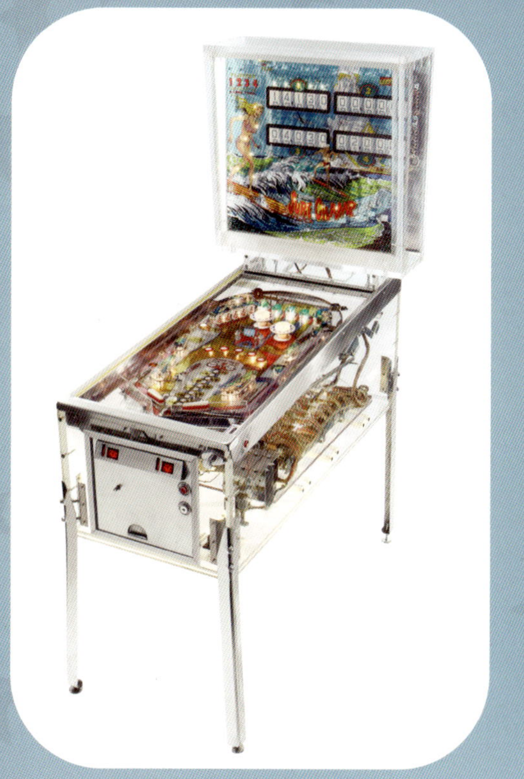

DROP A QUARTER IN THE SLOT

Good luck! Write or email me if you have a question, or don't understand something in this manual.

> A clean machine is a good machine.
>
> —David Gottlieb, founder of the Gottlieb Pinball Company

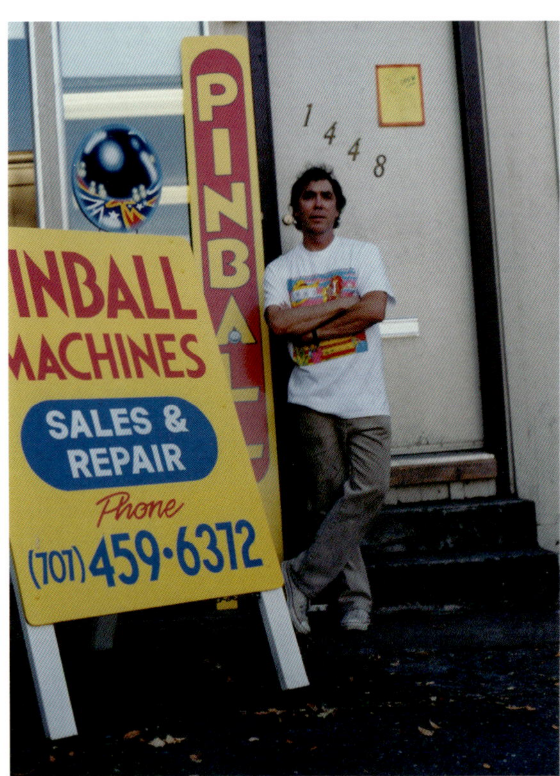

B.B. Kamoroff owns and operates Bell Springs Pinball Alley, a pinball sales and repair shop in Willits, California. He is the author of several how-to books. He and his wife, Sharon, live in Willits, in Mendocino County. (*Photo by Julia Jones*)

B.B. Kamoroff, PO Box 1240, Willits, CA 95490

kamoroff@bellsprings.com

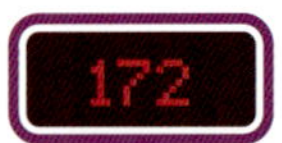

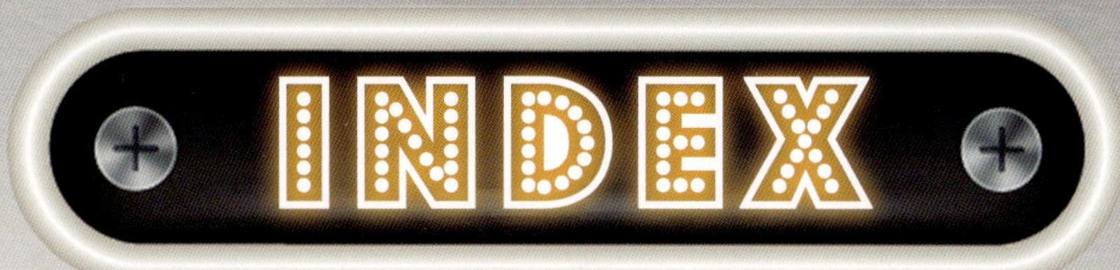

ILLUSTRATIONS

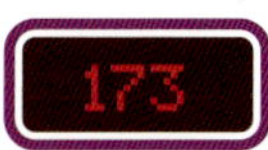

PINBALL MACHINES

INDEX

A

B

C

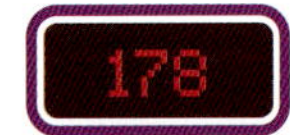

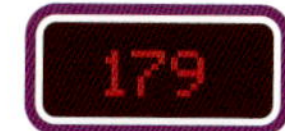

PRAISE FOR THE PREVIOUS EDITION:

A wealth of valuable information and step-by-step help, this manual is packed with tips and clear instructions.
—Play Meter Magazine

This should be the first book every new pinball owner buys. It answers all the basic questions, even the simple things like how to remove a glass, open a game, what types of light bulbs are used. What to do later, what maintenance is required, what basic repairs can one do, how assemblies work. Really, this book is what you need.
www.Flippers.de

A hugely useful guidebook for novice and professional alike. What Kamoroff has accomplished here is the best of both worlds. By explaining the simple things, and offering helpful ideas for the seasoned folk, he's handed the hobby an all-around silverball guide.
Game Room Magazine

It shines in its information on repairs and insights. Kamoroff has done an excellent job of making the technical information easy to understand.
Silverball News and Views

An excellent book, highly recommended.
—Pin Game Journal

A good troubleshooting guide.
—Multiball Magazine

The Go To manual for any novice pinball machine owner. There's gold in this here book.
—Pacific Pinball Museum

There isn't any other book that puts pinball repair in layman's terms. All the repairs in the book can be performed by just about anyone. It is truly an outstanding manual.
—Al Wang Pinball Repair, San Francisco

Packed full of useful information on about everything you wanted to know about pinball machines.
—Coin Tech Amusements

Just wanted to say thanks for a great book. Just finished my tenth pin and still find it helpful.
—Paul Starrider

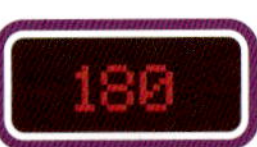